Birkhäuser

# Computer Science Foundations and Applied Logic



*Computer Science Foundations and Applied Logic* is a growing series that focuses on the foundations of computing and their interaction with applied logic, including how science overall is driven by this. Thus, applications of computer science to mathematical logic, as well as applications of mathematical logic to computer science, will yield many topics of interest. Among other areas, it will cover combinations of logical reasoning and machine learning as applied to AI, mathematics, physics, as well as other areas of science and engineering. The series (previously known as *Progress in Computer Science and Applied Logic*, https://www.springer.com/series/4814) welcomes proposals for research monographs, textbooks and polished lectures, and professional text/references. The scientific content will be of strong interest to both computer scientists and logicians.

Alexander Leitsch · David Michael Cerna ·
Anela Lolic

# First-Order Schemata and Inductive Proof Analysis

Birkhäuser

Alexander Leitsch
Institute of Logic and Computation
Technischen Universität Wien
Vienna, Austria

David Michael Cerna
Institute of Computer Science
Czech Academy of Sciences
Prague, Czech Republic

Anela Lolic
Institute of Logic and Computation
Technischen Universität Wien
Vienna, Austria

ISSN 2731-5754 ISSN 2731-5762 (electronic)
Computer Science Foundations and Applied Logic
ISBN 978-3-032-05740-2 ISBN 978-3-032-05741-9 (eBook)
https://doi.org/10.1007/978-3-032-05741-9



This book is published under the imprint Birkhäuser, www.birkhauser-science.com by the registered company Springer Nature Switzerland AG
The registered company address is: Gewerbestrasse 11, 6330 Cham, Switzerland

# Preface

This book is based on research of the authors on the method CERES (Cut-Elimination by RESolution) over the last 15 years. The main aim of this work is to demonstrate the potential of cut-elimination in proof analysis. In the method CERES, which was developed around 2000, cut elimination is based on automated theorem proving and thus provides a bridge between automated deduction and proof theory. The original method was applicable only to first-order logic without induction. But, as the induction principle is absolutely vital to real mathematics, further extensions of the method were necessary to make CERES applicable in mathematical practice. It turned out that *proof schemata* are most suited for integrating induction into the CERES method. The first part of the book serves the purpose to define and analyze schematic first-order theories on which proof schemata are based and shows the connections to first-order arithmetic, while the second part is building a schematic CERES method on this formalism. Finally applications and implementations are presented.

The authors did not develop their work in empty space. Many other scientists contributed to the development of CERES who deserve to be mentioned, though providing a complete list of those persons is hardly possible. It was Matthias Baaz with whom one of the authors developed the original CERES method. Stefan Hetzl, Daniel Weller, Nicolas Peltier, Giselle Reis and Martin Riener made important contributions to the theoretical analysis and extensions of the CERES method. In the development of the system GAPT (General Architecture for Proof Transformation) and in the implementation of CERES also Tomer Libal, Mikheil Rukhaia, Cvetan Dunchev and Gabriel Ebner made fundamental contributions. Finally we thank Dominik Pichler for finding a bug in some of our papers (which is corrected in this book). We also thank Stella Mahler for her cooperation in the investigations on schemata and primitive recursive arithmetic, a topic which is presented in this book, and for her recent contributions to GAPT.

Vienna, Austria Alexander Leitsch
Prague, Czech Republic David Michael Cerna
Vienna, Austria Anela Lolic

**Competing Interests** The authors have no competing interests to declare that are relevant to the content of this manuscript.

# Contents

# Chapter 1
# Introduction

What is a proof? Is a proof just a verification that a statement is true or is it also a key for understanding the theory in which the corresponding theorem is stated? The first to address the questions above was Georg Kreisel; in [26, 27] he developed the first ideas about extracting constructive information from classical proofs, and in [28] he discussed philosophical implications of formal proofs and the mathematical information they provide. In proof theory, which is a formal discipline for investigating proofs as syntactic objects, the forms and types of mathematical proofs are of central importance. In particular, proofs can contain information about mathematical objects, such as explicit information on bounds or algorithms, which are not visible in the statement itself. Therefore, analyzing mathematical proofs and extracting explicit (possibly hidden) information is at the very heart of mathematical reasoning. This process is generally known as *proof mining*.

In mathematical proofs of higher complexity arguments are typically structured by auxiliary statements, so-called *lemmas*. While lemmas are important in defining theories and making proofs comprehensible, they can be also considered as a form of detour; in fact, the information in lemma-based proofs is only implicit. To make it *explicit* the proof has to be transformed into specific normal forms. Perhaps the most famous normal form of formal proofs, obtained by eliminating lemmas, is the *cut-free proof*. It can be obtained by a method called *cut-elimination* which was introduced by Gerhard Gentzen in his famous paper [20]. In [21] Gentzen used his cut-elimination method in proving the consistency of Peano arithmetic. It was Georg Kreisel [26, 27] who recognized the potential of cut-elimination beyond proving consistency of theories. He suggested to apply cut-elimination in the investigation of real mathematical proofs formulated originally in "traditional" mathematics outside of formal logic. For mathematical proofs the result of cut-elimination is a purely (elementary) combinatorial proof, which can be used to extract explicit mathematical information, like bounds or algorithms. As a consequence, cut-elimination became a key technique in the analysis of mathematical proofs, Girard [22] being a prominent example: Girard's analysis of Fürstenberg and Weiss' topological proof, see [18], of



A. Leitsch et al., *First-Order Schemata and Inductive Proof Analysis*, Computer Science Foundations and Applied Logic, https://doi.org/10.1007/978-3-032-05741-9_1

van der Waerden's theorem [40]. Cut-elimination, applied to the proof of Fürstenberg and Weiss, resulted in van der Waerden's original elementary proof. Girard's analysis was carried out within mathematical meta-language by hand. Actually, his analysis is an application of cut-elimination to the proof of Fürstenberg and Weiss with the intention to obtain van der Waerden's combinatorial proof. Indeed, as any other complex proof transformation carried out by humans also cut-elimination requires a goal oriented strategy. Though such a strategy is essential in proof analysis by humans, other interesting elementary proofs might be missed by following a fixed goal. But how can we obtain different elementary proofs than those we were trying to get and which kind of information can be hidden in these proofs? The idea of *automated proof analysis* originates from such questions and indeed, a fully automated proof analysis may yield new and unexpected results. In order to automate the process of proof analysis *formal proofs* are required. In fact, is is essential to work with *correct proofs* formalized in specific calculi rather than with mathematical proofs in meta-language. The reason is obvious: although proofs in meta-language can be understood by humans very easily, computers (and algorithms) in general cannot process natural language; instead they deal with formal languages. The discipline of *proof checking* is concerned with the process of using software for checking formal (or better formalized) proofs for correctness. The proof checkers typically use powerful proof assistants, as *Isabelle*[1] and *Coq*.[2] Proof checkers are capable of checking also very complicated mathematical proofs, e.g. the proof of the famous four color theorem has been formally checked in Coq [23]. While proof checkers verify the correctness of proofs, automated proof analysis goes one step further: here not only the construction of formal proofs is required, but also their analysis and transformation and, in the end, an interpretation of the final results by humans. Therefore, automated proof analysis consists of formalizing mathematical proofs, applying algorithmic proof transformation (e.g. cut-elimination) and, finally, interpreting the resulting formal proof.

Unlike traditional methods of cut-elimination (which are typically based on proof rewriting a la Gentzen [20]) the method CERES (Cut-Elimination by RESolution, see [6, 8]) analyzes the structure of a given proof, extracts a so-called characteristic formula and uses a resolution refutation of this formula to construct a proof with only atomic cuts. As the core of this method is first-order resolution, CERES defines a bridge between automated deduction and proof theory. The first application of CERES to an interesting mathematical proof by Fürstenberg [17] can be found in [5]; the original proof (using topological arguments) was analyzed by resolution and it was shown that via cut-elimination Euclid's famous proof can be obtained. As the formalization of the proof (necessarily) contained an application of induction, the original CERES method was no longer applicable directly; in order to avoid application of the induction rule, the proof of Fürstenberg was represented as a *proof schema*. The computation of instances of the proof schema provided a (finite) sequence of characteristic formulas which supported the definition of a general resolution refutation

[1] http://isabelle.in.tum.de/index.html.

[2] https://coq.inria.fr/.

schema yielding Euclid's argument. At that time the necessary formal framework of first-order schemata was yet undeveloped and thus the main part of the proof analysis was based on traditional mathematics carried out by hand. In particular this proof analysis reveiled the need for a general formal theory of first-order schemata as a basis for an automation of this kind of proof analysis.

Mathematical induction is one of the most important principles in real mathematics, thus any substantial and relevant approach to proof analysis has to take into account induction. But in systems with induction rules, essential proof theoretic concepts and transformations become problematic. In particular Gentzen's method of cut-elimination fails for general induction proofs. However, there are methods for performing cut-elimination in presence of induction, but the resulting proofs do not have the subformula property and Herbrand's theorem cannot be realized [11, 34]. But if induction is represented via *schemata of proofs* [15, 31] schematic cut-elimination methods can be defined which allow the extraction of so-called *Herbrand systems*. A (partially) automated proof analysis using schemata was performed on Fürstenberg's proof of the infinitude of primes using topological concepts [5, 17]. Fürstenberg's proof by contradiction was formalized as a sequence of **LK**-proofs indexed by the (finite) number of primes assumed to exist; then the method CERES was applied to the entire sequence. The analysis was performed in a semi-automated way; in fact, major parts of the analysis had to be performed by hand. Nevertheless, the analysis showed that from Fürstenberg's proof Euclid's elementary proof could be obtained by applying a *formal* cut-elimination procedure. Though a fully automated analysis of this proof is not yet within reach, this example reveils the need for the development of *formal schematic proof systems* for handling proofs with induction. Recent developments based on schematic CERES can be considered as a step in this direction. Originally, existing schematic calculi with a variant of Herbrand's theorem were either defined for a weak induction principle [31], or did not guarantee a fully automated transformation and analysis of the given proof [15].

The first step towards a formal method was made in [15] where formula schemata, proof schemata, characteristic formula schema (at that time still in the format of clause set schemata), schematic refutations and schematic normal forms were formally defined; algorithms were developed to extract a characteristic formula schema. Schemata were only defined for one recursion parameter and the formalism was not well suited for practical use. However, the method served as a basis for a proof analysis published in [12]. In [31] the methods were substantially improved resulting in a fully automatable version of schematic CERES; still the new method was only capable of handling just one recursion parameter and was not well suited for interactive use (in case the automated theorem prover based on $n$-clause logic fails to produce a refutation). In the Ph.D., thesis [32] the range of schematic methods was extended substantially; schemata of terms, formulas and proofs were defined for arbitrary many parameters and a schematic CERES method algorithmically producing so-called Herbrand schemata as a final result. This schematic method is based on a novel formulation of the CERES method [29] avoiding the construction of proof normal forms and computing directly the set of Herbrand instances (and the expansion trees) of the resulting cut-free proof. All steps in the CERES method in [32], except the definition of a refutation schema for the characteristic formula schema which requires interactive reasoning, are fully automatable. The concept of resolution refutation schema (which forms the core of the schematic CERES method) was further generalized and improved in [13] using the general concept of point transition systems which increases the power of inductive definitions in [32]; this paper also addresses the problem of schematic unification in more depth and layed the ground for schematic unification algorithms (see also http://arxiv.org/abs/2201.10298).

The purpose of this book is the completion and unification of the existing material on first-order schemata described above. Large parts of the book are based on the thesis [32] which has not been published so far. Here, however, we develop a new improved general formalism which is clearer and more powerful than the original one; additionally it contains essential novel material unpublished so far (which is also not contained in [32]). Many important features of schematic proofs, like the relation to subsystems of arithmetic were published quite recently (see [30]). Indeed, there is a close connection between proof schemata and primitive recursive arithmetic which is analyzed in Chap. 8. New is also the general definition of schematic substitutions and unification of term schemata in Chap. 6, and the characterization of the range and the limits of schematic unification methods. Besides this new material the book provides a general formalism of handling schemata, by unifying and comparing all approaches to this topic. Schemata are capable of representing infinite sequences of objects. In first-order logic these objects can be terms or formulas. Proof schemata define infinite sequences of proofs in first-order logic and are tools for modeling inductive reasoning. In particular, proof schemata form a basis for computational cut-elimination and thus for an automated analysis of inductive proofs.

In Chap. 4 of this book we introduce a uniform formal language of term schemata and formula schemata which is based on the principle of primitive recursive definitions. In Chap. 5 we generalize primitive recursive definitions of formal objects to a powerful abstract framework (so-called point transition systems) allowing a more flexible specification of recursions. In Chap. 6 we develop a new formalism for defining schematic substitutions and define a schematic unification algorithm UAL (which forms the computational basis for schematic resolution calculi introduced in Sect. 7.2). As the problem of schematic unification is $\Pi_1$-complete (shown in Theorem 6.2.1) and thus undecidable, UAL—which is terminating—is necessarily incomplete but shown to be useful in practice, as illustrated by various applications described in this book. In Chap. 7 the schematic formalism for term- and formula schemata is extended to proofs; while in Sect. 7.1 schemata of **LK**-proofs are formalized, Sect. 7.2 is devoted to a formalization of resolution proof schemata. In Sect. 7.3 we decribe various extensions and variants of resolution refutation schemata. In Chap. 8 we compare proof schemata with inductive proofs (based on induction rules) and show that proof schemata simulate primitive recursive arithmetic. In Chap. 3 we switch from the analysis of schemata to cut-elimination; in Sect. 3.1 we first describe the role of cut-free proofs and proofs with quantifier-free cuts, the midsequent theorem and Herbrand's theorem. In Sect. 3.2 we introduce the method CERES for first-order logic and define a so-called CERES normal form, a proof with only quantifier-free cuts (from which a Herbrand sequent can be extracted); we show that the construction of a CERES normal form can be avoided in constructing a Herbrand sequent directly from a refutation of the characteristic formula and the proof projection of the input proof. In Chap. 9 we combine classical proof theory with the theory of first-order schemata; we extend the CERES method for first-order proofs to a CERES *method for first-order proof schemata* and (via the results in Chap. 8) obtain a method for extracting schemata of Herbrand sequents (Herbrand systems) for proofs with induction. This combination allows a formulation of Herbrand's theorem for inductive proofs and thus yields a powerful tool for the *analysis of inductive proofs*. In Chap. 10 we build a bridge from theory to applications by first describing an implementation of the schematic CERES method in the system GAPT (General Architecture for Proof Transformation) and then an analysis of a real mathematical proof in this system. Finally, in Chap. 11 we present a list of unsolved problems and discuss potentials and perspectives of future research.

# Chapter 2
# Preliminaries

In this chapter we will introduce all basic concepts that will be used throughout this work. Most of the concepts will be extended later on.

## 2.1 The Calculus LK

We denote predicate symbols by $P, Q, R, \ldots$, function symbols by $f, g, h, \ldots$, and constant symbols by $a, b, c, \ldots$. Moreover, we distinguish a countably infinite set of free variables $V_f$ and a countably infinite set of bound variables $V_b$. The distinction between free and bound variables is important for different proof-transformations, like cut-elimination. We use $\alpha, \beta, \gamma, \ldots$ for free variables and $x, y, z, \ldots$ for bound ones. Terms are defined as usual with the restriction that they may contain bound variables.

**Definition 2.1.1** (*semi-term, term* [8]) We define a set of semi-terms inductively:

- bound and free variables are semi-terms,
- constants are semi-terms,
- if $t_1, \ldots, t_n$ are semi-terms and $f$ is a $n$-place function symbol then $f(t_1, \ldots, t_n)$ is a semi-term.

Terms are semi-terms which do not contain bound variables.

**Definition 2.1.2** (*substitution* [8]) A substitution is a mapping from $V_f \cup V_b$ to the set of semi-terms s.t. $\sigma(v) \neq v$ for only finitely many $v \in V_f \cup V_b$. If $\sigma$ is a substitution with $\sigma(x_i) = t_i$ for $x_i \neq t_i$ $(1 \leq i \leq n)$ and $\sigma(v) = v$ for $v \notin \{x_1, \ldots, x_n\}$, then we denote $\sigma$ by $\{x_1 \leftarrow t_1, \ldots, x_n \leftarrow t_n\}$.

Substitutions can be extended to terms, atoms and formulas in a straightforward way. Some substitutions are more general than others, and some substitutions are unifiers of sets of atoms.



A. Leitsch et al., *First-Order Schemata and Inductive Proof Analysis*, Computer Science Foundations and Applied Logic, https://doi.org/10.1007/978-3-032-05741-9_2

**Definition 2.1.3** A substitution $\sigma$ is called *more general* than a substitution $\theta$ ($\sigma \leq_s \theta$) if there exists a substitution $\mu$ s.t. $\theta = \sigma\mu$.

**Definition 2.1.4** (*unifier* [8]) Let $\mathcal{A}$ be a nonempty set of atoms and $\sigma$ be a substitution. $\sigma$ is a unifier of $\mathcal{A}$ if the set $\mathcal{A}\sigma$ contains only one element. $\sigma$ is a most general unifier (m.g.u.) of $\mathcal{A}$ if $\sigma$ is a unifier of $\mathcal{A}$ and for all unifiers $\lambda$ of $\mathcal{A}$ $\sigma \leq_s \lambda$.

Note that we can also unify several sets of atoms simultaneously.

**Definition 2.1.5** (*simultaneous unifier* [8]) Let

$$W = (\mathcal{A}_1, \ldots, \mathcal{A}_n),$$

where the $\mathcal{A}_i$ are nonempty sets of atoms for $i = 1, \ldots, n$. A substitution $\sigma$ is called a simultaneous unifier of $W$ if $\sigma$ unifies all $\mathcal{A}_i$. $\theta$ is called a most general simultaneous unifier of $W$ if $\theta$ is a simultaneous unifier of $W$ and $\theta \leq_s \sigma$ for all simultaneous unifiers $\sigma$ of $W$.

**Definition 2.1.6** (*semi-formula, formula* [8]) $\top$ and $\bot$ are formulas. If $t_1, \ldots, t_n$ are terms and $P$ is a $n$-place predicate symbols then $P(t_1, \ldots, t_n)$ is an atomic formula.

- If $A$ is a formula then $\neg A$ is a formula.
- If $A, B$ are formulas, then $A \wedge B$, $A \vee B$ and $A \rightarrow B$ are formulas.
- If $A\{x \leftarrow a\}$ is a formula, then $\forall x A$ and $\exists x A$ are formulas.

Semi-formulas differ from formulas in containing free variables in $V_b$.

**Definition 2.1.7** (*prenex formula*) A formula is prenex (or in prenex form) if it is of the form $Q_1 x_1 Q_2 x_2 \ldots Q_n x_n F$, where $n \geq 0$ and $Q_i \in \{\forall, \exists\}$ for $1 \leq i \leq n$ and $F$ is quantifier-free.

By $V(t)$ we denote the the set of variables occurring freely in a term or formula $t$. The sequent calculus **LK** is based on sequents.

**Definition 2.1.8** (*sequent*) Let $\Gamma$ and $\Delta$ be two multi-sets of formulas and $\vdash$ be a symbol not belonging to the logical language. Then $\Gamma \vdash \Delta$ is called a sequent.

If $S_1 : \Gamma \vdash \Delta$ and $S_2 : \Pi \vdash \Lambda$ are sequents we define the *concatenation* of $S_1$ and $S_2$ (notation $S_1 \circ S_2$) as $\Gamma, \Pi \vdash \Delta, \Lambda$.

**Definition 2.1.9** (*semantics of a sequent*) Semantically a sequent

$$S : A_1, \ldots, A_n \vdash B_1, \ldots, B_m$$

stands for

$$F(S) : \bigwedge_{i=1}^{n} A_i \rightarrow \bigvee_{j=1}^{m} B_j.$$

In particular, we define $\mathcal{M}$ to be an interpretation of $S$ if $\mathcal{M}$ is an interpretation of $F(S)$. If $n = 0$ we assign $\top$ to $\bigwedge_{i=1}^{n} A_i$, if $m = 0$ we assign $\bot$ to $\bigvee_{j=1}^{m} B_j$. The empty sequent is represented by $\top \rightarrow \bot$, which is equivalent to $\bot$ and represents falsum. $S$ is true in $\mathcal{M}$ if $F(S)$ is true in $\mathcal{M}$. S is called valid if $F(S)$ is valid.

Substitutions can be extended to sequents.

**Definition 2.1.10** If $S : A_1, \ldots, A_n \vdash B_1, \ldots, B_m$ is a sequent and $\sigma$ is a substitution, then

$$S\sigma : A_1\sigma, \ldots, A_n\sigma \vdash B_1\sigma, \ldots, B_m\sigma.$$

**Definition 2.1.11** (*prenex sequent*) A sequent

$$S\colon A_1, \ldots, A_n \vdash B_1, \ldots, B_m$$

is prenex (or in prenex form) if the formulas $A_i$ and $B_j$, where $1 \leq i \leq n$ and $1 \leq j \leq m$, are prenex formulas.

In this book, we will use the notion of strong and weak quantifiers. This notion is defined via the polarity of a formula occurrence in a sequent.

**Definition 2.1.12** (*polarity* [8]) Let $\lambda$ be an occurrence of a formula $A$ in $B$.

- If $A \equiv B$ then $\lambda$ is a positive occurrence in $B$.
- If $B \equiv \neg C$ and $\lambda$ is a positive (negative) occurrence of $A$ in $C$ then the corresponding occurrence $\lambda'$ of $A$ in $B$ is negative (positive).
- If $B \equiv C \wedge D$, $B \equiv C \vee D$, $B \equiv \forall x C$ or $B \equiv \exists x C$ and $\lambda$ is a positive (negative) occurrence of $A$ in $C$ (or in $D$) then the corresponding occurrence $\lambda'$ of $A$ in $B$ is positive (negative).
- If $B \equiv C \rightarrow D$ and $\lambda$ is a positive (negative) occurrence of $A$ in $D$ then the corresponding occurrence $\lambda'$ in $B$ is positive (negative). If $\lambda$ is a positive (negative) occurrence of $A$ in $C$ then the corresponding occurrence $\lambda'$ of $A$ in $B$ is negative (positive).

If there exists a positive (negative) occurrence of a formula $A$ in $B$ we say that $A$ is of positive (negative) polarity in $B$.

**Definition 2.1.13** (*strong and weak quantifiers* [8]) If $\forall x$ occurs positively (or negatively) in $B$ then $\forall x$ is called a strong (or weak) quantifier. If $\exists x$ occurs positively (or negatively) in $B$ then $\exists x$ is called a weak (or strong) quantifier.

This definition can be extended to sequents: A sequent is called weakly (or strongly) quantified if all quantifier occurrences in $S$ are weak (or strong).

For the proof-theoretic transformations that will be presented in this book, in particular for the CERES method, we will use skolemized sequents.

**Definition 2.1.14** (*skolemized sequent* [8]) Let $sk$ be a function that maps closed formulas into closed ones. $sk(F)$ is the skolemization of $F$ and defined as

$$sk(F) = F \text{ if } F \text{ does not contain strong quantifiers,}$$

and otherwise assume that $Qy$ is the first strong quantifier in $F$ (in a tree ordering) in the scope of the weak quantifiers $Q_1x_1 Q_2x_2 \ldots Q_nx_n$. Let $f$ be an $n$-ary function symbol not occurring in $F$, then $sk(F)$ is inductively defined as

$$sk(F) = sk(F_{Qy}\{y \leftarrow f(x_1, \ldots, x_n)\}),$$

where $F_{Qy}$ is $F$ after removal of $Qy$.

Now let $S\colon A_1, \ldots, A_n \vdash B_1, \ldots, B_m$ be a sequent consisting of closed formulas only and

$$sk((A_1 \wedge \ldots \wedge A_n) \rightarrow (B_1 \vee \ldots \vee B_m)) = (A'_1 \wedge \ldots \wedge A'_n) \rightarrow (B'_1 \vee \ldots \vee B'_m).$$

Then the sequent $S'\colon A'_1, \ldots, A'_n \vdash B'_1, \ldots, B'_m$ is the skolemization of $S$.

Like most other calculi Gentzen's **LK** is based on axioms and rules.

**Definition 2.1.15** (*axiom set*) A (possibly infinite) set $\mathcal{A}$ of sequents is called an axiom set if it is closed under substitution, i.e. for all $S \in \mathcal{A}$ and for all substitutions $\theta$ we have $S\theta \in \mathcal{A}$. If $\mathcal{A}$ contains only atomic sequents we speak about an atomic axiom set.

**Definition 2.1.16** (*standard axiom set*) Let $\mathcal{A}_T$ be the smallest axiom set containing all sequents of the form $A \vdash A$ for arbitrary atomic formulas $A$. Then $\mathcal{A}_T$ is called the *standard axiom set*.

**Definition 2.1.17** (*sequent calculus* **LK** [8]) Here, we will use a variant of Gentzen's version of **LK** [19]. Since we consider multi-sets of formulas, we do not need exchange or permutation rules. There are two groups of rules, the logical and the structural ones. All rules except the cut have left and right versions, denoted by $l$ and $r$, respectively. The binary rules are of multiplicative type, i.e. no auto-contraction of the context is applied. Below, $A$ and $B$ denote formulas whereas $\Gamma$, $\Delta$, $\Pi$, $\Lambda$ denote multi-sets of formulas. In the rules there are introducing or auxiliary formulas in the premises and introduced or principal formulas in the conclusion. We indicate these formulas for all rules, auxiliary formula occurrences are marked by $+$ and principal formula occurrences are marked by $*$. We frequently say auxiliary (main) formula instead of auxiliary (main) formula occurrence.

**The logical rules**

$\wedge$-introduction

$$\frac{A^+, B^+, \Gamma \vdash \Delta}{(A \wedge B)^*, \Gamma \vdash \Delta}\ \wedge_l \qquad \frac{\Gamma_1 \vdash \Delta_1, A^+ \qquad \Gamma_2 \vdash \Delta_2, B^+}{\Gamma_1, \Gamma_2 \vdash \Delta_1, \Delta_2, (A \wedge B)^*}\ \wedge_r$$

$\vee$-introduction

$$\frac{\Gamma \vdash \Delta, A^+, B^+}{\Gamma \vdash \Delta, (A \vee B)^+} \vee_r \qquad \frac{A^+, \Gamma_1 \vdash \Delta_1 \qquad B^+, \Gamma_2 \vdash \Delta_2}{(A \vee B)^*, \Gamma_1, \Gamma_2 \vdash \Delta_1, \Delta_2} \vee_l$$

$\rightarrow$-introduction

$$\frac{\Gamma_1 \vdash \Delta_1, A^+ \qquad B^+, \Gamma_2 \vdash \Delta_2}{(A \rightarrow B)^*, \Gamma_1, \Gamma_2 \vdash \Delta_1, \Delta_2} \rightarrow_l \qquad \frac{A^+, \Gamma \vdash \Delta, B^+}{\Gamma \vdash \Delta, (A \rightarrow B)^*} \rightarrow_r$$

$\neg$-introduction

$$\frac{\Gamma \vdash \Delta, A^+}{\neg A^*, \Gamma \vdash \Delta} \neg_l \qquad \frac{A^+, \Gamma \vdash \Delta}{\Gamma \vdash \Delta, \neg A^*} \neg_r$$

$\forall$-introduction

$$\frac{A\{x \leftarrow t\}^+, \Gamma \vdash \Delta}{\forall x A^*, \Gamma \vdash \Delta} \forall_l \qquad \frac{\Gamma \vdash \Delta, A\{x \leftarrow \alpha\}^+}{\Gamma \vdash \Delta, \forall x A^*} \forall_r$$

where $t$ is an arbitrary term and $\alpha$ is a free variable which may not occur in $\Gamma$, $\Delta$, $A$. $\alpha$ is called an eigenvariable.

$\exists$-introduction

$$\frac{A\{x \leftarrow \alpha\}^+, \Gamma \vdash \Delta}{\exists x A^*, \Gamma \vdash \Delta} \exists_l \qquad \frac{\Gamma \vdash \Delta, A\{x \leftarrow t\}^+}{\Gamma \vdash \Delta, \exists x A^*} \exists_r$$

where the variable conditions for $\exists_l$ are the same as those for $\forall_r$ and similarly for $\exists_r$ and $\forall_l$. The quantifier-rules $\forall_l$, $\exists_r$ are called *weak*, the rules $\exists_l$, $\forall_r$ *strong*.

**The structural rules**

weakening

$$\frac{\Gamma \vdash \Delta}{A^*, \Gamma \vdash \Delta} w_l \qquad \frac{\Gamma \vdash \Delta}{\Gamma \vdash \Delta, A^*} w_r$$

contraction

$$\frac{A^+, A^+, \Gamma \vdash \Delta}{A^*, \Gamma \vdash \Delta} c_l \qquad \frac{\Gamma \vdash \Delta, A^+, A^+}{\Gamma \vdash \Delta, A^*} c_r$$

cut

$$\frac{\Gamma_1 \vdash \Delta_1, A^m \qquad A^n, \Gamma_2 \vdash \Delta_2}{\Gamma_1, \Gamma_2 \vdash \Delta_1, \Delta_2} cut(A, m, n)$$

where $m, n \geq 1$. The formulas in $A^m$ and $A^n$ are the auxiliary formulas of $cut(A, m, n)$ (they are also called cut-formulas) and there are no principal ones. We use this form of the cut rule as it can be easily transformed into the resolution rule, which will be used in the method CERES.

An **LK**-proof from a set of axioms $\mathcal{A}$ is a tree formed according to the rules of **LK** such that all leaves are in $\mathcal{A}$. In general, our **LK**-proofs will be proofs from $\mathcal{A}_T$.

When it is clear from the context, we will omit marking auxiliary and principal formula occurrences with $+$ and $*$.

We denote occurrences of sequents in a derivation with $\nu_0, \nu_1, \dots \nu_\alpha$. The sequent at node $\nu_\alpha$ is denoted by $Seq(\nu_\alpha)$. Formula occurrences in a sequent are denoted by $\mu_0, \mu_1, \dots, \mu_\beta$. Let $\mu$ be an occurrence of a formula $A$ in a sequent $S$, then we define $formula(\mu) = A$ in $S$.

**Definition 2.1.18** (*regularity*) An **LK**-derivation $\varphi$ is called regular if

- all eigenvariables of quantifier introductions $\forall_r$ and $\exists_l$ in $\varphi$ are mutually different, and
- if a variable $\alpha$ occurs as an eigenvariable in a proof node $\nu$ then $\alpha$ occurs only above $\nu$ in the proof tree.

There is a straightforward transformation from **LK**-derivations into regular ones, by just renaming the eigenvariables in different subderivations.

**Definition 2.1.19** (*prenex derivation*) A derivation is in prenex form (or prenex) if all its sequents are prenex.

**Definition 2.1.20** (*skolemized derivation*) A derivation is skolemized if its end-sequent is a skolemized sequent.

## 2.2 The Resolution Calculus $\mathbf{RPL_0}$

In the definition of the (schematic) CERES method we will make use of a calculus for refuting formulas in negation normal form. These formulas will be referred to as *characteristic formulas*. For refuting characteristic formulas we chose the calculus $\mathrm{RPL}_0$ for quantifier-free formulas, which combines dynamic normalization rules à la Andrews [1] with the resolution rule, but in contrast to [1] does not restrict the resolution rule to atomic formulas. The main motivation of the calculus $\mathrm{RPL}_0$ is that it can be extended to a schematic setting in a straightforward way, and that it is particularly suited for the extraction of Herbrand sequents.

The set of quantifier-free formulas in predicate logic will be denoted as $\mathrm{PL}_0$. As characteristic formulas do not contain $\to$, for simplicity we omit $\to$, but can represent it by $\neg$ and $\vee$ in the usual way. In this setting, as sequents we consider objects of the form $\Gamma \vdash \Delta$, where $\Gamma$ and $\Delta$ are multisets of formulas in $\mathrm{PL}_0$.

**Definition 2.2.1** ($\mathrm{RPL}_0$) The axioms of $\mathrm{RPL}_0$ are sequents $\vdash F$ for $F \in \mathrm{PL}_0$.

The rules are elimination rules for the connectives and the resolution rule.

$$\frac{\Gamma \vdash \Delta, A \wedge B}{\Gamma \vdash \Delta, A} \wedge\colon r_1 \quad \frac{\Gamma \vdash \Delta, A \wedge B}{\Gamma \vdash \Delta, B} \wedge\colon r_2 \quad \frac{A \wedge B, \Gamma \vdash \Delta}{A, B, \Gamma \vdash \Delta} \wedge\colon l$$

$$\frac{\Gamma \vdash \Delta, A \vee B}{\Gamma \vdash \Delta, A, B} \vee : r \quad \frac{A \vee B, \Gamma \vdash \Delta}{A, \Gamma \vdash \Delta} \vee : l_1 \quad \frac{A \vee B, \Gamma \vdash \Delta}{B, \Gamma \vdash \Delta} \vee : l_2$$

$$\frac{\Gamma \vdash \Delta, \neg A}{A, \Gamma \vdash \Delta} \neg : r \quad \frac{\neg A, \Gamma \vdash \Delta}{\Gamma \vdash \Delta, A} \neg : l$$

The resolution rule, where $\sigma$ is an m.g.u. of $\{A_1, \ldots, A_k, B_1, \ldots, B_l\}$ and $V(\{A_1, \ldots, A_k\}) \cap V(\{B_1, \ldots, B_l\}) = \emptyset$ is

$$\frac{\Gamma \vdash \Delta, A_1, \ldots, A_k \quad B_1, \ldots, B_m, \Pi \vdash \Lambda}{\Gamma\sigma, \Pi\sigma \vdash \Delta\sigma, \Lambda\sigma} \; res$$

A $\mathrm{RPL}_0$-derivation is a tree formed from axioms $\vdash F$ for $F \in \mathrm{PL}_0$ by application of the rules above.

**Proposition 2.2.1** $\mathrm{RPL}_0$ *is sound and refutationally complete, i.e.*

1. *all rules in* $\mathrm{RPL}_0$ *are sound and*
2. *for any unsatisfiable formula* $\forall F$ *and* $F \in \mathrm{PL}_0$ *there exists a* $\mathrm{RPL}_0$*-derivation of* $\vdash$ *from axioms of the form* $\vdash F\vartheta$ *where* $\vartheta$ *is a renaming of the free variables in* $F$.

***Proof*** 1. is trivial: if $\mathcal{M}$ is a model of the premise(s) of a rule then $\mathcal{M}$ is also a model of the conclusion.

For proving 2., first the standard clause set $\mathcal{C}$ of $F$ has to be derived. Therefore, the rules of $\mathrm{RPL}_0$ are applied to $\vdash F$, decomposing $F$ into its subformulas until no rule other than $res$ can be applied. The last subformula obtained in this way is atomic and a clause. The standard clause set $\mathcal{C}$ of $F$ is comprised of the clauses obtained in this way. As $\forall F$ is unsatisfiable, its standard clause set is refutable by resolution. Applying $res$ to the clauses results in a derivation of $\vdash$. The whole derivation is in $\mathrm{RPL}_0$.

**Example 2.2.1** Let

$$\begin{aligned} F &= Pa \wedge (Pf^2a \vee \neg Pf^2a) \wedge \neg Pf^4a \wedge (\neg P\alpha \vee Pf^2\alpha), \\ H &= (Pf^2a \vee \neg Pf^2a) \wedge \neg Pf^4a \wedge (\neg P\alpha \vee Pf^2\alpha), \\ G &= \neg Pf^4a \wedge (\neg P\alpha \vee Pf^2\alpha). \end{aligned}$$

We obtain the $\mathrm{RPL}_0$-refutation $\rho =$

$$\dfrac{\dfrac{\overset{(\rho_1)}{Pf^2a \vdash} \qquad \dfrac{\dfrac{\dfrac{\dfrac{\dfrac{\vdash F}{\vdash H}\,\wedge_{r_2}}{\vdash G}\,\wedge_{r_2}}{\vdash \neg P\alpha \vee Pf^2\alpha}\,\wedge_{r_2}}{\vdash \neg P\alpha, Pf^2\alpha}\,\vee_r}{P\alpha \vdash Pf^2\alpha}\,\neg_r}{Pa \vdash}\,res\{\alpha \leftarrow a\} \qquad \dfrac{\vdash F}{\vdash Pa}\,\wedge_{r_1}}{\vdash}\,res$$

where $\rho_1 =$

$$\dfrac{\dfrac{\dfrac{\dfrac{\dfrac{\vdash F}{\vdash H}\,\wedge_{r_2}}{\vdash G}\,\wedge_{r_2}}{\vdash \neg Pf^4a}\,\wedge_{r_1}}{Pf^4a \vdash}\,\neg_r \qquad \dfrac{\dfrac{\dfrac{\dfrac{\dfrac{\vdash F}{\vdash H}\,\wedge_{r_2}}{\vdash G}\,\wedge_{r_2}}{\vdash \neg P\alpha \vee Pf^2\alpha}\,\wedge_{r_2}}{\vdash \neg P\alpha, Pf^2\alpha}\,\vee_r}{P\alpha \vdash Pf^2\alpha}\,\neg_r}{Pf^2a \vdash}\,res\{\alpha \leftarrow f^2a\}$$

The set of unifiers in $\rho$ is $\{\{\alpha \leftarrow a\}, \{\alpha \leftarrow f^2a\}\}$.

As shown in the example above, several resolution rules may occur in a $\text{RPL}_0$-derivation, and hence several most general unifiers $\sigma_i$ need to be applied. A total unifier (or total m.g.u.) can be obtained by considering the most general simultaneous unifier of the unification problems given by the atoms in the premises of all resolution rules, after regularizing the derivation by renaming variables.

**Definition 2.2.2** (*regular* $\text{RPL}_0$*-derivation*) A $\text{RPL}_0$-derivation is regular if different occurrences of axioms are variable-disjoint. A regular $\text{RPL}_0$-derivation $\rho_r$ can always be obtained from a $\text{RPL}_0$-derivation $\rho$ (which is not regular) in the following way:

1. Let $\vdash F$ be an axiom in $\rho$. Replace $\vdash F$ by $\vdash F\sigma$, where $\sigma$ is a substitution replacing all variables $\alpha_1, \ldots, \alpha_n$ in $F$ by (new) fresh variables $\beta_1, \ldots, \beta_n$.
2. Starting from the axioms, apply the substitution $\sigma$ from top to bottom to each sequent that follows from a unary rule (note that the only binary rule in $\rho$ is the resolution rule).
3. For any resolution rule with premises $S_1$ and $S_2$ in $\rho$ and unifier $\sigma$, in $\rho_r$ we will have the premises $S_1\sigma_1$ and $S_2\sigma_2$ for some substitutions $\sigma_1$ and $\sigma_2$ resulting from regularizing the derivations leading to the premises. We apply the resolution rule as in $\rho$, but now with the unifier $\sigma\sigma_1$ and $\sigma\sigma_2$.

The resulting derivation is regular, as all axioms are variable-disjoint.

**Example 2.2.2** Let $\rho$ be the $\text{RPL}_0$-derivation in Example 2.2.1, then $\rho_r =$

$$
\begin{prooftree}
\AxiomC{$(\rho_{r_1})$}
\noLine
\UnaryInfC{$Pf^2a \vdash$}
\AxiomC{$\vdash F\{\alpha \leftarrow \beta\}$}
\RightLabel{$\wedge_{r_2}$}
\UnaryInfC{$\vdash H\{\alpha \leftarrow \beta\}$}
\RightLabel{$\wedge_{r_2}$}
\UnaryInfC{$\vdash G\{\alpha \leftarrow \beta\}$}
\RightLabel{$\wedge_{r_2}$}
\UnaryInfC{$\vdash \neg P\beta \vee Pf^2\beta$}
\RightLabel{$\vee_r$}
\UnaryInfC{$\vdash \neg P\beta, Pf^2\beta$}
\RightLabel{$\neg_r$}
\UnaryInfC{$P\beta \vdash Pf^2\beta$}
\RightLabel{$res\{\beta \leftarrow a\}$}
\BinaryInfC{$Pa \vdash$}
\AxiomC{$\vdash F$}
\RightLabel{$\wedge_{r_1}$}
\UnaryInfC{$\vdash Pa$}
\RightLabel{$res$}
\BinaryInfC{$\vdash$}
\end{prooftree}
$$

where $\rho_{r_1} =$

$$
\begin{prooftree}
\AxiomC{$\vdash F\{\alpha \leftarrow \delta\}$}
\RightLabel{$\wedge_{r_2}$}
\UnaryInfC{$\vdash H\{\alpha \leftarrow \delta\}$}
\RightLabel{$\wedge_{r_2}$}
\UnaryInfC{$\vdash G\{\alpha \leftarrow \delta\}$}
\RightLabel{$\wedge_{r_1}$}
\UnaryInfC{$\vdash \neg Pf^4a$}
\RightLabel{$\neg_r$}
\UnaryInfC{$Pf^4a \vdash$}
\AxiomC{$\vdash F\{\alpha \leftarrow \gamma\}$}
\RightLabel{$\wedge_{r_2}$}
\UnaryInfC{$\vdash H\{\alpha \leftarrow \gamma\}$}
\RightLabel{$\wedge_{r_2}$}
\UnaryInfC{$\vdash G\{\alpha \leftarrow \gamma\}$}
\RightLabel{$\wedge_{r_2}$}
\UnaryInfC{$\vdash \neg P\gamma \vee Pf^2\gamma$}
\RightLabel{$\vee_r$}
\UnaryInfC{$\vdash \neg P\gamma, Pf^2\gamma$}
\RightLabel{$\neg_r$}
\UnaryInfC{$P\gamma \vdash Pf^2\gamma$}
\RightLabel{$res\{\gamma \leftarrow f^2a\}$}
\BinaryInfC{$Pf^2a \vdash$}
\end{prooftree}
$$

is the $\text{RPL}_0$-derivation obtained by regularizing $\rho$. The set of substitutions in $\rho_r$ is $\{\{\beta \leftarrow a\}, \{\gamma \leftarrow f^2a\}\}$.

**Definition 2.2.3** (*total m.g.u.*) Let $\rho_r$ be a regular $\text{RPL}_0$-derivation containing $n$ resolution inferences. The unification problem of $\rho_r$ is defined as $W = (\mathcal{A}_1, \ldots, \mathcal{A}_n)$, where $\mathcal{A}_i$ ($i \in \{1, \ldots n\}$) is the set

$$\{A^i_1, \ldots, A^i_{k_i}, B^i_1, \ldots, B^i_{m_i}\}$$

in the resolution inferences

$$
\begin{prooftree}
\AxiomC{$\Gamma \vdash \Delta, A^i_1, \ldots, A^i_{k_i}$}
\AxiomC{$B^i_1, \ldots, B^i_{m_i}, \Pi \vdash \Lambda$}
\RightLabel{$res$}
\BinaryInfC{$\Gamma\sigma_i, \Pi\sigma_i \vdash \Delta\sigma_i, \Lambda\sigma_i$}
\end{prooftree}
$$

in $\rho_r$. If $\sigma$ is a most general simultaneous unifier of $W$ (see Definition 2.1.5), $\sigma$ is called a total m.g.u. of $\rho_r$.

**Example 2.2.3** The total m.g.u. of $\rho_r$ in Example 2.2.2 is

$$\{\{\beta \leftarrow a\}, \{\gamma \leftarrow f^2a\}\}.$$

$$\vdash B(x \leftarrow 1) \quad [illegible]$$

[illegible]

where $\theta_1$ is a most general simultaneous unifier of [illegible] (see Definition 2.[illegible]) [illegible] called a [illegible] of $\gamma$.

**Example 2.2.3** [illegible] in Example 2.[illegible]

[illegible]

# Chapter 3
# Cut-Elimination and the Method CERES

When Gentzen introduced the sequent calculus **LK** in [19] he also provided a method for the algorithmic elimination of cut inferences from proofs. While there are many important collaries to a cut-elimination theorem for a calculus such as the consistency of a system, we focus on one particular corollary with applications in proof minining, which Gentzen called the midsequent theorem. This theorem directly corresponds to Herbrand's theorem [24] and provides the basis for many methods of proof analysis.

## 3.1 Cut-Elimination

Cut-elimination is a proof transformation of essential importance to proof theory. The cut-elimination theorem, or Gentzen's Hauptsatz, can be formulated for **LK** as

**Theorem 3.1.1** *If a sequent is* **LK***-provable, then it is* **LK***-provable without a cut.*

The theorem can be proved by induction over the rank and grade of a formula, which are measures for a proof's complexity. In Gentzen's proof always the uppermost cut is eliminated, but there are other cut-elimination methods where, for instance, the cut formula with the highest complexity is eliminated first. These methods for cut-elimination can be described as reductive cut-elimination methods, as they follow certain reduction rules to eliminate the cuts step by step. It should be emphasized that this kind of cut-elimination is effective, in the sense that a new proof is constructed from a given proof.

An important application of the cut-elimination theorem lies in the analysis of proofs, often referred to as proof mining. The idea is that the cut-elimination theorem basically states that any theorem in **LK** can be proved without detours. Technically speaking, this can be described by the subformula property.

 
A. Leitsch et al., *First-Order Schemata and Inductive Proof Analysis*, Computer Science Foundations and Applied Logic, https://doi.org/10.1007/978-3-032-05741-9_3

**Definition 3.1.1** The set of subformulas of a formula is inductively defined by the number of logical symbols occurring in the formula.

1. The subformula of an atomic formula is the formula itself.
2. The subformulas of $\neg A$ are the subformulas of $A$ and $\neg A$ itself.
3. The subformulas of $A \circ B$ for $\circ \in \{\wedge, \vee, \rightarrow\}$ are the subformulas of $A$ and of $B$ and the formula $A \circ B$ itself.
4. The subformulas of $\forall x A(x)$ or $\exists x A(x)$ are the subformulas of any formula of the form $A(t)$, where $t$ is an arbitrary term, and the formulas $\forall x A(x)$ (or $\exists x A(x)$) themselves.

It is easy to see that for any inference rule in **LK** except for the cut rule, any formula that occurs in the upper sequent is a subformula of some formula occurring in the lower sequent. Thus, a proof without cuts contains only subformulas of formulas occurring in the end-sequent. This property is referred to as the subformula property. In other words, if a formula is provable in **LK**, it is provable by the use of its subformulas only, and hence, without detours.

When we consider mathematical proofs, cut-elimination corresponds to the removal of intermediate statements, i.e. lemmas, from proofs. These lemmas might contain mathematical concepts that do not occur in the theorem that is proved, thus removing these lemmas also removes these mathematical concepts and allows the computation of an elementary proof from an abstract one. An important application of the cut-elimination theorem was the analysis of the topological Fürstenberg-Weiss proof of van der Waerden's theorem by Girard in [22], where the original combinatorial proof was obtained.

The most important direct consequence of the cut-elimination theorem for proof mining is the midsequent theorem.

**Theorem 3.1.2** *Let $S$ be an* **LK***-provable sequent that consists of prenex formulas only. Then there is a cut-free proof of $S$ which contains a sequent (the midsequent) $S'$, which satisfies the following:*

1. *$S'$ is quantifier-free.*
2. *Every inference above $S'$ is either structural or propositional.*
3. *Every inference below $S'$ is either structural or a quantifier inference.*

*The midsequent splits the proof into an upper, propositional part, and a lower part, that contains the quantifier inferences.*

What makes this theorem so important for proof mining is its correspondence to Herbrand's theorem, which is a fundamental result in proof theory that connects first-order logic to propositional logic. In its simplest form it states that a formula $\exists x A(x)$ is provable only if there exist terms $t_1, \ldots, t_n$ such that the Herbrand disjunction $A(t_1) \vee \ldots \vee A(t_n)$ is provable. What separates first-order from propositional logic is the information about the terms $t_1, \ldots, t_n$. Given these terms, first-order and propositional provability coincide. A Herbrand sequent is a propositionally valid sequent computed from the end-sequent by instantiations, where quantified formulas from the end-sequent are replaced by their Herbrand instances. Herbrand sequents can be extracted directly from proofs of prenex end-sequents with at most quantifier-free cuts [25] ("at most quantifier-free cuts" means that the proof contains only cuts without quantifiers or no cuts at all). The method is based on the collection of instances of quantified formulas directly from the proof.

We will proceed with the formal definition of a Herbrand sequent, which is based on some preliminary concepts.

**Definition 3.1.2** (*thread*) Let $\varphi$ be an **LK**-proof. A thread in $\varphi$ is a path of sequent occurrences in the proof tree beginning at the end-sequent and ending in an initial sequent.

As a thread is a sequence of sequent occurrences, to every thread we can consider a corresponding sequence of formula occurrences in the sequents. Such sequences of occurrences starting in an occurrence of the end-sequent is called a trace.

**Definition 3.1.3** (*trace*) Let $\varphi$ be an **LK**-proof and

$$\theta \colon \nu_0, \ldots, \nu_\alpha$$

a thread in $\varphi$. Let $\mu_0$ be a formula occurrence in $Seq(\nu_0)$. For every $i \in \{0, \ldots, \alpha - 1\}$ let $\mu_{i+1}$ be an occurrence in $Seq(\nu_{i+1})$, which is an ancestor occurrence of $\mu_i$ in $Seq(\nu_i)$. Then the sequence

$$(\nu_0, \mu_0), \ldots, (\nu_\alpha, \mu_\alpha)$$

is called a trace of $\mu_0$ in $\theta$. A sequence $\mu_0, \ldots, \mu_\alpha$ is called a trace of $\mu_0$ in $\varphi$ if there exists a thread $\theta$ in $\varphi$ such that $(\nu_0, \mu_0), \ldots, (\nu_\alpha, \mu_\alpha)$ is a trace of $\mu_0$ in $\theta$.

Note that if $\mu_i$ in the definition above is the principal formula of a weakening, the occurrence $\mu_{i+1}$ does not exist. Then, we cannot define a trace of $\mu_0$ corresponding to a thread $\theta$. Thus, traces, like threads, start in the end-sequent and end in an initial sequent. As ancestor occurrences are not unique, a thread may define several traces.

Every trace defines a sequence of formulas, the yield.

**Definition 3.1.4** (*yield*) Let $\tau \colon \mu_0, \ldots, \mu_\alpha$ be a trace in an **LK**-proof $\varphi$. Then the sequence of formulas

$$formula(\mu_0), \ldots, formula(\mu_\alpha)$$

is called the yield of $\tau$.

**Example 3.1.1** Consider the **LK**-proof $\varphi =$

$$\dfrac{\dfrac{(\varphi_1)}{\nu_2 : B, B \vdash \mathbf{C}, C, Q(a)} \quad \dfrac{\dfrac{\dfrac{\dfrac{Q(a) \vdash Q(a)}{P(a), Q(a) \vdash Q(a)}\, w:l}{Q(a) \vdash P(a) \to Q(a)}\, \to:r}{Q(a) \vdash \exists y(P(a) \to Q(y))}\, \exists:r}{Q(a), B \vdash C}\, w:l}{\dfrac{\nu_1 : B, B, B \vdash \mathbf{C}, C, C}{\nu_0 : \forall x(P(x) \to Q(x)) \vdash \mathbf{\exists y(P(a) \to Q(y))}}\, c^*}\, cut$$

where $\varphi_1 =$

$$\dfrac{\dfrac{\dfrac{\dfrac{\dfrac{\nu_7 : \mathbf{P(a)} \vdash P(a)}{\nu_6 : \mathbf{P(a)} \vdash P(a), Q(a)}\, w:r}{\nu_5 : \vdash \mathbf{P(a) \to Q(a)}, P(a)}\, \to:r}{\nu_4 : \vdash \mathbf{\exists y(P(a) \to Q(y))}, P(a)}\, \exists:r}{\nu_3 : B \vdash \mathbf{C}, P(a)}\, w:l \quad \dfrac{\dfrac{\dfrac{P(a) \vdash P(a) \quad Q(a) \vdash Q(a)}{P(a), P(a) \to Q(a) \vdash Q(a)}\, \to:l}{P(a), \forall x(P(x) \to Q(x)) \vdash Q(a)}\, \forall:l}{P(a), B \vdash C, Q(a)}\, w:r}{\nu_2 : B, B \vdash \mathbf{C}, C, Q(a)}\, cut$$

and $B = \forall x(P(x) \to Q(x))$ and $C = \exists y(P(a) \to Q(y))$.

Then $\vartheta = \nu_0, \nu_1, \nu_2, \nu_3, \nu_4, \nu_5, \nu_6, \nu_7$ is a thread in $\varphi$. Let $\mu_0$ be the formula occurrence of $\exists y(P(a) \to Q(y))$ in $\mathrm{Seq}(\nu_0)$. Then the sequence

$$\tau = (\nu_0, \mu_0), (\nu_1, \mu_0), (\nu_2, \mu_0), (\nu_3, \mu_0), (\nu_4, \mu_0), (\nu_5, \mu_1), (\nu_6, \mu_2), (\nu_7, \mu_2)$$

is a trace of $\mu_0$ in $\vartheta$, where $\mu_1$ is the occurrence of $P(a) \to Q(a)$ in $\mathrm{Seq}(\nu_5)$ and $\mu_2$ the occurrence of $P(a)$ in $\mathrm{Seq}(\nu_6)$ and $\mathrm{Seq}(\nu_7)$. This trace is indicated in bold letters in the derivation $\varphi$. The sequence $\exists y(P(a) \to Q(y))$, $\exists y(P(a) \to Q(y))$, $\exists y(P(a) \to Q(y))$, $\exists y(P(a) \to Q(y))$, $\exists y(P(a) \to Q(y))$, $P(a) \to Q(a)$, $P(a)$, $P(a)$ is the yield of $\tau$.

**Proposition 3.1.1** *Let $\varphi$ be an* **LK***-proof with at most quantifier-free cuts of a skolemized prenex end-sequent and $\mu_0$ an occurrence of a formula*

$$Qx_1 \ldots Qx_\beta \cdot A(x_1, \ldots, x_\beta)$$

*in the end-sequent, where $Q \in \{\exists, \forall\}$ and $A$ quantifier-free. Let $\tau$ be a trace of $\mu_0$ in $\varphi$. Then there is a substitution $\sigma$ such that $A\sigma$ occurs in the yield of $\tau$.*

***Proof*** By induction on the number of quantifiers $Qx_1 \ldots Qx_\beta$ in the formula $Qx_1 \ldots Qx_\beta \cdot A(x_1, \ldots, x_\beta)$.

For the base case assume there is one quantifier, and the formula is of the form $QxA$. We follow the trace $\tau$ of occurrences $\mu_0$ of $QxA$ in the proof-tree. By the subformula property, and as axioms are atomic, the quantified variable $x$ in $QxA$ has to be replaced by a term $t$ in the proof-tree and thus, $\{x \leftarrow t\}$ is the desired substitution $\sigma$.

Now assume as an induction hypothesis that for $n$ quantifiers in the formula $Qx_1 \dots Qx_n \cdot A(x_1, \dots, x_n)$ there exists a substitution $\sigma$ such that $A\sigma$ occurs in the yield of $\tau$.

Consider the case of $n+1$ quantifiers in $Qx_1 \dots Qx_{n+1} \cdot A(x_1, \dots, x_{n+1})$. We follow the trace $\tau$ of the occurrence of $\mu_0$ of $Qx_1 \dots Qx_{n+1} \cdot A(x_1, \dots, x_{n+1})$ in the proof-tree. By the subformula property, and as axioms are atomic, the first quantified variable that will be eliminated is $Qx_1$. Therefore, $x_1$ will be replaced by a term $t_1$ and a substitution $\sigma_1 = \{x_1 \leftarrow t_1\}$ is obtained. The formula $Qx_1 \dots Qx_{n+1} \cdot A(x_1, \dots, x_{n+1})$ changes to a formula $A'$: $Qx_2 \dots Qx_{n+1} \cdot A\sigma_1$. In $A'$ there are only $n$ quantifiers left and we apply the induction hypothesis, i.e. there is a substitution $\sigma$ such that the formula $A\sigma_1\sigma$ occurs in the yield of $\tau$. $\sigma_1\sigma$ is the desired substitution.

Now we are ready to define a Herbrand substitution.

**Definition 3.1.5** (*Herbrand substitution*) Let $\varphi$ be an **LK**-proof with at most quantifier-free cuts of a skolemized prenex end-sequent and $\mu_0$ an occurrence of a formula

$$Qx_1 \dots Qx_\beta \cdot A(x_1, \dots, x_\beta)$$

in the end-sequent, where $Q \in \{\exists, \forall\}$ and $A$ quantifier-free. Let $\tau$ be a trace of $\mu_0$ in $\varphi$ such that $A(t_1, \dots, t_\beta)$ occurs in the yield of $\tau$. Then

$$S(\varphi, \tau) = \{x_1 \leftarrow t_1, \dots, x_\beta \leftarrow t_\beta\}$$

is called a Herbrand substitution for $Qx_1 \dots Qx_\beta \cdot A(x_1, \dots, x_\beta)$ in $\varphi$.

The formula $A(t_1, \dots, t_\beta)$ and the substitution $\{x_1 \leftarrow t_1, \dots, x_\beta \leftarrow t_\beta\}$ are uniquely defined by the trace $\tau$.

**Definition 3.1.6** Let $\varphi$ be an **LK**-proof with at most quantifier-free cuts of a skolemized prenex end-sequent and $\mu_0$ an occurrence of a formula

$$Qx_1 \dots Qx_\beta \cdot A(x_1, \dots, x_\beta)$$

in the end-sequent, where $Q \in \{\exists, \forall\}$ and $A$ quantifier-free. Then $T(\varphi, \mu_0)$ is the set of all traces of $\mu_0$ in $\varphi$.

**Definition 3.1.7** (*Herbrand instances*) Let $\varphi$ be an **LK**-proof with at most quantifier-free cuts of a skolemized prenex end-sequent and $\mu_0$ an occurrence of a formula

$$Qx_1 \dots Qx_\beta \cdot A(x_1, \dots, x_\beta)$$

in the end-sequent, where $Q \in \{\exists, \forall\}$ and $A$ quantifier-free. Let $\tau$: $\mu_0, \dots, \mu_\alpha$ be a trace in $T(\varphi, \mu_0)$. Then the set of all Herbrand instances for $\mu_0$ is defined as

$$S(\varphi, \mu_0) = \bigcup \{S(\varphi, \tau) \mid \tau \in T(\varphi, \mu_0)\}.$$

The sequence of formulas obtained by substituting all Herbrand instances for the quantified variables in a formula is called the Herbrand expansion.

**Definition 3.1.8** (*Herbrand expansion*) Let $\varphi$ be an **LK**-proof with at most quantifier-free cuts of a skolemized prenex end-sequent and $\mu_0$ an occurrence of a formula

$$Qx_1 \dots Qx_\beta \cdot A(x_1, \dots, x_\beta)$$

in the end-sequent, where $Q \in \{\exists, \forall\}$ and $A$ quantifier-free. Let

$$S(\varphi, \mu_0) = \{\{x_1 \leftarrow t_1^1, \dots, x_\beta \leftarrow t_\beta^1\}, \dots, \{x_1 \leftarrow t_1^\alpha, \dots, x_\beta \leftarrow t_\beta^\alpha\}\}$$

be the set of Herbrand instances for $\mu_0$. Then the Herbrand expansion for $\mu_0$ is

$$H(\varphi, \mu_0)\colon\ A(t_1^1, \dots, t_\beta^1), \dots, A(t_1^\alpha, \dots, t_\beta^\alpha).$$

**Example 3.1.2** Consider the **LK**-proof $\varphi$ from Example 3.1.1. Let $\mu_0$ be the occurrence of the formula

$$\forall x(P(x) \to Q(x))$$

in the end-sequent. The set of all Herbrand instances for $\mu_0$ is

$$S(\varphi, \mu_0) = \{x \leftarrow a\}.$$

The Herbrand expansion for $\mu_0$ is $H(\varphi, \mu_0) = P(a) \to Q(a)$.

Let $\mu_0'$ be the occurrence of the formula

$$\exists y(P(a) \to Q(y))$$

in the end-sequent. The set of all Herbrand instances for $\mu_0'$ is

$$S(\varphi, \mu_0') = \{y \leftarrow a\}.$$

The Herbrand expansion for $\mu_0'$ is $H(\varphi, \mu_0') = P(a) \to Q(a)$.

Herbrand sequents can be obtained constructively from **LK**-proofs with quantifier-free cut-formulas. They are constructed by replacing quantified formulas occurring in an end-sequent by their Herbrand expansions.

**Definition 3.1.9** (*Herbrand sequents*) Let $\varphi$ be an **LK**-proof with at most quantifier-free cuts of a skolemized prenex end-sequent

$$\Gamma, A_0, \dots, A_n \vdash B_0, \dots, B_m, \Delta$$

such that $\Gamma, \Delta$ are multisets of quantifier-free formulas and

$$A_0, \dots, A_n, B_0, \dots, B_m$$

are formulas containing quantifiers. Let $\mu_0, \ldots, \mu_n$ be the occurrences of $A_0, \ldots, A_n$ and $\nu_0, \ldots, \nu_m$ the occurrences of $B_0, \ldots, B_m$ in the end-sequent. Then

$$H(\varphi)\colon \Gamma, H(\varphi, \mu_0), \ldots, H(\varphi, \mu_n) \vdash H(\varphi, \nu_0), \ldots, H(\varphi, \nu_m), \Delta$$

is called the Herbrand sequent of $\varphi$.

**Theorem 3.1.3** *Let $\varphi$ be an* **LK***-proof with at most quantifier-free cuts of a skolemized prenex end-sequent, then*

1. *the formulas of $H(\varphi)$ are substitution instances of the formulas of the end-sequent without their quantifiers.*
2. *$H(\varphi)$ is valid.*

***Proof*** 1. follows by the construction of $H(\varphi)$, as quantified formulas are replaced by their substitution instances. 2. follows as every Herbrand expansion (and thus every substitution instance) of a quantified formula in the end-sequent occurs in the corresponding yield. Therefore, every substitution instance is provable, and thus, the Herbrand sequent is provable.

**Example 3.1.3** Let $\varphi$ be the **LK**-proof in Example 3.1.1. Its Herbrand sequent is

$$P(a) \to Q(a) \vdash P(a) \to Q(a).$$

## 3.2 The Method CERES

In contrast to the reductive cut-elimination methods à la Gentzen or Tait, where cuts are eliminated step by step following certain rewrite rules, the method CERES was developed with a focus on efficiency, by eliminating all cuts simultaneously. Indeed, in [7] it was shown that CERES outperforms reductive methods of cut-elimination in computational complexity. There are infinite sequences of proofs where the computing time of CERES is nonelementarily faster than that of the reductive methods, on the other hand a nonelementary speed-up of CERES via reductive methods is shown impossible.

The CERES method to be presented here was first introduced in [31], and differs from the original CERES method developed in [6], which we will refer to as clausal CERES, by constructing a characteristic formula rather than a characteristic clause set, and only one proof projection instead of one proof projection per clause in the characteristic clause set. However, the clausal CERES method can be considered as a special case of the CERES method. Towards the end of Sect. 3.2.1 we will show that the clausal CERES method from [6] can be simulated with the CERES method developed in [31].

CERES can be described by considering a proof $\varphi$ of a skolemized end-sequent $S$ that contains cuts, and splitting $\varphi$ in two parts. The first part of $\varphi$, let's call it

$\varphi_1$, contains all inferences that go into a cut, i.e. all formulas in sequents that are ancestors of a cut-formula, and all the inferences on these formulas. By construction, as all formulas in $\varphi_1$ go into a cut, they derive the empty sequent. Thus, the axioms of $\varphi_1$ derive a contradiction. This fact is a crucial part of CERES, and indeed, the formulas in the axioms of $\varphi_1$ are used to construct the characteristic formula. More precisely, all atomic ancestors of the cuts which occur in the axioms of the proof are considered, and depending on how these atoms are related via binary inferences in the proof the characteristic formula is formed. The definition of the characteristic formula is based on the cut-status of the formula occurrences in a proof, i.e. whether a given formula occurrence is a cut-ancestor or not. In the definition below we will use a set $\Omega$ describing the set of all cut-ancestors in a given proof $\varphi$.

What is important is that the characteristic formula is always an unsatisfiable formula, and thus can be refuted by resolution using any theorem prover. The refutation of the characteristic formula is then combined with the proof projection, which can be obtained from the second part of $\varphi$.

The second part of $\varphi$, let's call it $\varphi_2$, basically is all that is left in $\varphi$ when deleting $\varphi_1$. By construction, $\varphi_2$ is a cut-free proof. There are some additional proof transformations performed on $\varphi_2$, which will be presented in the following definitions. What is obtained from $\varphi_2$ is the proof projection, a proof of a sequent composed of the original end-sequent $S$ and the characteristic formula. This derivation can then be combined with the refutation of the characteristic formula in such a way, that the resulting proof is a proof of the same end-sequent $S$, and moreover, it contains quantifier-free cuts only, thus allows the extraction of Herbrand sequents.

From now on we will only consider proofs of closed end-sequents, i.e. sequents of formulas that do not contain free variables. The following definition is from [31].

**Definition 3.2.1** (*characteristic formula and proof projection*) Let $\varphi$ be a skolemized **LK**-proof and $\Omega$ the set containing all the occurrences of cut formulas in $\varphi$. We define a formula $F_\nu$ and a proof $\varphi^*(\nu)$ inductively for all nodes $\nu$ in $\varphi$. With $S(\nu, \Omega)$ we will denote the subsequent of $S(\nu)$ consisting of ancestors of $\Omega$ and with $\bar{S}(\nu, \Omega)$ the subsequent of $S(\nu)$ consisting of ancestors of of the end-sequent of $\varphi$.

- Let $\nu$ be a leaf node in $\varphi$. Then $Seq(\nu) = A \vdash A$ for some atom $A$. We distinguish the cases:

  1. $S(\nu, \Omega) = \vdash$.
     Then $F_\nu = \bot$ and $\varphi^*(\nu) = \nu\colon A \vdash A$.
  2. $S(\nu, \Omega) = \vdash A$.
     Then $F_\nu = A$ and $\varphi^*(\nu) = \nu\colon A \vdash A$.
  3. $S(\nu, \Omega) = A \vdash$.
     Then $F_\nu = \neg A$ and $\varphi^*(\nu) =$

$$\frac{A \vdash A}{\vdash A, \neg A}\ \neg : r$$

  4. $S(\nu, \Omega) = A \vdash A$.
     Then $F_\nu = A \vee \neg A$ and $\varphi^*(\nu) =$

$$\dfrac{\dfrac{A \vdash A}{\vdash A, \neg A}\ \neg: r}{\vdash A \vee \neg A}\ \vee: r$$

- Now let us assume that $\xi$ is a unary inference with premise $\nu'$ and conclusion $\nu$ in $\varphi$. Inductively we assume that $F_{\nu'}$ and $\varphi^*(\nu')$ have been defined and $\varphi^*(\nu')$ is a proof of $\Gamma \vdash \Delta, F_{\nu'}$ where $\Gamma \vdash \Delta = \bar{S}(\nu', \Omega)$.
  We define $F_\nu = F_{\nu'}$. For defining $\varphi^*(\nu)$ we distinguish two cases:

  1. the principal formula of $\xi$ is an ancestor of $\Omega$ then $\varphi^*(\nu) = \varphi^*(\nu')$.
  2. The principal formula of $\xi$ is an ancestor of the end-sequent. Then $\varphi^*(\nu) =$

$$\dfrac{\begin{matrix}(\varphi^*(\nu'))\\ \Gamma \vdash \Delta, F_{\nu'}\end{matrix}}{\Gamma' \vdash \Delta', F_{\nu'}}\ \xi$$

- Assume that $\xi$ is a binary inference with premises $\nu_1, \nu_2$ and conclusion $\nu$. Assume further that $F_{\nu_1}, F_{\nu_2}$ are defined and $\varphi^*(\nu_i)$ are derivations of $\Gamma_i \vdash \Delta_i, F_{\nu_i}$ (for $i = 1, 2$) such that $\Gamma_i \vdash \Delta_i = \bar{S}(\nu_i, \Omega)$. We distinguish two cases:

  1. the auxiliary formulas in $\xi$ are ancestors of $\Omega$. Then we define $\varphi^*(\nu) =$

$$\dfrac{\begin{matrix}(\varphi^*(\nu_i))\\ \Gamma_1 \vdash \Delta_1, F_{\nu_1}\end{matrix} \quad \begin{matrix}(\varphi^*(\nu_2))\\ \Gamma_2 \vdash \Delta_2, F_{\nu_2}\end{matrix}}{\Gamma_1, \Gamma_2 \vdash \Delta_1, \Delta_2, F_{\nu_1} \wedge F_{\nu_2}}\ \wedge: r$$

  and $F_\nu = F_{\nu_1} \wedge F_{\nu_2}$.

  2. the auxiliary formulas in $\xi$ are ancestors of the end-sequent. Then we define $\varphi^*(\nu) =$

$$\dfrac{\dfrac{\begin{matrix}(\varphi^*(\nu_i))\\ \Gamma_1 \vdash \Delta_1, F_{\nu_1}\end{matrix} \quad \begin{matrix}(\varphi^*(\nu_2))\\ \Gamma_2 \vdash \Delta_2, F_{\nu_2}\end{matrix}}{\Gamma_1', \Gamma_2' \vdash \Delta_1', \Delta_2', F_{\nu_1}, F_{\nu_2}}\ \xi}{\Gamma_1', \Gamma_2' \vdash \Delta_1', \Delta_2', F_{\nu_1} \vee F_{\nu_2}}\ \vee: r$$

  and $F_\nu = F_{\nu_1} \vee F_{\nu_2}$.

Finally, the characteristic formula of $\varphi$ is defined as $F = F_{\nu_0}$ and the proof projection as $\varphi^* = \varphi^*(\nu_0)$ for the root node (end-sequent) $\nu_0$ of $\varphi$.

**Example 3.2.1** Let $\varphi$ be the following **LK**-proof:

$$\dfrac{\begin{matrix}(\varphi_1)\\ \forall x(Px \to Pfx) \vdash \forall \mathbf{x}(\mathbf{Px} \to \mathbf{Pf^2x})\end{matrix} \qquad \begin{matrix}(\varphi_2)\\ Pa, \forall \mathbf{x}(\mathbf{Px} \to \mathbf{Pf^2x}) \vdash \exists z Pf^4 z\end{matrix}}{Pa, \forall x(Px \to Pfx) \vdash \exists z Pf^4 z}\ cut$$

where $\varphi_1 =$

$$
\dfrac{\dfrac{\dfrac{\dfrac{\dfrac{\mathbf{P}\alpha \vdash P\alpha \qquad \dfrac{Pf\alpha \vdash Pf\alpha \qquad Pf^2\alpha \vdash \mathbf{Pf^2}\alpha}{Pf\alpha, Pf\alpha \to Pf^2\alpha \vdash \mathbf{Pf^2}\alpha}\;{\to_l}}{\mathbf{P}\alpha, P\alpha \to Pf\alpha, Pf\alpha \to Pf^2\alpha \vdash \mathbf{Pf^2}\alpha}\;{\to_l}}{\mathbf{P}\alpha, \forall x(Px \to Pfx), \forall x(Px \to Pfx) \vdash \mathbf{Pf^2}\alpha}\;{\forall_l, \forall_l}}{\forall x(Px \to Pfx), \forall x(Px \to Pfx) \vdash \mathbf{P}\alpha \to \mathbf{Pf^2}\alpha}\;{\to_r}}{\forall x(Px \to Pfx) \vdash \mathbf{P}\alpha \to \mathbf{Pf^2}\alpha}\;{c_l}}{\forall x(Px \to Pfx) \vdash \forall \mathbf{x}(\mathbf{Px} \to \mathbf{Pf^2x})}\;{\forall_r}
$$

and $\varphi_2 =$

$$
\dfrac{\dfrac{\dfrac{\dfrac{Pa \vdash \mathbf{Pa} \qquad \dfrac{\mathbf{Pf^2a} \vdash \mathbf{Pf^2a} \qquad \mathbf{Pf^4a} \vdash Pf^4a}{\mathbf{Pf^2a}, \mathbf{Pf^2a} \to \mathbf{Pf^4a} \vdash Pf^4a}\;{\to_l}}{Pa, \mathbf{Pa} \to \mathbf{Pf^2a}, \mathbf{Pf^2a} \to \mathbf{Pf^4a} \vdash Pf^4a}\;{\to_l}}{Pa, \forall \mathbf{x}(\mathbf{Px} \to \mathbf{Pf^2x}), \forall \mathbf{x}(\mathbf{Px} \to \mathbf{Pf^2x}) \vdash Pf^4a}\;{\forall_l, \forall_l}}{Pa, \forall \mathbf{x}(\mathbf{Px} \to \mathbf{Pf^2x}) \vdash Pf^4a}\;{c_l}}{Pa, \forall \mathbf{x}(\mathbf{Px} \to \mathbf{Pf^2x}) \vdash \exists z Pf^4z}\;{\exists_r}
$$

Following Definition 3.2.1 we obtain the proof projection $\pi^* =$

$$
\dfrac{(\pi_1^*) \qquad (\pi_2^*)}{Pa, \forall x(Px \to Pfx) \vdash \exists z Pf^4z, F}\;{\wedge : r}
$$

where $F = Pa \wedge (Pf^2a \vee \neg Pf^2a) \wedge \neg Pf^4a \wedge (\neg P\alpha \vee Pf^2\alpha)$ and $\pi_1^* =$

$$
\dfrac{\dfrac{\dfrac{\dfrac{\dfrac{P\alpha \vdash P\alpha}{\vdash P\alpha, \neg P\alpha}\;{\neg : r} \qquad \dfrac{Pf\alpha \vdash Pf\alpha \qquad Pf^2\alpha \vdash Pf^2\alpha}{Pf\alpha, Pf\alpha \to Pf^2\alpha \vdash Pf^2\alpha}\;{\to : l}}{P\alpha \to Pf\alpha, Pf\alpha \to Pf^2\alpha \vdash \neg P\alpha, Pf^2\alpha}\;{\to : l}}{P\alpha \to Pf\alpha, Pf\alpha \to Pf^2\alpha \vdash \neg P\alpha \vee Pf^2\alpha}\;{\vee : r}}{\forall x(Px \to Pfx), \forall x(Px \to Pfx) \vdash \neg P\alpha \vee Pf^2\alpha}\;{\forall_l, \forall_l}}{\forall x(Px \to Pfx) \vdash \neg P\alpha \vee Pf^2\alpha}\;{c_l}
$$

and $\pi_2^* =$

$$
\dfrac{\dfrac{Pa \vdash Pa \qquad \dfrac{\dfrac{\dfrac{Pf^2a \vdash Pf^2a}{\vdash Pf^2a, \neg Pf^2a}\;{\neg : r}}{\vdash Pf^2a \vee \neg Pf^2a}\;{\vee : r} \qquad \dfrac{Pf^4a \vdash Pf^4a}{\vdash Pf^4a, \neg Pf^4a}\;{\neg : r}}{\vdash Pf^4a, (Pf^2a \vee \neg Pf^2a) \wedge \neg Pf^4a}\;{\wedge : r}}{Pa \vdash Pf^4a, Pa \wedge (Pf^2a \vee \neg Pf^2a) \wedge \neg Pf^4a}\;{\wedge : r}}{Pa \vdash \exists z Pf^4z, Pa \wedge (Pf^2a \vee \neg Pf^2a) \wedge \neg Pf^4a}\;{\exists_r}
$$

and the characteristic formula is

$$F = Pa \wedge (Pf^2a \vee \neg Pf^2a) \wedge \neg Pf^4a \wedge (\neg P\alpha \vee Pf^2\alpha).$$

Let $\mu_0$ be the occurrence of the formula $\forall x(Px \to Pfx)$ and $\mu_1$ the occurrence of the formula $\exists z Pf^4 z$ in the end-sequent of $\pi^*$. Then

$$S(\pi^*, \mu_0) = \{\{x \leftarrow \alpha\}, \{x \leftarrow f\alpha\}\}, \ S(\pi^*, \mu_1) = \{\{z \leftarrow a\}\}.$$

The Herbrand expansion of the formula $\forall x(Px \to Pfx)$ is $H(\pi^*, \mu_0) = P\alpha \to Pf\alpha, Pf\alpha \to Pf^2\alpha$ and the Herbrand expansion of the formula $\exists z Pf^4 z$ is $H(\pi^*, \mu_1) = Pf^4a$. The Herbrand sequent of $\pi^*$ is

$$H(\pi^*) = Pa, P\alpha \to Pf\alpha, Pf\alpha \to Pf^2\alpha$$

$$\vdash Pf^4a, Pa \wedge (Pf^2a \vee \neg Pf^2a) \wedge \neg Pf^4a \wedge (\neg P\alpha \vee Pf^2\alpha).$$

The universal closure of a characteristic formula is always unsatisfiable. As already mentioned, a refutation of the characteristic formula serves as a skeleton for a proof in normal form, from which Herbrand sequents can be extracted. However, in general the construction of a refutation of the characteristic formula is the most problematic step of the CERES method. In Sect. 3.2.1 we will show how the resolution calculus $\text{RPL}_0$ can be used for CERES. To move on with the method CERES we only need to know that the universal closure of the characteristic formula is unsatisfiable.

**Proposition 3.2.1** ([31]) Let $\varphi$ be a skolemized **LK**-proof and $F$ its characteristic formula. Then $\forall F$ is unsatisfiable.

***Proof*** In [6] a set of clauses $\text{CL}(\varphi)$ (the characteristic clause set) is defined and shown unsatisfiable. It is easy to check that $\text{CL}(\varphi)$ can be obtained from $F$ (the conjuncts of $F$ define the clauses in $\text{CL}(\varphi)$, see Sect. 3.2.1), hence $\forall F$ is unsatisfiable as well.

As $\forall F$ is unsatisfiable there exists a cut-free proof (or a proof with quantifier-free cuts) $\chi$ of $\forall F \vdash$. By Herbrand's theorem there exist ground substitutions $\theta_1, \ldots, \theta_\alpha$ on the variables in $F$ s.t. $F\sigma_1, \ldots, F\sigma_\alpha \vdash$ is a propositionally valid sequent. Therefore, there exists an **LK**-proof $\psi$ deriving $F\sigma_1, \ldots, F\sigma_\alpha \vdash$. This proof can be used to construct a CERES normal form.

**Definition 3.2.2** (CERES *normal form*, [31]) Let $\varphi$ be a skolemized proof of a sequent $S\colon \Gamma \vdash \Delta$, $F$ its characteristic formula and $\pi^*$ the proof projection. As $\forall F$ is unsatisfiable there exists a cut-free proof (or a proof with quantifier-free cuts) $\chi$ of $\forall F \vdash$. By Herbrand's theorem there exist ground substitutions $\theta_1, \ldots, \theta_\alpha$ on the variables in $F$ s.t. $S_\alpha\colon F\theta_1, \ldots, F\theta_\alpha \vdash$ is propositionally valid. Therefore, there exists an **LK**-proof $\psi$ deriving $S_\alpha$. We abbreviate $F\theta_\beta$ by $F_\beta$ and $\pi^*\theta_\beta$ by $\pi^*_\beta$. The CERES normal form is

$$
\dfrac{\begin{matrix}(\pi_n^*)\\ \Gamma \vdash \Delta, F_\alpha\end{matrix} \qquad \dfrac{\dfrac{\begin{matrix}(\pi_2^*)\\ \Gamma \vdash \Delta, F_2\end{matrix} \qquad \dfrac{\begin{matrix}(\pi_1^*)\\ \Gamma \vdash \Delta, F_1\end{matrix} \qquad \begin{matrix}(\psi)\\ F_1, \ldots, F_\alpha \vdash\end{matrix}}{\Gamma, F_2, \ldots, F_\alpha \vdash \Delta}\ cut}{\Gamma, F_3, \ldots, F_\alpha \vdash \Delta}\ cut + c^*}{\vdots}}{\Gamma \vdash \Delta}\ cut + c^*
$$

The CERES normal form is not a cut-free proof, but a proof containing at most quantifier-free cuts, which suffices for the extraction of Herbrand sequents.

### 3.2.1 *Refuting the Characteristic Formula*

As the characteristic formula is always in NNF, we can use the calculus $\mathrm{RPL}_0$ to refute it. In order to define a CERES normal form, the $\mathrm{RPL}_0$-refutation of a characteristic formula $F$ has to be transformed into an **LK**-derivation of $F\theta_1, \ldots, F\theta_n \vdash$, for Herbrand substitutions $\theta_1, \ldots, \theta_n$ of $F$. To this aim we work with grounded $\mathrm{RPL}_0$-derivations, by first applying a total m.g.u., and then grounding the remaining variables. Note that by doing so, the resolution inferences in the resulting derivations do not define substitutions. The obtained derivation is clearly not in $\mathrm{RPL}_0$, but the resolution rules can now be replaced by cuts. We call the obtained calculus $\mathrm{RPL}_0^c$.

**Definition 3.2.3** ($\mathrm{RPL}_0^c$-*derivations*) Let $\rho$ be an $\mathrm{RPL}_0$-derivation, $\rho_r$ its regularization with total m.g.u. $\sigma$, and $X = V(\rho_r)$ the variables in $\rho_r$. Let $X = \{x_1, \ldots, x_k\}$, $c$ a constant symbol and $\theta = \{x_1 \leftarrow c, \ldots, x_k \leftarrow c\}$. An $\mathrm{RPL}_0^c$-derivation $\rho^*$ is obtained from $\rho_r$ by first grounding, i.e. applaying $\sigma\theta$ to $\rho_r$, and then replacing all resolution inferences by cuts.

**Example 3.2.2** Let $\rho_r$ be the regular $\mathrm{RPL}_0$-refutation in Example 2.2.2, then $\rho^* =$

$$
\dfrac{\dfrac{\begin{matrix}(\rho_{r_1}^*)\\ Pf^2a \vdash\end{matrix} \qquad \dfrac{\dfrac{\dfrac{\dfrac{\dfrac{\dfrac{\vdash F\{\alpha \leftarrow \beta\}\{\beta \leftarrow a\}}{\vdash H\{\alpha \leftarrow \beta\}\{\beta \leftarrow a\}}\ \wedge_{r_2}}{\vdash G\{\alpha \leftarrow \beta\}\{\beta \leftarrow a\}}\ \wedge_{r_2}}{\vdash \neg Pa \vee Pf^2a}\ \wedge_{r_2}}{\vdash \neg Pa, Pf^2a}\ \vee_r}{Pa \vdash Pf^2a}\ \neg_r}{Pa \vdash}\ cut \qquad \dfrac{\vdash F\{\alpha \leftarrow a\}}{\vdash Pa}\ \wedge_{r_1}}{\vdash}\ cut
$$

where $\rho_{r_1}^* =$

$$
\dfrac{
\dfrac{\dfrac{\dfrac{\dfrac{\vdash F\{\alpha \leftarrow \delta\}\{\delta \leftarrow a\}}{\vdash H\{\alpha \leftarrow \delta\}\{\delta \leftarrow a\}}\,\wedge_{r_2}}{\vdash G\{\alpha \leftarrow \delta\}\{\delta \leftarrow a\}}\,\wedge_{r_2}}{\vdash \neg Pf^4a}\,\wedge_{r_1}}{Pf^4a \vdash}\,\neg_r
\qquad
\dfrac{\dfrac{\dfrac{\dfrac{\dfrac{\vdash F\{\alpha \leftarrow \gamma\}\{\gamma \leftarrow f^2a\}}{\vdash H\{\alpha \leftarrow \gamma\}\{\gamma \leftarrow f^2a\}}\,\wedge_{r_2}}{\vdash G\{\alpha \leftarrow \gamma\}\{\gamma \leftarrow f^2a\}}\,\wedge_{r_2}}{\vdash \neg P\gamma \vee Pf^4a}\,\wedge_{r_2}}{\vdash \neg Pf^2a, Pf^4a}\,\vee_r}{Pf^2a \vdash Pf^4a}\,\neg_r
}{Pf^2a \vdash}\,cut
$$

By construction, $\rho^*$ is a $\mathrm{RPL}_0^c$-refutation with axioms of the form $\vdash F\sigma_i$, where $\sigma_i$ is a substitution obtained by grounding. Appending $F\sigma_i$ to the antecedent of these axioms, results in **LK**-axioms $F\sigma_i \vdash F\sigma_i$. Propagating these formulas in the antecedents of the axioms to the end-sequent results in a derivation of $F\sigma_1, \ldots F\sigma_n \vdash$. An **LK**-proof of $F\sigma_1, \ldots F\sigma_n \vdash$ is then obtained by additional cuts.

**Proposition 3.2.2** *Let $\rho$ be a* $\mathrm{RPL}_0$*-refutation of a characteristic formula $F$. Then there exists an* **LK***-proof $\rho'$ of $F\sigma_1, \ldots, F\sigma_n \vdash$, where $\sigma_1, \ldots, \sigma_n$ are the substitutions of the axiom $\vdash F$ in the corresponding* $\mathrm{RPL}_0^c$ *refutation $\rho^*$.*

***Proof*** Start with the $\mathrm{RPL}_0^c$ refutation $\rho^*$. Append $F\sigma_i$ to the antecedent of the axioms in $\rho^*$, resulting in **LK** axioms $F\sigma_i \vdash F\sigma_i$. Propagate the antecedents of the axioms to the end-sequent, resulting in a derivation $\rho'^*$ of $F\sigma_1, \ldots, F\sigma_n \vdash$ from **LK** axioms but $\mathrm{RPL}_0$ inferences. To obtain an **LK** derivation, $\mathrm{RPL}_0$ inferences are inductively transformed into **LK** derivations starting from the axioms in $\rho'^*$. Case distinction on the top-most $\mathrm{RPL}_0$ inference $x$:

- $x$ is $\wedge\colon r_1$. Then we replace $\wedge\colon r_1$ with

$$
\dfrac{\begin{matrix}\vdots\\ \Gamma \vdash \Delta, A \wedge B\end{matrix}}{\Gamma \vdash \Delta, A}\;\wedge\colon r_1^*
$$

which denotes the **LK** derivation

$$
\dfrac{\begin{matrix}\vdots\\ \Gamma \vdash \Delta, A \wedge B\end{matrix} \quad \dfrac{A \vdash A}{A \wedge B \vdash A}\,\wedge : l_1}{\Gamma \vdash \Delta, A}\;cut
$$

- $x$ is $\wedge\colon r_2$. Then we replace $\wedge\colon r_2$ with

$$
\dfrac{\begin{matrix}\vdots\\ \Gamma \vdash \Delta, A \wedge B\end{matrix}}{\Gamma \vdash \Delta, B}\;\wedge\colon r_2^*
$$

which denotes the **LK** derivation

$$\dfrac{\dfrac{\vdots}{\Gamma \vdash \Delta, A \wedge B} \quad \dfrac{B \vdash B}{A \wedge B \vdash B}\wedge : l_2}{\Gamma \vdash \Delta, B}\ cut$$

- $x$ is $\wedge : l$. Then we replace $\wedge : l$ with

$$\dfrac{\dfrac{\vdots}{A \wedge B, \Gamma \vdash \Delta}}{A, B, \Gamma \vdash \Delta}\wedge : l^*$$

which denotes the **LK** derivation

$$\dfrac{\dfrac{A \vdash A \quad B \vdash B}{A, B \vdash A \wedge B}\wedge : r \quad \dfrac{\vdots}{A \wedge B, \Gamma \vdash \Delta}}{B, \Gamma \vdash \Delta}\ cut$$

- $x$ is $\vee : r$. Then we replace $\vee : r$ with

$$\dfrac{\dfrac{\vdots}{\Gamma \vdash \Delta, A \vee B}}{\Gamma \vdash \Delta, A, B}\vee : r^*$$

which denotes the **LK** derivation

$$\dfrac{\dfrac{\vdots}{\Gamma \vdash \Delta, A \vee B} \quad \dfrac{A \vdash A \quad B \vdash B}{A \vee B \vdash A, B}\vee : l}{\Gamma \vdash \Delta, A, B}\ cut$$

- $x$ is $\vee : l_1$. Then we replace $\vee : l_1$ with

$$\dfrac{\dfrac{\vdots}{A \vee B, \Gamma \vdash \Delta}}{A, \Gamma \vdash \Delta}\vee : l_1^*$$

which denotes the **LK** derivation

$$\dfrac{\dfrac{A \vdash A}{A \vdash A \vee B}\vee : r_1 \quad \dfrac{\vdots}{A \vee B, \Gamma \vdash \Delta}}{A, \Gamma \vdash \Delta}\ cut$$

- $x$ is $\vee : l_2$. Then we replace $\vee : l_2$ with

$$\frac{\begin{array}{c}\vdots\\ A \lor B, \Gamma \vdash \Delta\end{array}}{B, \Gamma \vdash \Delta}\ \lor\colon l_2^*$$

which denotes the **LK** derivation

$$\frac{\dfrac{B \vdash B}{B \vdash A \lor B}\ \lor : r_2 \qquad \begin{array}{c}\vdots\\ A \lor B, \Gamma \vdash \Delta\end{array}}{B, \Gamma \vdash \Delta}\ cut$$

- $x$ is $\neg\colon r$. Then we replace $\neg\colon r$ with

$$\frac{\begin{array}{c}\vdots\\ \Gamma \vdash \Delta, \neg A\end{array}}{A, \Gamma \vdash \Delta}\ \neg\colon r^*$$

which denotes the **LK** derivation

$$\frac{\begin{array}{c}\vdots\\ \Gamma \vdash \Delta, \neg A\end{array} \qquad \dfrac{A \vdash A}{\neg A, A \vdash}\ \neg : l}{A, \Gamma \vdash \Delta}\ cut$$

- $x$ is $\neg\colon l$. Then we replace $\neg\colon l$ with

$$\frac{\begin{array}{c}\vdots\\ \neg A, \Gamma \vdash \Delta\end{array}}{\Gamma \vdash \Delta, A}\ \neg\colon l^*$$

which denotes the **LK** derivation

$$\frac{\dfrac{A \vdash A}{\vdash A, \neg A}\ \neg : r \qquad \begin{array}{c}\vdots\\ \neg A, \Gamma \vdash \Delta\end{array}}{\Gamma \vdash \Delta, A}\ cut$$

After all $\mathrm{RPL}_0^c$ inferences $x$ have been replaced by their corresponding **LK**-derivations $x^*$, an **LK**-proof $\rho'$ of $F\sigma_1, \ldots, F\sigma_n \vdash$ is obtained. Double occurrences in the end-sequent are contracted.

**Example 3.2.3** Consider the grounded refutation $\rho^*$ from Example 2.2.1. Following the construction in the proof of Proposition 3.2.2 we obtain an **LK**-proof $\rho' =$

$$\dfrac{\dfrac{\begin{array}{c}(\rho_1')\\ F\{\alpha \leftarrow a\}, F\{\alpha \leftarrow a\}, F\{\alpha \leftarrow f^2a\}, Pa \vdash\end{array} \qquad \dfrac{F\{\alpha \leftarrow a\} \vdash F\{\alpha \leftarrow a\}}{F\{\alpha \leftarrow a\} \vdash Pa}\,\wedge^*_{r_1}}{F\{\alpha \leftarrow a\}, F\{\alpha \leftarrow a\}, F\{\alpha \leftarrow a\}, F\{\alpha \leftarrow f^2a\} \vdash}\,cut}{F\{\alpha \leftarrow a\}, F\{\alpha \leftarrow f^2a\} \vdash}\,c_l$$

where $\rho_1' =$

$$\dfrac{\begin{array}{c}(\rho_2')\\ F\{\alpha \leftarrow a\}, F\{\alpha \leftarrow f^2a\}, Pf^2a \vdash\end{array} \qquad \dfrac{\dfrac{\dfrac{\dfrac{\dfrac{F\{\alpha \leftarrow a\} \vdash F\{\alpha \leftarrow a\}}{F\{\alpha \leftarrow a\} \vdash H\{\alpha \leftarrow a\}}\,\wedge^*_{r_2}}{F\{\alpha \leftarrow a\} \vdash G\{\alpha \leftarrow a\}}\,\wedge^*_{r_2}}{F\{\alpha \leftarrow a\} \vdash \neg Pa \vee Pf^2a}\,\wedge^*_{r_2}}{F\{\alpha \leftarrow a\} \vdash \neg Pa, Pf^2a}\,\vee^*_r}{F\{\alpha \leftarrow a\}, Pa \vdash Pf^2a}\,\neg^*_r}{F\{\alpha \leftarrow a\}, F\{\alpha \leftarrow a\}, F\{\alpha \leftarrow f^2a\}, Pa \vdash}\,cut$$

and $\rho_2' =$

$$\dfrac{\dfrac{\dfrac{\dfrac{\dfrac{F\{\alpha \leftarrow a\} \vdash F\{\alpha \leftarrow a\}}{F\{\alpha \leftarrow a\} \vdash H\{\alpha \leftarrow a\}}\,\wedge^*_{r_2}}{F\{\alpha \leftarrow a\} \vdash G\{\alpha \leftarrow a\}}\,\wedge^*_{r_2}}{F\{\alpha \leftarrow a\} \vdash \neg Pf^4a}\,\wedge^*_{r_1}}{F\{\alpha \leftarrow a\}, Pf^4a \vdash}\,\neg^*_r \qquad \dfrac{\dfrac{\dfrac{\dfrac{\dfrac{F\{\alpha \leftarrow f^2a\} \vdash F\{\alpha \leftarrow f^2a\}}{F\{\alpha \leftarrow f^2a\} \vdash H\{\alpha \leftarrow f^2a\}}\,\wedge^*_{r_2}}{F\{\alpha \leftarrow f^2a\} \vdash G\{\alpha \leftarrow f^2a\}}\,\wedge^*_{r_2}}{F\{\alpha \leftarrow f^2a\} \vdash \neg Pf^2a \vee Pf^4a}\,\wedge^*_{r_2}}{F\{\alpha \leftarrow f^2a\} \vdash \neg Pf^2a, Pf^4a}\,\vee^*_r}{F\{\alpha \leftarrow f^2a\}, Pf^2a \vdash Pf^4a}\,\neg^*_r}{F\{\alpha \leftarrow a\}, F\{\alpha \leftarrow f^2a\}, Pf^2a \vdash}\,cut$$

Now we can also give an example of a CERES normal form.

**Example 3.2.4** Let $\varphi$ be the **LK**-proof of Example 3.2.1. Then the CERES normal form of $\varphi$ is $\pi =$

$$\dfrac{\begin{array}{c}(\pi^*\sigma_2)\\ Pa, A \vdash \exists z Pf^4z, F\sigma_2\end{array} \qquad \begin{array}{c}(\pi_1^*)\\ Pa, A, F\sigma_2 \vdash \exists z Pf^4z\end{array}}{Pa, \forall x(Px \to Pfx) \vdash \exists z Pf^4z}\,cut, c^*$$

where $A = \forall x(Px \to Pfx)$ and $\pi_1^* =$

$$\dfrac{\begin{array}{c}(\pi^*\sigma_1)\\ Pa, \forall x(Px \to Pfx) \vdash \exists z Pf^4z, F\sigma_1\end{array} \qquad \begin{array}{c}(\delta^*)\\ F\sigma_1, F\sigma_2 \vdash\end{array}}{Pa, \forall x(Px \to Pfx), F\sigma_2 \vdash \exists z Pf^4z}\,cut$$

and where $\sigma_1 = \{\alpha \leftarrow a\}$, $\sigma_2 = \{\alpha \leftarrow f^2a\}$, $\delta^*$ is the **LK**-proof in Example 3.2.3, $\pi^*$ is the proof projection in Example 3.2.1 and $F$ is the characteristic formula $Pa \wedge (Pf^2a \vee \neg Pf^2a) \wedge \neg Pf^4a \wedge (\neg P\alpha \vee Pf^2\alpha)$.

We can also extract a Herbrand sequent of $\pi$. Let $\mu_0$ be the occurrence of the formula $\forall x(Px \to Pfx)$ and $\mu_1$ the occurrence of the formula $\exists z Pf^4z$ in the end-sequent of $\pi$. Then $S(\pi, \mu_0) = S(\pi^*\sigma_1, \mu_0) \cup S(\pi^*\sigma_2, \mu_0) = S(\pi^*, \mu_0)\sigma_1 \cup S(\pi^*, \mu_0)\sigma_2$. From Example 3.2.1 we know that $S(\pi^*, \mu_0) = \{\{x \leftarrow \alpha\}, \{x \leftarrow f\alpha\}\}$. Therefore,

$$\begin{aligned} S(\pi, \mu_0) &= \{\{x \leftarrow \alpha\}, \{x \leftarrow f\alpha\}\}\sigma_1 \cup \{\{x \leftarrow \alpha\}, \{x \leftarrow f\alpha\}\}\sigma_2 \\ &= \{\{x \leftarrow a\}, \{x \leftarrow fa\}\} \cup \{\{x \leftarrow f^2a\}, \{x \leftarrow f^3a\}\}. \end{aligned}$$

$S(\pi, \mu_1) = S(\pi^*\sigma_1, \mu_1) \cup S(\pi^*\sigma_2, \mu_1) = S(\pi^*, \mu_1)\sigma_1 \cup S(\pi^*, \mu_1)\sigma_2$. From Example 3.2.1 we know that $S(\pi^*, \mu_1) = \{\{z \leftarrow a\}\}$. Therefore,

$$S(\pi, \mu_1) = \{\{z \leftarrow a\}\}\sigma_1 \cup \{\{z \leftarrow a\}\}\sigma_2 = \{\{z \leftarrow a\}\}.$$

The Herbrand expansion of the formula $\forall x(Px \to Pfx)$ is

$$H(\pi, \mu_0) = Pa \to Pfa, Pfa \to Pf^2a, Pf^2a \to Pf^3a, Pf^3a \to Pf^4a$$

and the Herbrand expansion of the formula $\exists z Pf^4z$ is $H(\pi, \mu_1) = Pf^4a$.

The Herbrand sequent of $\pi$ is $H(\pi) =$

$$Pa, Pa \to Pfa, Pfa \to Pf^2a, Pf^2a \to Pf^3a, Pf^3a \to Pf^4a \vdash Pf^4a.$$

Having defined $\text{RPL}_0$ we are now ready to show that the clausal CERES method can be simulated by the CERES method used in this book. The crucial part is the construction of the characteristic clause set, as the refutation of the characteristic clause set can be performed with any theorem prover, and the proof projections are constructed for the clauses in the characteristic clause set.

The clause set extraction in clausal CERES consists in collecting all atomic ancestors of the cuts which occur in the axioms of the proof. The clauses are formed depending on how these atoms are related via binary inferences in the proof. The formal definition is given below.

**Definition 3.2.4** (*characteristic clause-set* [6]) Let $\varphi$ be a proof of a skolemized sequent. The characteristic clause set is built recursively from the leaves of the proof to the end-sequent. Let $\nu$ be the occurrence of a sequent in this proof, then

- If $\nu$ is an axiom, then $\text{CL}(\nu)$ contains the sub-sequent of $\nu$ composed only of cut-ancestors.
- If $\nu$ is the result of the application of a unary rule on a sequent. $\mu$, then $\text{CL}(\nu) = \text{CL}(\mu)$
- If $\nu$ is the result of the application of a binary rule on sequents $\mu_1$ and $\mu_2$, then we distinguish two cases:

– If the rule is applied to ancestors of the cut formula, then

$$\mathrm{CL}(\nu) = \mathrm{CL}(\mu_1) \cup \mathrm{CL}(\mu_2).$$

– If the rule is applied to ancestors of the end-sequent, then

$$\mathrm{CL}(\nu) = \mathrm{CL}(\mu_1) \times \mathrm{CL}(\mu_2).$$

where

$$\mathrm{CL}(\mu_1) \times \mathrm{CL}(\mu_2) = \{C \circ D \mid C \in \mathrm{CL}(\mu_1), D \in \mathrm{CL}(\mu_2)\}.$$

If $\nu_0$ is the root node $\mathrm{CL}(\nu_0)$ is called the characteristic clause set of $\varphi$.

Below we show that the clauses of characteristic clause set can be derived from the characteristic clause set and thus the refutation of the characteristic clause set can be simulated in $\mathrm{RPL}_0$.

**Definition 3.2.5** A derivation $\gamma$ in $\mathrm{RPL}_0$ is called *resolution-free* if only the elimination rules but no resolution rule appears in $\gamma$.

Note that, as all elimination rules in $\mathrm{RPL}_0$ are unary, resolution free $\mathrm{RPL}_0$-derivations have only one leaf node of the form $\vdash F$ where $F$ is the characterisic formula. We give a proof for **LK**-proofs with tautological axioms of the form $A \vdash A$ which can easily be generalized to arbitrary atomic axioms.

**Lemma 3.2.1** *Let $\varphi$ be an* **LK**-*proof, $F$ the characteristic formula of $\varphi$ and* $\mathrm{CL}(\varphi)$ *the characteristic clause set of $\varphi$. Then every clause in* $\mathrm{CL}(\varphi)$ *is derivable from $\vdash F$ by a resolution-free* $\mathrm{RPL}_0$*-derivation.*

***Proof*** By the construction of characteristic clause sets and characteristic formulas, to every node in the proof $\varphi$ corresponds a characteristic clause set and a characteristic formula. We prove the lemma by induction on the proof nodes in $\varphi$ where $\mathcal{C}(\nu)$ is the characteristic clause set at node $\nu$ and $F(\nu)$ the characteristic formula at node $\nu$. (IB) Let $\nu$ be an axiom node $A \vdash A$ in $\varphi$. Then

1. If $A$ to the left is ancestor of a cut then $\mathcal{C}(\nu) = \{A \vdash\}$ and $F(\nu) = \neg A$.
2. If $A$ to the right is ancestor of a cut then $\mathcal{C}(\nu) = \{\vdash A\}$ and $F(\nu) = A$.
3. If both formula occurrences are ancestors of a cut then $\mathcal{C}(\nu) = \{A \vdash A\}$ and $F(\nu) = \neg A \vee A$.

It is easy to see that in all three cases the (single) clause in $\mathcal{C}(\nu)$ is derivable from $F(\nu)$ by a resolution-free $\mathrm{RPL}_0$-derivation; we only show the case 3:

$$\cfrac{\cfrac{\vdash \neg A \vee A}{\vdash \neg A, A}\ \vee : r}{A \vdash A}\ \neg : r$$

(IH) Now let $\nu$ be a conclusion of an inference $\xi$ in $\varphi$ and assume that for the premises of $\xi$ the lemma holds.
We show that the lemma holds for $\nu$. If $\xi$ is a unary rule then the lemma holds for $\nu$ as unray rules neither change the characteristic clause set nor the characteristic formula. So let us assume that $\xi$ is a binary rule of the form

$$\frac{\nu_1 \colon S_1 \quad \nu_2 \colon S_2}{\nu \colon S}\ \xi$$

We distinghish two cases:

a. The auxiliary formulas of $\xi$ are ancestors of a cut in $\varphi$, and
b. the auxiliary formulas of $\xi$ are ancestors of the end-sequent of $\varphi$.

In case a we have

- $\mathcal{C}(\nu) = \mathcal{C}(\nu_1) \cup \mathcal{C}(\nu_2)$ and $F(\nu) = F(\nu_1) \land F(\nu_2)$.

Now let $C \in \mathcal{C}(\nu)$ then $C \in \mathcal{C}(\nu_1)$ or $C \in \mathcal{C}(\nu_2)$. Let us assume $C \in \mathcal{C}(\nu_1)$; then by (IH) there exists a resolution-free $\mathrm{RPL}_0$-derivation $\gamma$ of $C$ from $\vdash F(\nu_1)$. The corresponding $\mathrm{RPL}_0$-derivation from $F(\nu)$ is then

$$\cfrac{\cfrac{\vdash F(\nu_1) \land F(\nu_2)}{\vdash F(\nu_1)}\ \land\colon r}{C}\ \gamma$$

The case $C \in \mathcal{C}(\nu_2)$ is analogous.

In case b we have

- $\mathcal{C}(\nu) = \mathcal{C}(\nu_1) \times \mathcal{C}(\nu_2)$ and $F(\nu) = F(\nu_1) \lor F(\nu_2)$.

Let $C \in \mathcal{C}(\nu)$ then $C = \Gamma_1, \Gamma_2 \vdash \Delta_1, \Delta_2$ where $\Gamma_1 \vdash \Delta_1 \in \mathcal{C}(\nu_1)$ and $\Gamma_2 \vdash \Delta_2 \in \mathcal{C}(\nu_2)$. By (IH) there exists a resolution-free $\mathrm{RPL}_0$-derivation $\gamma_1$ of $\Gamma_1 \vdash \Delta_1$ from $F(\nu_1)$ and a resolution-free $\mathrm{RPL}_0$-derivation $\gamma_2$ of $\Gamma_2 \vdash \Delta_2$ from $F(\nu_2)$. Then the corresponding $\mathrm{RPL}_0$-derivation of $C$ from $F(\nu)$ is of the form

$$\cfrac{\cfrac{\cfrac{\vdash F(\nu_1) \lor F(\nu_2)}{\vdash F(\nu_1), F(\nu_2)}\ \lor\colon r}{\Gamma_1 \vdash \Delta_1, F(\nu_2)}\ \gamma_1(+)}{\Gamma_1, \Gamma_2 \vdash \Delta_1, \Delta_2}\ \gamma_2(*)$$

where $\gamma_1(+)$ is the extension of $\gamma_1$ by the context $F(\nu_2)$ and $\gamma_2(*)$ the extension of $\gamma_2$ by the context $\Gamma_1 \vdash \Delta_1$.

The lemma is proven when we select $\nu_0$, the root node of $\varphi$.

**Example 3.2.5** Let $F = Pa \land (Pf^2a \lor \neg Pf^2a) \land \neg Pf^4a \land (\neg P\alpha \lor Pf^2\alpha)$ be the characteristic formula in Example 3.2.1. We obtain the clauses of $\mathrm{CL}(\varphi)$ by:

$$\frac{\vdash F}{\vdash Pa} \wedge_{r_1}$$

$$\cfrac{\cfrac{\cfrac{\cfrac{\vdash F}{\vdash (Pf^2a \vee \neg Pf^2a) \wedge \neg Pf^4a \wedge (\neg P\alpha \vee Pf^2\alpha)} \wedge_{r_2}}{\vdash (Pf^2a \vee \neg Pf^2a)} \wedge_{r_1}}{\vdash Pf^2a, \neg Pf^2a} \vee_r}{Pf^2a \vdash Pf^2a} \neg_r$$

$$\cfrac{\cfrac{\cfrac{\cfrac{\vdash F}{\vdash (Pf^2a \vee \neg Pf^2a) \wedge \neg Pf^4a \wedge (\neg P\alpha \vee Pf^2\alpha)} \wedge_{r_2}}{\vdash \neg Pf^4a \wedge (\neg P\alpha \vee Pf^2\alpha)} \wedge_{r_2}}{\vdash \neg Pf^4a} \wedge_{r_1}}{Pf^4a \vdash} \neg_r$$

$$\cfrac{\cfrac{\cfrac{\cfrac{\cfrac{\vdash F}{\vdash (Pf^2a \vee \neg Pf^2a) \wedge \neg Pf^4a \wedge (\neg P\alpha \vee Pf^2\alpha)} \wedge_{r_2}}{\vdash \neg Pf^4a \wedge (\neg P\alpha \vee Pf^2\alpha)} \wedge_{r_2}}{\vdash \neg P\alpha \vee Pf^2\alpha} \wedge_{r_2}}{\vdash \neg P\alpha, Pf^2\alpha} \vee_r}{P\alpha \vdash Pf^2\alpha} \neg_r$$

**Proposition 3.2.3** *Let $F$ be the characteristic formula of an* **LK**-*proof $\varphi$, then there exists an* $\mathrm{RPL}_0$-*refutation of* $\vdash F$.

***Proof*** By [6] a characteristic clause set $\mathcal{C}$ of a proof $\varphi$ is always refutable by resolution. Let $\gamma$ be a resolution refutation tree with leaves in $\mathcal{C}$. From $\gamma$ we can obtain a $\mathrm{RPL}_0$-refutation $\delta$ of $\vdash F$ by applying Lemma 3.2.1 and appending the corresponding $\mathrm{RPL}_0$-derivation to every leaf in $\gamma$.

### 3.2.2 *Herbrand Sequent Extraction Without the Construction of a* CERES *Normal Form*

Usually, Herbrand sequents are extracted from a CERES normal form after the application of CERES. However, in [29] it was shown that for the extraction of Herbrand sequents the construction of such a normal form can be omitted. Indeed, the crucial information about the CERES normal form is already contained in the proof projection and the resolution refutation of the characteristic formula. Therefore, instances of quantified formulas occurring in the end-sequent of the CERES normal form can be extracted from the proof projection. Substituting these instances with the substitutions of the characteristic formula $F$ obtained by the total unifier defined by a resolution refutation of $\vdash F$ results in the Herbrand sequent of the CERES normal form. This improvement yields a gain in asymptotic complexity for the clausal CERES method [29].

The modified CERES method especially suited for Herbrand sequent extraction is as we describe below. For each quantified formula occurrence $\mu_i$ in a prenex end-sequent of a skolemized **LK**-proof $\varphi$ a structure $\overline{H}(\varphi, \mu_i, \rho)$ is constructed. First, the projection $\pi^*$ of $\varphi$ is computed in the usual way. It is a crucial observation that for a proof $\varphi$ and the corresponding proof projection $\pi^*$ their end-sequents are the same, except that the end-sequent of $\pi^*$ contains the characteristic formula $F$ in the succedent. This means that the quantified formula occurrence $\mu_i$ in the end-sequent of $\varphi$ also occurs in the end-sequent of $\pi^*$. Therefore, we can extract Herbrand instances of the quantified formula occurrence $\mu_i$ in the end-sequent of $\pi^*$ by collecting instances from $\pi^*$. Then, a refutation $\rho$ of the characteristic formula $F$ of $\varphi$ is constructed, from which Herbrand instances of $F$ are obtained. The Herbrand instances obtained from the projection are then further instantiated with the Herbrand instances of $F$. Merging these substituted Herbrand instances results in $\overline{H}(\varphi, \mu_i, \rho)$. It is shown that $\overline{H}(\varphi, \mu_i, \rho)$ defines the set of Herbrand instances for the quantified formula occurrence $\mu_i$ of the original **LK**-proof $\varphi$.

**Definition 3.2.6** Let $\varphi$ be an **LK**-proof of a skolemized prenex end-sequent with characteristic formula $F$, $\rho$ a $\mathrm{RPL}_0^c$-refutation of $\vdash F$, $\pi^*$ the projection of $\varphi$ and for each quantified formula occurrence $\mu_i$ in the end-sequent of $\pi^*$ let $S(\pi^*, \mu_i)$ be the set of Herbrand instances defined by $\pi^*$. Then we define a set

$$\overline{H}(\varphi, \mu_i, \rho) = \bigcup_{j=1}^{m} S(\pi^*, \mu_i)\sigma_j$$

for $\sigma_1, \ldots, \sigma_m \in \Sigma(\rho)$, where $\Sigma(\rho)$ is the set of F-substitutions defined by $\rho$.

**Example 3.2.6** Let $\pi$ be the CERES normal form in Example 3.2.4 and let $\mu_0$ be the occurrence of the formula $\forall x(Px \rightarrow Pfx)$ and $\mu_1$ the occurrence of the formula $\exists z Pf^4 z$ in the end-sequent of $\pi$. $\Sigma(\rho) = \{\{\alpha \leftarrow a\}, \{\alpha \leftarrow f^2 a\}\}$ as in Example 2.2.1.

$S(\pi^*, \mu_0) = \{\{x \leftarrow \alpha\}, \{x \leftarrow f\alpha\}\}$ and $S(\pi^*, \mu_1) = \{\{z \leftarrow a\}\}$ as in Example 3.2.1.

Then,

$$\begin{aligned}
\overline{H}(\varphi, \mu_0, \rho) &= S(\pi^*, \mu_0)\{\alpha \leftarrow a\} \cup S(\pi^*, \mu_0)\{\alpha \leftarrow f^2 a\} \\
&= \{\{x \leftarrow a\}, \{x \leftarrow fa\}\} \cup \{\{x \leftarrow f^2 a\}, \{x \leftarrow f^3 a\}\}, \\
&= \{\{x \leftarrow a\}, \{x \leftarrow fa\}, \{x \leftarrow f^2 a\}, \{x \leftarrow f^3 a\}\},
\end{aligned}$$

$$\begin{aligned}\overline{H}(\varphi, \mu_1, \rho) &= S(\pi^*, \mu_1)\{\alpha \leftarrow a\} \cup S(\pi^*, \mu_1)\{\alpha \leftarrow f^2 a\} \\ &= \{\{z \leftarrow a\}\} \cup \{\{z \leftarrow a\}\} = \{\{z \leftarrow a\}\}.\end{aligned}$$

The obtained sets $\overline{H}(\varphi, \mu_0, \rho)$ and $\overline{H}(\varphi, \mu_1, \rho)$ in the example above are equal to the sets $S(\pi, \mu_0)$ and $S(\pi, \mu_1)$ obtained in Example 3.2.4 from the CERES normal form. This is not a coincidence but a generally valid observation. It can be shown that for any quantified formula occurrence $\mu_i$ the set $\overline{H}(\varphi, \mu_i, \rho)$ is equal to the set of Herbrand instances of the quantified formula occurrence $\mu_i$ in the end-sequent of the CERES normal form $\pi$.

**Theorem 3.2.1** *Let $\varphi$ be an* **LK**-*proof of a skolemized prenex end-sequent $\Gamma \vdash \Delta$, $F$ its characteristic formula and $\pi^*$ the projection of $\varphi$ to $F$. Let $\rho$ be a $\mathrm{RPL}_0^c$-refutation of $\vdash F$ and $\pi$ the* CERES *normal form of $\varphi$. Then for each quantified formula occurrence $\mu_i$ in $\varphi$, $\overline{H}(\varphi, \mu_i, \rho) = S(\pi, \mu_i)$.*

***Proof*** By induction on the number of substitutions in $\rho$. For the base case assume there is only one substitution $\sigma$. Then the CERES normal form $\pi$ is of the form

$$\dfrac{\dfrac{\pi^*\sigma}{\Gamma \vdash \Delta, F\sigma} \quad \dfrac{\rho^*}{F\sigma \vdash}}{\Gamma \vdash \Delta}\ (cut)$$

$\rho^*$ is the **LK**-proof of the Herbrand instances of $F$, in this case of $F\sigma \vdash$, obtained from $\rho$.

$S(\pi, \mu_i)$ is constructed by collecting instances of the formula at occurrence $\mu_i$. By construction of $\pi$, the quantified formula occurrence $\mu_i$ in the end-sequent of $\pi$ occurs also in the sequent $\Gamma \vdash \Delta, F\sigma$. Therefore,

$$S(\pi, \mu_i) = S(\pi^*\sigma, \mu_i) = S(\pi^*, \mu_i)\sigma = \overline{H}(\varphi, \mu_i, \rho).$$

The formula at occurrence $\mu_i$ is a closed formula and does not change after substitution with one of the substitutions in $\rho$.

Assume as an induction hypothesis that if there are $n$ substitutions

$$\sigma_1, \ldots, \sigma_n$$

in $\rho$, then the assumption holds, i.e. $S(\pi, \mu_i) = \{S(\pi_1^*, \mu_i), \ldots, S(\pi_n^*, \mu_i)\} = \bigcup_{j=1}^n S(\pi^*, \mu_i)\sigma_j = \overline{H}(\varphi, \mu_i, \rho)$, where $\pi_i^* = \pi^*\sigma_i$ and $\pi$ is of the form

$$\dfrac{\dfrac{\pi_n^*}{\Gamma \vdash \Delta, F_n} \quad \dfrac{\dfrac{\dfrac{\pi_2^*}{\Gamma \vdash \Delta, F_2} \quad \dfrac{\dfrac{\pi_1^*}{\Gamma \vdash \Delta, F_1} \quad \dfrac{\rho^*}{F_1, \ldots, F_n \vdash}}{\Gamma, F_2, \ldots, F_n \vdash \Delta}\ (cut)}{\Gamma, F_3, \ldots, F_n \vdash \Delta}\ (cut + c^*)}{\vdots}}{\Gamma \vdash \Delta}\ (cut + c^*)$$

and $F_i = F\sigma_i$.

Now assume that there are $n+1$ substitutions $\sigma_1, \ldots, \sigma_n, \sigma_{n+1}$ in $\rho$. Here we have to consider a proof $\rho^*$ of $F_1, \ldots, F_n, F_{n+1} \vdash$ and the additional substitution instance $\pi^*_{n+1}$ of the projection $\pi^*$:

$$
\dfrac{\dfrac{\pi^*_{n+1}}{\Gamma \vdash \Delta, F_{n+1}} \quad \dfrac{\dfrac{\pi^*_n}{\Gamma \vdash \Delta, F_n} \quad \dfrac{\begin{array}{c}\dfrac{\dfrac{\pi^*_1}{\Gamma \vdash \Delta, F_1} \quad \dfrac{\rho^*}{F_1, \ldots, F_n, F_{n+1} \vdash}}{\Gamma, F_2, \ldots, F_n, F_{n+1} \vdash \Delta}\,(cut) \\ \vdots \end{array}}{\Gamma, F_n, F_{n+1} \vdash \Delta}}{\nu\colon \Gamma, F_{n+1} \vdash \Delta}\,(cut + c^*)}{\Gamma \vdash \Delta}\,(cut + c^*)
$$

At node $\nu$ there are only $n$ instances of the proof projection $\pi^*$, only that $F_{n+1} \vdash$ is propagated through the proof tree starting from $\rho^*$. By the induction hypothesis the Herbrand instances for $\mu_i$ at node $\nu$ are

$$S(\pi_\nu, \mu_i) = \{S(\pi^*_1, \mu_i), \ldots, S(\pi^*_n, \mu_i)\} = \bigcup_{j=1}^{n} S(\pi^*, \mu_i)\sigma_j.$$

The Herbrand instances $S(\pi, \mu_i)$ are by construction

$$
\begin{aligned}
S(\pi, \mu_i) &= S(\pi^*_{n+1}, \mu_i) \cup S(\pi_\nu, \mu_i) \\
&= S(\pi^*_{n+1}, \mu_i) \cup \bigcup_{j=1}^{n} S(\pi^*, \mu_i)\sigma_j \\
&= S(\pi^*, \mu_i)\sigma_{n+1} \cup \bigcup_{j=1}^{n} S(\pi^*, \mu_i)\sigma_j \\
&= \bigcup_{j=1}^{n+1} S(\pi^*, \mu_i)\sigma_j = \overline{H}(\varphi, \mu_i, \rho).
\end{aligned}
$$

# Chapter 4
# Term Schemata and Formula Schemata

A computable function on the natural numbers is typically specified by a finite expression using defined function symbols and operators; a syntactic expression $F$ specifying a function $f: \mathbb{N} \to \mathbb{N}$ can be considered as a finite representation of the infinite set $D(f): \{f(n) \mid n \in \mathbb{N}\}$. We can say that $F$ is a "schema" for the function $f$. In a similar way there are schemata for infinite sets of terms and infinite sets of formulas. In this chapter we define such schemata based on primitive recursion and investigate their properties. These schemata will be the backbone for proof schemata and their application in inductive proof analysis.

## 4.1 Term Schemata

Term schemata are syntactic objects describing infinite sets of terms via parameters, i.e. variables ranging over natural numbers. These syntactic objects are based on inductive definitions. In this section we will present the classes of numeric terms and of schematic individual terms. Given a parameter assignment mapping parameters to numerals, the objects of the first class will evaluate to numerals, these of the second class to first-order terms.

### *4.1.1 Numeric Terms*

We start with simple numeric terms of type $\omega$, the type of the natural numbers. The terms are based on the constant symbol $\mathbf{0}: \omega$, and the function symbols successor $s: \omega \to \omega$ and predecessor $p: \omega \to \omega$. Let $\mathcal{N}$ be a countably infinite set of variables of type $\omega$ which we call *parameters*.



A. Leitsch et al., *First-Order Schemata and Inductive Proof Analysis*, Computer Science Foundations and Applied Logic, https://doi.org/10.1007/978-3-032-05741-9_4

**Definition 4.1.1** The set of basic terms, denoted by $T_0$ is defined inductively below:
• $\mathbf{0} \in T_0$,
• $\mathcal{N} \subseteq T_0$,
• if $t \in T_0$ then $s(t) \in T_0$,
• if $t \in T_0$ then $p(t) \in T_0$.
The set of ground terms in $T_0$ (terms not containing parameters) is denoted by $T_0^G$.

The definition above is unsatisfactory as, besides $s$, we also have the function symbol $p$ and ground the terms $\mathbf{0}$ and $p(s(\mathbf{0}))$ in $T_0^G$ are considered as (semantically) different. But when we define $p$ as predecessor then every ground term is equal to a term of the form $s^\alpha(\mathbf{0})$ for $\alpha \in \mathbb{N}$; we call such terms *numerals*.

**Definition 4.1.2** (*numerals*) The subset of $T_0^G$ in which only the symbols $\mathbf{0}$ and $s$ occur is called the set of numerals and is denoted by *Num*.

We define now an evaluation function $\downarrow: T_0^G \rightarrow Num$ assigning a numeral to every term in $T_0^G$.

**Definition 4.1.3** (*evaluation of* $T_0^G$) Let $t$ be a term in $T_0^G$. We distinguish the following cases:
(a) $t = \mathbf{0}$: $t \downarrow = t$.
(b) $t = s(t')$ for $t' \in T_0^G$: $t \downarrow = s(t' \downarrow)$.
(c) $t = p(\mathbf{0})$: $t \downarrow = \mathbf{0}$.
(d) $t = p(s(t'))$ for a term $t' \in T_0^G$: $t \downarrow = t' \downarrow$.
(e) $t = p(p(t'))$ for a term $t' \in T_0^G$: $t \downarrow = p(p(t') \downarrow) \downarrow$.

It is easy to see that the cases (a)–(e) above cover all terms in $T_0^G$ and we have a complete and unique definition; but we still have to show that $t \downarrow \in Num$ for every $t \in T_0^G$. We first give an example of an evaluation of a ground term.

**Example 4.1.1**

$$p(s(s(p(\mathbf{0})))) \downarrow = s(p(\mathbf{0})) \downarrow = s(p(\mathbf{0}) \downarrow) = s(\mathbf{0}).$$

Numerals evaluate to themselves.

**Lemma 4.1.1** *Let $r \in Num$, then $r \downarrow = r$.*

***Proof*** Straightforward by induction on term depth of $r$ using cases (a) and (b) of Definition 4.1.3.

**Proposition 4.1.1** *Let $t \in T_0^G$, then $t \downarrow \in Num$.*

***Proof*** By induction on the term depth of $t$ (denoted by $\tau(t)$).

- $\tau(t) = 0$: then $t = \mathbf{0}$ and by Definition 4.1.3 case (a) $t \downarrow = \mathbf{0}$ and $\mathbf{0} \in Num$.
- $\tau(t) = 1$: then either $t = s(\mathbf{0})$ or $t = p(\mathbf{0})$.
  If $t = s(\mathbf{0})$ then by (b) $t \downarrow = s(\mathbf{0} \downarrow)$ and by (a) $\mathbf{0} \downarrow = \mathbf{0}$, so $t \downarrow = s(\mathbf{0})$ and $s(\mathbf{0}) \in Num$.
  If $t = p(\mathbf{0})$ then by (c) $t \downarrow = \mathbf{0} \in Num$.

- (IH) assume that for an $n \in \mathbb{N}$ and for all $t$ with $\tau(t) \leq n$ we have $t \downarrow \in Num$.

Now let $t \in T_0^G$ with $\tau(t) = n + 1$. If $n = 0$ then we are done; so we may assume $n > 0$ and $\tau(t) \geq 2$. We distinguish the following cases:

- $t = s(t')$: by Definition 4.1.3 case (b) $t \downarrow = s(t' \downarrow)$. By (IH) $t' \downarrow = r$ for an $r \in Num$. So $t \downarrow = s(r) \in Num$.
- $t = p(s(t'))$: by Definition 4.1.3 case (d) $t \downarrow = t' \downarrow$ and by (IH) $t' \downarrow \in Num$ and so $t \downarrow \in Num$.
- $t = p(p(t'))$: by Definition 4.1.3 case (e) $t \downarrow = p(p(t') \downarrow) \downarrow$. By (IH) $p(t') \downarrow = r$ for an $r \in Num$. Then either $r = \mathbf{0}$ or $r = s(r')$ for an $r' \in Num$.

  - $r = \mathbf{0}$. Then $t \downarrow = p(\mathbf{0}) \downarrow$ and by case (c) $p(\mathbf{0}) \downarrow = \mathbf{0}$ resulting in $t \downarrow = \mathbf{0}$.
  - $r = s(r')$. Then $t \downarrow = p(s(r')) \downarrow$ and by case (d) $p(s(r')) \downarrow = r' \downarrow$. By Lemma 4.1.1 $r' \downarrow = r'$ and so $t \downarrow = r' \downarrow = r' \in Num$.

We are now ready to define the semantics of $T_0$ itself.

**Definition 4.1.4** (*parameter assignment*) A function $\sigma\colon \mathcal{N} \to Num$ is called a *parameter assigment*. The set of all parameter assignments is denoted by $\mathcal{S}$. Any parameter assignment $\sigma$ can be extended to a function $T_0 \to \mathcal{N}$ in the following way:

- $\sigma(\mathbf{0}) = \mathbf{0}$,
- $\sigma(s(t)) = s(\sigma(t))$,
- $\sigma(p(t)) = p(\sigma(t))$.

**Lemma 4.1.2** *Let $\sigma$ be a parameter assignment, then for all terms $t \in T_0$ $\sigma(t) \in T_0^G$.*

***Proof*** Immediate by induction on the depth of the term $t$.

**Proposition 4.1.2** *If $\sigma$ is a parameter assigment and $t \in T_0$ then $\sigma(t) \downarrow \in Num$.*

***Proof*** By Lemma 4.1.2 $\sigma(t) \in T_0^G$. Therefore, by Proposition 4.1.1, $\sigma(t) \downarrow \in Num$.

For better reading we frequently write $n + 1$ for $s(n)$ and $n - 1$ for $p(n)$ (but note that, due to this notation, $0 - 1 = 0$). The expressivity of the class $T_0$ is quite low: when the expressions of $T_0$ are interpreted as functions with respect to parameter assignments not even the function $+$ is expressible. We choose an extension of $T_0$ via *primitive recursive definitions* which extends the expressivity considerably: the functions expressible in this extension are the primitive recursive functions. The extension is based on the introduction of a class of function symbols and corresponding defining equations.

**Definition 4.1.5** For every $n \geq 1$ we define a countably infinite class $\mathbf{F}_n^\omega$ containing function symbols of type $\omega^n \to \omega$ (the numeric function symbols of arity $n$) and $\mathbf{F}^\omega = \bigcup_{n=1}^{\infty} \mathbf{F}_n^\omega$ (the numeric function symbols). For the basic functions $s$ and $p$ we require $s, p \notin \mathbf{F}^\omega$.

Below we extend the set of terms $T_0$ to the set of *numeric terms.*

**Definition 4.1.6** ($\mathbf{T}^\omega$) The set $\mathbf{T}^\omega$ is defined inductively as follows:

- $0 \in \mathbf{T}^\omega$,
- $\mathcal{N} \subseteq \mathbf{T}^\omega$,
- if $t \in \mathbf{T}^\omega$ then $s(t) \in \mathbf{T}^\omega$ and $p(t) \in \mathbf{T}^\omega$,
- if $t_1, \ldots, t_m \in \mathbf{T}^\omega$ and $f \in \mathbf{F}^\omega_m$ then $f(t_1, \ldots, t_m) \in \mathbf{T}^\omega$.

So far the terms show no difference to ordinary first-order terms of type $\omega$. What adds additional structure is an assignment of *defining equations* to numeric function symbols. Instead of a fixed assignment of defining equations to every symbol in $\mathbf{F}^\omega$ we select a finite subset $\mathcal{F}$ of $\mathbf{F}^\omega$ and define a "theory" corresponding to $\mathcal{F}$. Such a theory consists of assigments of equations to every element of $\mathcal{F}$, a selected element of $\mathcal{F}$ and some ordering condition on the elements of $\mathcal{F} \cup \{s, p\}$. For the set of numeric function symbols occurring in a term $t \in \mathbf{T}^\omega$ we write $\mathbf{F}^\omega(t)$.

**Definition 4.1.7** *($\omega$-Theory)* A tupel $(f, \mathcal{F}, D(\mathcal{F}), <)$ is an $\omega$-theory of $f$ if

- $f \in \mathcal{F}$,
- $\mathcal{F}$ is a finite subset of $\hat{F}$,
- $<$ is an irreflexive, transitive relation on $\mathcal{F} \cup \{s, p\}$ such that for all $g \in \mathcal{F} \backslash \{f\}$: $g < f$ ($f$ is maximal in $\mathcal{F}$) and for all $g \in \mathcal{F}$ $s < g$ and $p < g$ ($s$ and $p$ are minimal).
- $D(\mathcal{F}) = \bigcup_{g \in \mathcal{F}} D(g)$ where $D(g)$ is a set containing the following equation (we assume $g \in \mathbf{F}^\omega_{m+1}$): $g(x_1, \ldots, x_m, y) =_d$

$$\underline{\text{if}}\ y = \mathbf{0}\ \underline{\text{then}}\ t^g_B\ \underline{\text{else}}\ t^g_S\{z \leftarrow g(x_1, \ldots, x_m, p(y))\},$$

where $t^g_S, t^g_B \in \mathbf{T}^\omega$, $V\mathcal{N}(t^g_S) \subseteq \{x_1, \ldots, x_m, y, z\}$, $\mathcal{N}(t^g_B) \subseteq \{x_1, \ldots, x_m\}$, $\mathcal{G}$: $\mathbf{F}^\omega(t_S) \cup \mathbf{F}^\omega(t_B) \subseteq \mathcal{F}$, and for all $h \in \mathcal{G} : h < g$.

**Example 4.1.2** Let $f, g \in \mathbf{F}^\omega_2$, $\mathcal{F} = \{f, g\}$, $g < f$, $s < g$, $p < g$. Then $\mathcal{D}$: $(f, \mathcal{F}, D(\{f, g\}), <)$ is an $\omega$-theory for

$$D(f) = \{f(x, y) =_d \underline{\text{if}}\ y = \mathbf{0}\ \underline{\text{then}}\ \mathbf{0}\ \underline{\text{else}}\ g(x, z)\{z \leftarrow f(x, p(y))\}\}.$$

and

$$D(g) = g(x, y) =_d \underline{\text{if}}\ y = \mathbf{0}\ \underline{\text{then}}\ x\ \underline{\text{else}}\ s(z)\{z \leftarrow g(x, p(y))\}\}.$$

Here we have $t^f_S = g(x, z)$, $t^f_B = \mathbf{0}$ and $t^g_S = s(z)$, $t^g_B = x$. We see that all conditions of Definition 4.1.7 are fulfilled. Note that $g$ is minimal in $\mathcal{F}$, so $t^g_S, t^g_B$ do not contain symbols in $\mathcal{F}$-these terms belong to $T_0$.

When we replace $f$ by $*$ and $g$ by $+$ and use the usual infix notation we obtain the definitions

$$D(*) = \{x * y =_d \underline{\text{if}}\ y = \mathbf{0}\ \underline{\text{then}}\ \mathbf{0}\ \underline{\text{else}}\ (x + (x * p(y)))\}.$$

and

$$D(+) = \{x + y =_d \underline{\text{if}}\ y = \mathbf{0}\ \underline{\text{then}}\ x\ \underline{\text{else}}\ s(x + p(y))\}.$$

which corresponds to the recursive definitions of $*$ and $+$ via $p$ and $s$:

$$\begin{aligned} x * y &= x + x * p(y) \text{ for } y > 0, \\ &= 0 \text{ for } y = 0, \\ x + y &= s(x + p(y)) \text{ for } y > 0, \\ &= x \text{ for } y = 0. \end{aligned}$$

The restriction of $\mathbf{T}^\omega$ to terms with function symbols in $\mathcal{F} \cup \{s, p\}$ is denoted by $\mathbf{T}^\omega(\mathcal{F})$. Given an $\omega$-theory $T\colon (f, \mathcal{F}, D(\mathcal{F}), <)$ and $g \in \mathcal{F}$ we write $\mathcal{F}_{<g}$ for the set $\{h \mid h \in \mathcal{F} \cup \{s, p\}, h < g\}$; with $\mathcal{F}_{\leq g}$ we denote the set $\mathcal{F}_{<g} \cup \{g\}$.

The semantics of numeric terms is defined via parameter assignments: Let $T$ be an $\omega$-theory $(f, \mathcal{F}, D(\mathcal{F}), <)$; then for any $t \in \mathbf{T}^\omega(\mathcal{F})$ and any $\sigma \in \mathcal{S}$ we have $\sigma(t) \downarrow \in \mathit{Num}$. The formal definitions are given below. We first extend $\sigma \in \mathcal{S}$ to terms in $\mathbf{T}^\omega(\mathcal{F})$ and to definitions $D(g)$ for $g \in \mathcal{F}$:

**Definition 4.1.8** The extension of $\sigma$ to terms in $\mathbf{T}^\omega(\mathcal{F})$ is the defined as:

- $\sigma(0) = 0$,
- if $t \in \mathbf{T}^\omega(\mathcal{F})$ then $\sigma(s(t)) = s(\sigma(t))$, $\sigma(p(t)) = p(\sigma(t))$,
- if $f(t_1, \ldots, t_m) \in \mathbf{T}^\omega(\mathcal{F})$ then $\sigma(f(t_1, \ldots, t_m)) = f(\sigma(t_1), \ldots, \sigma(t_m))$.

It is easy to see that, for $t \in \mathbf{T}^\omega(\mathcal{F})$, $\sigma(t)$ is a ground term—typically containing function symbols in $\mathcal{F}$. These terms have still to be evaluated to numerals. The set of ground terms in $\mathbf{T}^\omega(\mathcal{F})$ is denoted by $\mathbf{T}^\omega_G(\mathcal{F})$.

**Definition 4.1.9** Let $T\colon (f, \mathcal{F}, D(\mathcal{F}), <)$ be an $\omega$-theory and $t \in \mathbf{T}^\omega_G(\mathcal{F})$.

- If $t \in T_0^G$ then $t \downarrow$ is a numeral as defined in Definition 4.1.3.
- If $t \in \mathbf{T}^\omega_G(\mathcal{F}) \backslash T_0^G$ then $s(t) \downarrow = s(t \downarrow)$.
- If $t \in \mathbf{T}^\omega_G(\mathcal{F}) \backslash T_0^G$ then $p(t) \downarrow = p(t \downarrow) \downarrow$.
- If $g \in \mathcal{F}_{m+1}$ where $D(g) = \{g(x_1, \ldots, x_m, y) =_d$

  $$\underline{\text{if}}\ y = \mathbf{0}\ \underline{\text{then}}\ t^g_B\ \underline{\text{else}}\ t^g_S\{z \leftarrow g(x_1, \ldots, x_m, p(y))\}\}$$

  and $s_1, \ldots, s_m, r \in \mathit{Num}$ then

  $$\begin{aligned} g(s_1, \ldots, s_m, r) \downarrow &= t^g_B\{x_1 \leftarrow s_1, \ldots, x_m \leftarrow s_m\} \downarrow \text{ for } r = \mathbf{0}, \\ &= t^g_S\{x_1 \leftarrow s_1, \ldots, x_m \leftarrow s_m, y \leftarrow p(r) \downarrow, \\ &\quad z \leftarrow g(s_1, \ldots, s_m, p(r) \downarrow) \downarrow\} \downarrow \\ &\text{for } r > \mathbf{0}. \end{aligned}$$

Furthermore let $t_1, \ldots, t_{m+1} \in \mathbf{T}_G^\omega(\mathcal{F})$ then

$$g(t_1, \ldots, t_{m+1}) \downarrow = g(t_1 \downarrow, \ldots, t_{m+1} \downarrow) \downarrow.$$

By Definition 4.1.9 we see that the normalization of a term remains the same when first a subterm is normalized and then the remaining term.

**Lemma 4.1.3** *Let $t \in \mathbf{T}_G^\omega$ and $\Lambda$ be an occurrence in $t$ such that $t = t[s]_\Lambda$. Then $t \downarrow = t[s \downarrow]_\Lambda \downarrow$.*

***Proof*** By induction on the depth of the occurrence $\Lambda$.

**Lemma 4.1.4** *Let $T\colon (f, \mathcal{F}, D(\mathcal{F}), <)$ be an $\omega$-theory and $\mathcal{F}' \subseteq \mathcal{F}$; assume that for all $g \in \mathcal{F}'_m$ (for all $m$) and for all numerals $s_1, \ldots, s_m$ we have $g(s_1, \ldots, s_m) \downarrow \in Num$. Then, for all terms $t \in \mathbf{T}_G^\omega(\mathcal{F}')$, $t \downarrow \in Num$ (where $\mathbf{T}_G^\omega(\mathcal{F}')$ is the subset of $\mathbf{T}^\omega(\mathcal{F})$ containing only function symbols in $\mathcal{F}' \cup \{s, p\}$).*

***Proof*** By induction on the number of occurrences of symbols in $\mathcal{F}'$ in the term $t$ ($occ_{\mathcal{F}'}(t)$).

(IB) If $occ_{\mathcal{F}'}(t) = 0$ then $t \in T_0^G$ and, therefore, $t \downarrow \in Num$.

Now assume that $occ_{\mathcal{F}'}(t) > 0$. Then there exists a subterm $g(t_1, \ldots, t_m)$ such that $g \in \mathcal{F}_m$ and $t_1, \ldots, t_m \in T_0^G$ (note that all function symbols in $\mathcal{F}$ are at least unary, so $m \geq 1$). By definition of the evaluation we have

$$g(t_1, \ldots, t_m) \downarrow = g(t_1 \downarrow, \ldots, t_m \downarrow) \downarrow.$$

As $t_1, \ldots, t_m \in T_0^G$ we have $t_i \downarrow = s_i$ for numerals $s_i$. Hence

$$g(t_1, \ldots, t_m) \downarrow = g(s_1, \ldots, s_m) \downarrow.$$

By assumption $g(s_1, \ldots, s_m) \downarrow = s_0$ for some $s_0 \in Num$. Now assume $t = t[g(t_1, \ldots, t_m)]_\lambda$. By Lemma 4.1.3 we get $t \downarrow = t[s_0]_\lambda \downarrow$. But $occ_{\mathcal{F}'}(t[s_0]_\lambda) < occ_{\mathcal{F}'}(t)$ and, by (IH), we obtain $t[s_0]_\lambda \downarrow \in Num$ and, eventually, $t \downarrow \in Num$.

**Definition 4.1.10** Let $T = (f, \mathcal{F}, D(\mathcal{F}), <)$ be an $\omega$-theory and $g \in \mathcal{F}$. By $\mathcal{F}_{<g}$ we denote the set of numeric function symbols in $\mathcal{F}$ which are smaller than $g$ and by $\mathcal{F}_{\leq g}$ the set $\mathcal{F}_{<g} \cup \{g\}$.

**Lemma 4.1.5** *Let $T = (f, \mathcal{F}, D(\mathcal{F}), <)$ be an $\omega$-theory and $g \in \mathcal{F}$. Then, for any term $t \in \mathbf{T}^\omega(\mathcal{F}_{\leq g})$ and for any $\sigma \in \mathcal{S}$, $\sigma(t) \downarrow \in Num$.*

***Proof*** We proceed by induction on $<$.

(IB) Let $g \in \mathcal{F}_{\leq g}$ and $g$ minimal in $\mathcal{F}$. Then $\mathcal{F}_{<g} = \emptyset$ and $\mathbf{T}^\omega(\mathcal{F}_{<g}) = T_0$, hence for all $t \in T(\mathcal{F}_{<g})$ and for all $\sigma$ $\sigma(t) \downarrow \in Num$ by Proposition 4.1.1.

(IH) assume that for all $t \in \mathbf{T}^\omega(\mathcal{F}_{<g})$ and all $\sigma \in \mathcal{S}$ $\sigma(t) \downarrow \in Num$. We have to show that $\sigma(t) \downarrow \in Num$ for all $\sigma \in \mathcal{S}$ and for all $t \in \mathbf{T}^\omega(\mathcal{F}_{\leq g})$. By Lemma 4.1.4 it is

sufficient to show that (for $m+1$-ary $g$) $g(s_1, \ldots, s_m, s_0) \downarrow \in Num$ for $s_1, \ldots, s_m, r \in Num$. Note that (IH) implies that, for all $h \in \mathcal{F}_{<g}$ and $h$ $k$-ary, $h(s_1, \ldots, s_k) \downarrow \in Num$ for $s_1, \ldots, s_k \in Num$.

Let

$$D(g) = \{g(x_1, \ldots, x_m, y) =_d \underline{\text{if}}\ y = \mathbf{0}\ \underline{\text{then}}\ t^B\ \underline{\text{else}}\ t^S\{y \leftarrow g(x_1, \ldots, x_m, p(y))\}\}.$$

By Lemma 4.1.4 and $T = (g, \{g\}, D(\{g\}), <)$ it is sufficient to show that $g(s_1, \ldots, s_m, s_0) \downarrow \in Num$ (for $s_1, \ldots, s_n, s_0 \in Num$) to obtain $\sigma(t) \downarrow \in Num$ for all $t \in \mathcal{F}_{\leq g}$. We proceed by induction on $s_0$.

(IB2) If $s_0 = \mathbf{0}$ then, by Definition 4.1.9,

$$g(s_1, \ldots, s_m, s_0) \downarrow = t^g_B\{x_1 \leftarrow s_1, \ldots, x_m \leftarrow s_m\} \downarrow.$$

By Definition of $D(g)$ all function symbols in $t^g_B$ are in $\mathcal{F}_{<g}$ and so

$$t' \colon t^g_B\{x_1 \leftarrow s_1, \ldots, x_m \leftarrow s_m\} \in \mathbf{T}^\omega_G(\mathcal{F}_{<g});$$

by (IH) $t' \downarrow \in Num$.

(IH2) Now assume for all $s_0 < i$ and $s_1, \ldots, s_m \in Num$ that

$$g(s_1, \ldots, s_m, s_0) \downarrow \in Num.$$

By Definition 4.1.9 we have

$$g(s_1, \ldots, s_m, i) \downarrow = t^g_S\{x_1 \leftarrow s_1, \ldots, x_1 \leftarrow s_m, y \leftarrow g(s_1, \ldots, s_n, p(i) \downarrow) \downarrow\} \downarrow.$$

By induction hypothesis $g(s_1, \ldots, s_n, p(i) \downarrow) \downarrow \in Num$ and therefore we may consider (for $j = g(s_1, \ldots, s_n, p(i) \downarrow) \downarrow$)

$$t'' \colon t^g_S\{x_1 \leftarrow s_1, \ldots, x_1 \leftarrow s_m, y \leftarrow j\} \downarrow.$$

By Definition of $g$ we have $t'' \in \mathbf{T}^\omega_G(\mathcal{F}_{<g})$ and therefore $t'' \downarrow \in Num$ by (IH). So we have shown that for $g$ and all $t \in \mathbf{T}^\omega(\mathcal{F}_{\leq g})$ $\sigma(t) \downarrow \in Num$ for all $\sigma \in \mathcal{S}$. As a consequence , for all $h \in \mathcal{F}_{\leq g}$ and $h \in \hat{F}_m$, we obtain

$$h(s_1, \ldots, s_m) \downarrow \in Num.$$

By Lemma 4.1.4 $\sigma(t) \downarrow \in Num$ for all $\sigma \in \mathcal{S}$ and for all $t \in \mathbf{T}^\omega(\mathcal{F}_{\leq g})$.

**Corollary 4.1.1** *Let $T = (f, \mathcal{F}, D(\mathcal{F}), <)$ be an $\omega$-theory and $t \in \mathbf{T}^\omega(\mathcal{F})$. Then, for all $\sigma \in \mathcal{S}$, $\sigma(t) \downarrow \in Num$.*

***Proof*** Choose $g = f$ in Lemma 4.1.5.

Finally we analyze the expressivity of theories by the functions they are able to compute.

**Definition 4.1.11** Let $T: (f, \mathcal{F}, D(\mathcal{F}), <)$ be an $\omega$-theory and $f \in \mathbf{F}^{\omega}_{m}$. We define a function $Func(f)$ corresponding to $f$ by

$$Func(f) = \{(k_1, \ldots, k_m, l) \mid f(\mathbf{k}_1, \ldots, \mathbf{k}_m) \downarrow = \mathbf{l}, k_1, \ldots, k_m, l \in \mathbb{N}\}.$$

For the constant $\mathbf{0}$ we define $Func(\mathbf{0}) = 0$.

**Definition 4.1.12** A function $\varphi\colon \mathbb{N}^m \to \mathbb{N}$ is called *computable* in $\mathbf{F}^{\omega}$ if there exists an $\omega$-theory $T: (f, \mathcal{F}, D(\mathcal{F}), <)$ such that $Func(f) = \varphi$. The number $n$ is computable by the corresponding numeral $\bar{n}$.

We will show that all primitive recursive functions are computable in $\mathbf{F}^{\omega}$ which illustrates the strength and expressivity of our theories.

**Definition 4.1.13** The class of primitive recursive functions **PR** is defined as follows:

- $n \in \mathbf{PR}$ for $n \in \mathbb{N}$ (the number $n$ is a nullary function).
- $z \in \mathbf{PR}$ for $z(n) = 0$ for all $n \in \mathbb{N}$.
- For all $m \geq 1$ and $i \in \{1, \ldots, m\}$ $I^m_i \in \mathbf{PR}$ where

$$I^m_i(x_1, \ldots, x_m) = x_i \text{ for all } x_1, \ldots, x_n \in \mathbb{N}.$$

- $succ \in \mathbf{PR}$ for $succ(n) = n + 1$ for all $n \in \mathbb{N}$.

Furthermore **PR** is closed under the operators $C$ (composition) and Pr (primitive recursion) where

$$C(f; g_1, \ldots, g_k)(x_1, \ldots, x_m) = f(g_1(x_1, \ldots, x_m), \ldots, g_k(x_1, \ldots, x_m))$$

for $f\colon \mathbb{N}^k \to \mathbb{N}$ and $g\colon \mathbb{N}^m \to \mathbb{N}$ for $k, m \geq 1$. For $f = \Pr(g, h)$ we have

$$\begin{aligned} f(x_1, \ldots, x_m, y) &= g(x_1, \ldots, x_m) \text{ for } y = 0, \\ &= h(x_1, \ldots, x_m, y, f(x_1, \ldots, x_m, y - 1)) \text{ for } y > 0 \end{aligned}$$

for $g\colon \mathbb{N}^m \to \mathbb{N}$ and $h\colon \mathbb{N}^{m+2} \to \mathbb{N}$.

We introduce the following convention: if in the Definition $D(g)$ the terms $t^g_B$ and $t^g_S$ coincide (that means no real recursion is performed and $t^g_S$ does not contain the recursion variable $z$) then we write

- $\{g(x_1, \ldots, x_n) =_d t^g_B\}$

for $D(g)$ instead of

$$\{g(x_1, \ldots, x_n) =_d \underline{\text{if}}\ x_n = \mathbf{0}\ \underline{\text{then}}\ t^g_B\ \underline{\text{else}}\ t^g_S\{z \leftarrow g(x_1, \ldots, n_{n-1}, p(x_n))\}\}.$$

**Lemma 4.1.6** *The numbers $n \in \mathbb{N}$ and the functions $z$, $I_i^m$, succ are computable in* $\mathbf{F}^\omega$.

***Proof*** • $n$: by definition $n$ is computable by $\mathbf{n}$.
• $z$: we define $T = (f, \{f\}, D(\{f\}), <)$ where $f \in \mathbf{F}_1^\omega$,

$$D(f) = \{f(x) =_d \mathbf{0}\} \text{ and } s < f, p < f.$$

Then $Func(f) = z$.
• $I_i^m$: we define $T = (g, \{g\}, D(\{g\}), <)$ for $g \in \mathbf{F}_m^\omega$,

$$D(g) = \{g(x_1, \ldots, x_m) =_d x_i\} \text{ and } s < g, p < g.$$

Then $Func(g) = I_i^m$.
• *succ*: we choose $T = (h, \{h\}, D(\{h\}), <)$ for $h \in \mathbf{F}_1^\omega$,

$$D(h) = \{h(x) =_d s(x)\} \text{ and } s < h, p < h.$$

**Lemma 4.1.7** *Assume that $\varphi \colon \mathbb{N}^k \to \mathbb{N}$ is computable in $\mathbf{F}^\omega$ and $\psi_1, \ldots, \psi_k \colon \mathbb{N}^m \to \mathbb{N}$ are computable in $\mathbf{F}^\omega$ where $k, m \geq 1$; then*

$$C(\varphi; \psi_1, \ldots, \psi_k)$$

*is computable in* $\mathbf{F}^\omega$.

***Proof*** We assume that there are theories $T_0, T_1, \ldots, T_k$ such that

$$T_0 = (f, \mathcal{F}_0, D(\mathcal{F}_0), <_0),\ T_i = (g_i, \mathcal{F}_i, D(\mathcal{F}_i), <_i) \text{ for } i = 1, \ldots, k$$

and $Func(f) = \varphi$, $Func(g_i) = \psi_i$ for $i = 1, \ldots, k$. Without loss of generality we may assume that the $\mathcal{F}_i$ are pairwise disjoint (we can always rename the symbols in $\mathbf{F}^\omega$ accordingly). So the only function symbols these theories have in common are $s$ and $p$. Now we define

$$\mathcal{F} = \{h\} \cup \bigcup_{i=0}^{k} \mathcal{F}_i \text{ where } h \in \mathbf{F}_m^\omega \text{ and } h \notin \bigcup_{j=0}^{k} \mathcal{F}_i.$$

With $h$ we build the theory $T = (h, \mathcal{F}, D(\mathcal{F}), <)$ where

$$D(h) = \{h(x_1, \ldots, x_k) =_d f(g_1(x_1, \ldots, x_k), \ldots, g_k(x_1, \ldots, x_k))\}.$$

The $D(\mathcal{F}_i)$ remain unchanged for $i = 0, \ldots, k$ and $<$ is defined as the following extension of $\bigcup_{i=0}^k \{<_i\}$: $<$ is defined as the transitive closure of the relation

$$f < h,\ g_i < h \text{ for } i = 1, \ldots, k,\ r < s \text{ iff } r, s \in \mathcal{F}_i \text{ and } r <_i s.$$

Obviously $T$ is a theory and $Func(h) = C(\varphi; \psi_1, \ldots, \psi_k)$.

**Lemma 4.1.8** *Assume that* $\varphi\colon \mathbb{N}^m \to N$ *and* $\psi\colon \mathbb{N}^{m+2} \to \mathbb{N}$ *are computable in* $\mathbf{T}^\omega$ *where* $m \in \mathbb{N}$*; then* $\Pr(\varphi, \psi)$ *is computable in* $\mathbf{T}^\omega$.

***Proof*** As $\varphi$ is computable then either there exists a theory

$$T_1\colon (g, \mathcal{F}_1, D(\mathcal{F}_1), <_1) \text{ and } Func(g) = \varphi \text{ or } \varphi \in \mathbb{N};$$

in the second case $\varphi$ is computed by the corresponding numeral. For $\psi$ there exists a theory $T_2\colon (h, \mathcal{F}_2, D(\mathcal{F}_2), <_2)$ such that $Func(h) = \psi$.

Again we may assume without loss of generality that $\mathcal{F}_1 \cap \mathcal{F}_2 = \emptyset$. We define $\mathcal{F} = \{f\} \cup \mathcal{F}_1 \cup \mathcal{F}_2$ where $f \in \mathbf{F}^\omega_{m+1}$, $f \notin \mathcal{F}_1 \cup \mathcal{F}_2$, and a new theory $T\colon (f, \mathcal{F}, D(\mathcal{F}), <)$ where

$$D(f) = \{f(x_1, \ldots, x_m, y) =_d \underline{\text{if}}\ y = \mathbf{0}\ \underline{\text{then}}\ g(x_1, \ldots, x_m)\ \underline{\text{else}} \\ h(x_1, \ldots, x_m, y, f(x_1, \ldots, x_m, p(y)))\}.$$

If $m = 0$ the term $g(x_1, \ldots, x_m)$ is replaced by the numeral corresponding to the number constant. $<$ is an extension of $<_1 \cup <_2$ by taking the transitive closure of the relation

$$g < f, h < f,\ r < s \text{ if } r, s \in \mathcal{F}_i \text{ and } r <_i s.$$

Obviously $T$ is a theory. It remains to show that $Func(f) = \Pr(\varphi, \psi)$. We assume $m \geq 1$ as, for $m = 0$, the arguments are similar. We denote $\Pr(\varphi, \psi)$ by $\chi$. We have to show that, for all $k_1, \ldots, k_m, n \in \mathbb{N}$,

$$f(\mathbf{k}_1, \ldots, \mathbf{k}_m, \mathbf{n}) \downarrow= \mathbf{r} \text{ iff } \chi(k_1, \ldots, k_m, n) = r.$$

We proceed by induction on $n$:

- $n = 0$: then

$$f(\mathbf{k}_1, \ldots, \mathbf{k}_m, \mathbf{0}) \downarrow= g(\mathbf{k}_1, \ldots, \mathbf{k}_m) \downarrow.$$

  By assumption $Func(g) = \varphi$ and therefore $g(\mathbf{k}_1, \ldots, \mathbf{k}_m) \downarrow= \mathbf{r}$ iff $\varphi(k_1, \ldots, k_m) = r$. Therefore $f(\mathbf{k}_1, \ldots, \mathbf{k}_m, \bar{0}) \downarrow= \mathbf{r}$ iff $\varphi(k_1, \ldots, k_m) = r$ iff $\chi(k_1, \ldots, k_m, 0) = r$.
- (IH) assume that we have

$$f(\mathbf{k}_1, \ldots, \mathbf{k}_m, \bar{n}) \downarrow= \mathbf{r} \text{ iff } \chi(k_1, \ldots, k_m, n) = r$$

  Now

$$\begin{aligned} f(\mathbf{k}_1, \ldots, \mathbf{k}_m, s(\mathbf{n})) \downarrow &= h(\mathbf{k}_1, \ldots, \mathbf{k}_m, \mathbf{n}, f(\mathbf{k}_1, \ldots, \mathbf{k}_m, p(s(\mathbf{n})) \downarrow)) \downarrow \\ &= h(\mathbf{k}_1, \ldots, \mathbf{k}_m, \mathbf{n}, f(\mathbf{k}_1, \ldots, \mathbf{k}_m, \mathbf{n})) \downarrow. \end{aligned}$$

Now let us assume that $f(\mathbf{k}_1, \ldots, \mathbf{k}_m, \mathbf{n})) \downarrow = \mathbf{r}$; then, by (IH) we have that $\chi(k_1, \ldots, k_m, n) = r$. By assumption

$$h(\mathbf{k}_1, \ldots, \mathbf{k}_m, \mathbf{n}, \mathbf{r}) \downarrow = \bar{q} \text{ iff } \psi(k_1, \ldots, k_m, n, r) = q.$$

Putting things together we obtain

$$\begin{aligned} f(\bar{k}_1, \ldots, \bar{k}_m, s(\mathbf{n})) \downarrow = \mathbf{q} \;\; &\text{iff} \;\; \psi(k_1, \ldots, k_m, n, \chi(k_1, \ldots, k_m, n)) = q \\ &\text{iff} \;\; \chi(k_1, \ldots, k_m, n+1)) = q. \end{aligned}$$

**Theorem 4.1.1** *Let $\varphi$ be a primitive recursive function. Then $\varphi$ is computable in $\mathbf{F}^\omega$, i.e. there exists a theory $T : (f, \mathcal{F}, D(\mathcal{F}), <)$ such that*

$$Func(f) = \varphi.$$

***Proof*** Immediate by the Lemmas 4.1.6, 4.1.7 and 4.1.8.

### *4.1.2 Schemata of Individual Terms*

The numeric terms introduced in Sect. 4.1.1 are of type $\omega$ and evaluate to numerals under given parameter assignments, so—in some sense—numeric terms define a schema of numerals which can be interpreted as a function on the natural numbers. In this section we investigate expressions evaluating to first-order terms (to them we assign the type $\iota$) under given parameter assignments, which we call term schemata.

**Example 4.1.3** Let $f$ be a one-place function symbol of type $\iota \to \iota$ and $x$ be an individual (first-order) variable of type $\iota$. Then the terms

$$x, f(x), f(f(x)), \ldots$$

form an infinite sequence which, in an informal mathematical notation, could be denoted by $f^n(x)_{n \in \mathbb{N}}$. In analogy to numeric terms we describe this sequence recursively via a new (defined) function symbol of type $\iota \times \omega \to \iota$; let us call this symbol $\hat{f}$ and define it as

$$\begin{aligned} \hat{f}(x, n) &= x \text{ for } n = \mathbf{0}, \\ &= f(\hat{f}(x, p(n))) \text{ for } n > \mathbf{0}. \end{aligned}$$

Let $\sigma \in \mathcal{S}$ and $\sigma(n) = \bar{3}$. Then, as for the numeric terms, $\hat{f}(x, \bar{3}) \downarrow$ is computed as follows

$$\hat{f}(x, \bar{3}) \downarrow = f(\hat{f}(x, \bar{2}) \downarrow) = f(f(\hat{f}(x, \bar{1}) \downarrow) = f(f(f(\hat{f}(x, \mathbf{0}) \downarrow))) = f(f(f(x))).$$

By evaluating $\hat{f}(x, n)$ successively under increasing values of $\sigma$ we obtain the desired sequence.

In the example above all terms in the specified sequence contain the variable $x$ and only the variable $x$. But what about sequences of terms with an increasing number of variables?

**Example 4.1.4** Let us consider the problem of computing the maximum of arbitrary sequences of objects (which can be natural numbers, integers, rationals, lattices etc.). Then a recursive definition of a maximum would look like this

$$m(x_0) = x_0,\ m(x_0, \ldots, x_{n+1}) = \max(m(x_0, \ldots, x_n), x_{n+1}),$$

where $\max(x, y)$ denotes the maximum of $x$ and $y$. We see that $m(x, n)$ contains $n + 1$ variables thus defining an infinite sequence of terms with an increasing number of variables. In our approach we use defined symbols and—instead of an individual variable $x$—a second-order variable $X$ of type $\omega \to \iota$. The expression $X(n)$ (of type $\iota$) then evaluates under a parameter assignment $\sigma$ for $\sigma(n) = \bar{k}$ to $X(\bar{k})$, an object behaving like a first-order variable . Let us assume that max is an ordinary two-place function symbol and $\hat{m}$ a defined symbol having the following definition:

$$\begin{aligned} \hat{m}(X, n) &= X(\mathbf{0}) \text{ for } n = \mathbf{0}, \\ &= \max(\hat{m}(X, p(n)), X(n)) \text{ for } n > \mathbf{0}. \end{aligned}$$

Let $\sigma(n) = \bar{2}$. Then the evaluation of $\hat{m}(X, \bar{2})$ is computed as

$$\begin{aligned} \hat{m}(X, \bar{2}) \downarrow &= \max(\hat{m}(X, \bar{1}) \downarrow, X(\bar{2})) = \max(\max(\hat{m}(X, \mathbf{0}), X(\bar{1})), X(\bar{2})) \\ &= \max(\max(X(\mathbf{0}), X(\bar{1})), X(\bar{2})). \end{aligned}$$

It is easy to see that, for $\sigma(n) = \bar{k}$, $\hat{m}(X, \sigma(n))$ evaluates to a term with $k + 1$ different variables. Therefore, in contrast to the schema in Example 4.1.3, $\hat{m}$ cannot be considered as the specification of a $k$-ary function for fixed $k$; in fact the list of arguments for $\hat{m}$ increases with the value of $\sigma$.

The examples above illustrate the formal difference between term schemata with fixed and others with "variable" variables. Therefore we will distinguish the class of *simple* schemata (with fixed variables) and the class of $\upsilon$-schemata, which together will form the general class of term schemata. Simple term schemata are defined similar to numeric terms, with the difference that we also include tuples of individual variables, in $\upsilon$-schemata we have to work with classes of individual variables as illustrated in Example 4.1.4). In the syntax we have to define the set of first-order terms built from ordinary (theory-free) function symbols and terms containing schematic symbols.

**Definition 4.1.14** (*function symbols*) For every $n \geq 0$ we define an infinite set of function symbols of type $\iota^n \to \iota$ which we denote by $\mathbf{F}^\iota_n$, the set of all function symbols is defined as $\mathbf{F}^\iota = \bigcup_{n \in \mathbb{N}} \mathbf{F}^\iota_n$.

In Example 4.1.4 we have seen that the first-order variables obtained from the evaluation can be of the form $X(\bar{2})$ which suggests that we have to generalize the concept of first-order variables.

**Definition 4.1.15** (*variables and variable expressions*) We define as $\mathbf{V}^\iota_n$ for $n \geq 0$ as an infinite set variables of type $\omega^n \to \iota$; the set $\mathbf{V}^\iota_0$ represents the set of first-order variables for which we generally use the symbols $x, y, z, x_1, \ldots$. Let $X \in \mathbf{V}^\iota_k$ and $t_1, \ldots, t_k$ be terms in $T_0$ (see Definition 4.1.1) then the expression $X(t_1, \ldots, t_k)$ of type $\iota$ is called a *variable expression*. If $t_1, \ldots, t_k \in Num$ then we call the variable expression $X(t_1, \ldots, t_k)$ *a variable in* $X$. The set of all variables (in some $X$) is denoted by $\mathrm{var}^\iota$.

**Example 4.1.5** Let $X \in \mathbf{V}^\iota_3$ then $X(\mathbf{0}, \bar{2}, \bar{1})$ is a variable in $X$ and thus $X(\mathbf{0}, \bar{2}, \bar{1}) \in \mathrm{var}^\iota$ . $X(p(n), m, n)$ for $n, m \in \mathcal{N}$ is a variable expression. Note that for $\sigma(n) = \bar{1}, \sigma(m) = \bar{2}$ and $\sigma(X(p(n), m, n) = X(p(\sigma(n)) \downarrow, \sigma(m), \sigma(n))$ we obtain $\sigma(X(p(n), m, n)) = X(\mathbf{0}, \bar{2}, \bar{1})$, the latter being a variable in $X$. We see that variable expressions are, in some sense, *schematic variables* as they evaluate to variables (in some $X \in \mathbf{V}^\iota_k$ under parameter assigments).

**Remark** Though the $X \in \mathbf{V}^\iota_k$ for $k > 0$ are second-order variables we actually do not make use of the higher-order properties of $X$; e.g. given $X(n)$ we will not build expressions of the form $\lambda n.X(n)$; neither will be substitute X itself by expressions of the form $\lambda n.t$. Instead, for $X \in \mathbf{V}^\iota_k$, we will substitute only the variable expressions $X(t_1, \ldots, t_k)$ (which are of type $\iota$ and are somehow schematic first-order variables) by schematic terms. In Sect. 4.1.3 we will describe what happens if we use the full power of the higher-order variables.

**Definition 4.1.16** (*individual terms*) The set $\mathbf{T}^\iota$ of individual terms is defined inductively as follows.

- $\mathbf{F}^\iota_0 \subseteq \mathbf{T}^\iota$,
- all variable expressions are in $\mathbf{T}^\iota$,
- if $t_1, \ldots, t_k \in \mathbf{T}^\iota$ and $f \in \mathbf{F}^\iota_k$ then $f(t_1, \ldots, t_k) \in \mathbf{T}^\iota$.

The subset of $\mathbf{T}^\iota$ containing only variable expressions in $\mathrm{var}^\iota$ is denoted by $\mathbf{T}^\iota_0$.

**Example 4.1.6** Let $f \in \mathbf{F}^\iota_2, g \in \mathbf{F}^\iota_1, X \in \mathbf{V}^\iota_2, y \in \mathbf{V}^\iota_0$. Then

$$f(X(n, m), f(X(s(n), m), g(y))) \text{ is a term in } \mathbf{T}^\iota,$$
$$f(X(\bar{1}, \bar{1}), f(X(\bar{2}, \bar{1}), g(y))) \text{ is a term in } \mathbf{T}^\iota_0.$$

The term in $\mathbf{T}^\iota_0$ rewritten in ordinary first-order syntax would be of the form $f(x, f(z, g(y)))$.

We obtained the numeric terms $\mathbf{T}^\omega$ by extending the basic language $T_0$ (built only by parameters and the function symbols $s$ and $p$) by infinitely many theory symbols, where the corresponding theories consist of primitive recursive definitions of functions. In case of individual terms the role of $T_0$ is taken over by $\mathbf{T}^\iota$—with the difference that all the function symbols are theory-free (in contrast to $s$ and $p$ only). Schemata of individual terms are designed in a way that they evaluate to terms in $\mathbf{T}_0^\iota$ under any parameter assignment. In $\mathbf{T}^\iota$ we have variable classes of different arity and to each schematic term symbol we will assign variables in a class $\mathbf{V}_k^\iota$ for some $k$. The combination of arities of variable classes, numbers of occurrences and numbers of parameters we call a *profile*.

**Definition 4.1.17** (*profile*) A *profile* is a type of the form $(\omega^i \to \iota)^j \times \omega^k \to \iota$ for $i, j, k \in \mathbb{N}$ and $j, k \geq 1$ and is represented by the triple $(i, j, k)$.

**Example 4.1.7** The profile $(1, 2, 3)$ stands for the type $(\omega \to \iota)^2 \times \omega^3 \to \iota$, where 1 denotes a tuple of variables in $\mathbf{V}_1^\iota$, the number 2 encodes the arity of the tuple of those variables, and the last element 3 represents a tuple $(n_1, n_2, n_3)$ for $n_i \in \mathcal{N}$.

Let us consider the expression $\hat{m}(X, n)$ in Example 4.1.4; then the profile of $\hat{m}$ is $(1, 1, 1)$, i.e. $\hat{m}$ is of the type $(\omega \to \iota) \times \omega \to \iota$.

**Definition 4.1.18** (*schematic term symbols*) To each profile $\pi$ we assign an infinite set of symbols $\hat{F}(\pi)$ which we call *term symbols*. The set obtained by taking a union over all $\hat{F}(\pi)$ for profiles $\pi$ is denoted by $\hat{F}$.

Term symbols take over the role the symbols in $\mathbf{F}^\omega$ played in the definition of $\mathbf{T}^\omega$.

**Definition 4.1.19** (*schematic individual terms*) We define the set $\hat{\mathbf{T}}^\iota$ (the set of *schematic individual terms*) inductively as follows:

- $\mathbf{T}^\iota \subseteq \hat{\mathbf{T}}^\iota$,
- If $t_1, \ldots, t_k \in \hat{\mathbf{T}}^\iota$ and $f \in \mathbf{F}_k^\iota$ then $f(t_1, \ldots, t_k) \in \hat{\mathbf{T}}^\iota$,
- if $\pi = (i, j, k)$ is a profile and $\hat{t} \in \hat{F}(\pi)$, $X_l \in \mathbf{V}_i^\iota$ for $l = 1, \ldots, j, t_1, \ldots, t_k \in \mathbf{T}^\omega$ then $\hat{t}(X_1, \ldots, X_k, t_1, \ldots, t_j) \in \hat{\mathbf{T}}^\iota$.

A term of the form $\hat{t}(X_1, \ldots, X_k, t_1, \ldots, t_j)$ is called a *schematic head term.*

**Example 4.1.8** Let $f \in \mathbf{F}_2^\iota$, $t_1, t_2 \in \mathbf{T}^\iota$, $\pi = (2, 2, 1)$, $\hat{t} \in \hat{F}(\pi)$, $\dot{g} \in \mathbf{F}_2^\omega$, $X, Y \in \mathbf{V}_2^\iota$, $n \in \mathcal{N}$ then

$$f(t_1, f(t_2, \hat{t}(X, Y, \dot{g}(\mathbf{0}, n)))) \in \hat{\mathbf{T}}^\iota.$$

$\hat{t}(X, Y, \dot{g}(\mathbf{0}, n))$ is a schematic head term.

If all schematic individual terms in a term have a profile of the form $(0, j, k)$ (the variables are nullary and are just ordinary first-order variables) we call the term simple (such a simple term appears in Example 4.1.3). In contrast to variables in $\mathbf{V}_i^\iota$ for $i \geq 1$ we can substitute the individual variables by other simple schematic terms; note that in Definition 4.1.19 schematic terms cannot be nested, i.e. no schematic term appears inside $\hat{t}(X_1, \ldots, X_k, t_1, \ldots, t_j)$. We will illustrate in Sect. 4.1.3 that substitution of variable classes would correspond to higher-order substitutions which we want to avoid. The price we have to pay is that simple schematic terms are not just a subset of schematic individual terms and we have to define them separately.

**Definition 4.1.20** We define the set $\hat{\mathbf{T}}^\iota_s$ (the set of *simple schematic terms*) inductively as follows:

- $\mathbf{T}^\iota \subseteq \hat{\mathbf{T}}^\iota_s$,
- If $t_1, \ldots, t_k \in \hat{\mathbf{T}}^\iota_s$ and $f \in \mathbf{F}^\iota_k$ then $f(t_1, \ldots, t_k) \in \hat{\mathbf{T}}^\iota_s$,
- if $\pi = (0, j, k)$ is a profile and $\hat{t} \in \hat{F}(\pi)$, $s_l \in \hat{\mathbf{T}}^\iota_s$ for $l = 1, \ldots, j, t_1, \ldots, t_k \in \mathbf{T}^\omega$ then $\hat{t}(s_1, \ldots, s_j, t_1, \ldots, t_k) \in \hat{\mathbf{T}}^\iota_s$.

**Example 4.1.9** Let $\pi = (0, 1, 1)$, $\hat{t} \in \hat{F}(\pi)$, $f \in \mathbf{F}^\iota_1$, $X \in \mathbf{V}^\iota_1$, $x \in \mathbf{V}^\iota_0$, $n \in \mathcal{N}$ and $g(n) \in \mathbf{T}^\omega$. Then

$$f(\hat{t}(x, g(n))) \in \hat{\mathbf{T}}^\iota_s \text{ and } \hat{t}(\hat{t}(X(n), g(n)), g(n)) \in \hat{\mathbf{T}}^\iota_s.$$

Note that the first term is also in $\hat{\mathbf{T}}^\iota$, but the second is not.

What is missing is the "meaning" of the expressions $\hat{t}(X_1, \ldots, X_k, t_1, \ldots, t_j)$ in $\hat{\mathbf{T}}^\iota$ and of $\hat{t}(x_1, \ldots, x_k, t_1, \ldots, t_j)$ in $\hat{\mathbf{T}}^\iota_s$ (note that the last expression is contained also in $\hat{\mathbf{T}}^\iota$). Like for the symbols in $\mathbf{F}^\omega$ we have to assign an equation to every such symbol $\hat{t}$ taking into account existing $\omega$-theories for symbols in $\mathbf{F}^\omega$. So we have to define an $\iota$-theory for the symbols in $\hat{F}(\pi)$ together with an $\omega$-theory for the symbols in $\mathbf{F}^\omega$. We write $\hat{F}(T)$ for the set of symbols in $\hat{F}$ appearing in a set of terms $T \subseteq \hat{\mathbf{T}}^\iota$.

**Definition 4.1.21** *($\omega\iota$-theory)* A tupel $(T, T')$ is called an $\omega\iota$-theory of $\hat{t}$ if

- $T = (\hat{t}, \mathcal{F}, D(\mathcal{F}), V, <)$,
- $T'$: $(\hat{f}, \mathcal{F}', D(\mathcal{F}'), <')$ is an $\omega$-theory,
- $\hat{t} \in \mathcal{F}$,
- $\mathcal{F}$ is a finite subset of $\hat{F}$,
- $V$ is a finite subset of $\mathbf{V}^\iota_k$ for some $k \in \mathbb{N}$,
- $<$ is an irreflexive, transitive relation on $\mathcal{F} \cup \mathbf{T}^\iota$ such that for all $\hat{s} \in \mathcal{F}\backslash\{\hat{t}\}$: $\hat{s} < \hat{t}$ ($\hat{t}$ is maximal in $\mathcal{F}$) and for all $\hat{s} \in \mathcal{F}$ and $t \in \mathbf{T}^\iota$ we have $t < \hat{s}$ (the terms in $\mathbf{T}^\iota$ are smaller than all schematic term symbols).
- $D(\mathcal{F}) = \bigcup_{\hat{s} \in \mathcal{F}} D(\hat{s})$ where each $D(\hat{s})$ is a set containing an equation. We assume that $\pi(\hat{s}) = (k, i, j)$, $X_l \in V$ for $l = 1, \ldots, i$ and $n_1, \ldots, n_{j-1}, n_j \in \mathcal{N}$. Then $D(\hat{s})$ is defined as

$$\hat{s}(X_1, \ldots, X_i, n_1, \ldots, n_{j-1}, n_j) =_d \underline{\text{if}}\ n_j = \mathbf{0}\ \underline{\text{then}}\ \hat{s}_B\ \underline{\text{else}}$$
$$\hat{s}_S\{z \leftarrow \hat{s}(X_1, \ldots, X_i, n_1, \ldots, n_{j-1}, p(n_j))\}\}.$$

we require

- $\hat{s}_S, \hat{s}_B \in \hat{\mathbf{T}}^\iota(T, T')$, where $\hat{\mathbf{T}}^\iota(T, T')$ is the subset of all terms $t$ in $\hat{\mathbf{T}}^\iota$ such that $\hat{F}(t) \subseteq \mathcal{F}$ and $\mathbf{F}^\omega(t) \subseteq \mathcal{F}'$,
- $\mathbf{V}^\iota(\hat{s}_S) \subseteq \{X_1, \ldots, X_i, z\}$, $\mathcal{N}(\hat{s}_S) = \{n_1, \ldots, n_j\}$, and $z \in \mathbf{V}^\iota_0\backslash\{X_1, \ldots, X_i\}$; $\mathbf{V}^\iota(\hat{s}_B) \subseteq \{X_1, \ldots, X_i\}$, $\mathcal{N}(\hat{s}_B) \subseteq \{n_1, \ldots, n_{j-1}\}$,
- Let $\mathcal{G} = \hat{F}(\hat{s}_S) \cup \hat{F}(\hat{s}_B)$; then, for all $t \in \mathcal{G} : t < \hat{s}$.

If $V$ is a finite subset of $\mathbf{V}_0^\iota$ then $(T, T')$ is called a *simple* $\omega\iota$-theory. For simple theories we replace the condition $\hat{s}_S, \hat{s}_B \in \hat{\mathbf{T}}^\iota(T, T')$ by $\hat{s}_S, \hat{s}_B \in \hat{\mathbf{T}}^\iota_s(T, T')$ where $\hat{\mathbf{T}}^\iota_s(T, T')$ is the subset of all terms $t$ in $\hat{\mathbf{T}}^\iota_s$ such that $\hat{F}(t) \subseteq \mathcal{F}$ and $\mathbf{F}^\omega(t) \subseteq \mathcal{F}'$.

**Example 4.1.10** Let $T^* = (T_1, T_2)$ where

$$T_1 = (\hat{t}, \{\hat{t}, \hat{s}\}, D(\{\hat{t}, \hat{s}\}), \{X_1, X_2\}, <)$$

and

- $\pi(\hat{t}) = (2, 2, 2), \pi(\hat{s}) = (2, 1, 1)$,
- $X_1, X_2 \in \mathbf{V}_2^\iota$,
- $\hat{s} < \hat{t}$,

and

$$D(\hat{t}) = \{\hat{t}(X_1, X_2, n_1, n_2) =_d \underline{\text{if}}\ n_2 = \mathbf{0}\ \underline{\text{then}}\ \hat{t}_B\ \underline{\text{else}}\ \hat{t}_S\{z \leftarrow \hat{t}(X_1, X_2, n_1, p(n_2))\}\},$$

$$D(\hat{s}) = \{\hat{s}(X_1, n_1) =_d \underline{\text{if}}\ n_1 = \mathbf{0}\ \underline{\text{then}}\ \hat{s}_B\ \underline{\text{else}}\ \hat{s}_S\{z \leftarrow \hat{s}(X_1, p(n_1))\}$$

where

$$\hat{t}_S = g(X_1(n_1, n_2), z, \hat{s}(X_1, n_1)) \text{ and } \hat{t}_B = g(X_1(n_1, \mathbf{0}), X_2(n_1, \mathbf{0}), c),$$
$$\hat{s}_S = h(X_1(\mathbf{0}, n_1), z) \text{ and } \hat{s}_B = X_1(\mathbf{0}, \mathbf{0}).$$

Here we have $g \in \mathbf{F}_3^\iota$, $h \in \mathbf{F}_2^\iota$.

$T_2 = (\dot{f}, \{\dot{f}, \dot{g}\}, D(\{\dot{f}, \dot{g}\}), <')$ where $\dot{f}, \dot{g} \in \mathbf{F}_2^\omega$, $\dot{g} < \dot{f}$ and

$$D(\dot{f}) = \{\dot{f}(x, y)) =_d \underline{\text{if}}\ y = \mathbf{0}\ \underline{\text{then}}\ \mathbf{0}\ \underline{\text{else}}\ \dot{g}(x, \dot{f}(x, p(y))),$$
$$D(\dot{g}) = \{\dot{g}(x, y) =_d \underline{\text{if}}\ y = \mathbf{0}\ \underline{\text{then}}\ x\ \underline{\text{else}}\ s(\dot{g}(x, p(y)))\}.$$

When we have a look on the definition $D(\hat{s})$ we see (without having the formal definition of evaluations yet) that

$$\hat{s}(X_1, \bar{2}) = h(X1(\mathbf{0}, \bar{2}), \hat{s}(X_1, \bar{1})) = h(X_1(\mathbf{0}, \bar{2}), h(X_1(\mathbf{0}, \bar{1}), \hat{s}(X_1, \mathbf{0}))) = h(X_1(\mathbf{0}, \bar{2}), h(X_1(\mathbf{0}, \bar{1}), X_1(\mathbf{0}, \mathbf{0}))).$$

The same term schema (just rename the variables) could be defined via a variable $Y \in \mathbf{V}_1^\iota$ and

$$D(\hat{s}_1) = \{\hat{s}_1(Y, n_1) =_d \underline{\text{if}}\ n_1 = \mathbf{0}\ \underline{\text{then}}\ Y(\mathbf{0})\ \underline{\text{else}}\ h(Y(n_1), \hat{s}_1(Y, p(n_1)))\}.$$

We see that there is never a problem in increasing the arity of variable classes. This also justifies the uniform arity of variable classes in the definition of term schemata.

**Example 4.1.11** Let $T = (T_1, T_2)$ be a simple $\omega\iota$-theory where $T_2$ is defined as in Example 4.1.10 and $T_1 = (\hat{t}, \{\hat{t}, \hat{s}\}, D(\{\hat{t}, \hat{s}\}), \{x, y\}, <)$ where

- $\pi(\hat{t}) = (0, 2, 2)$, $\pi(\hat{s}) = (0, 1, 1)$,
- $x, y \in \mathbf{V}_0^\iota$,
- $\hat{s} < \hat{t}$,
- $g \in \mathbf{F}_2^\iota$, $h \in \mathbf{F}_1^\iota$.

Moreover,

$$D(\hat{t}) = \{\hat{t}(x, y, n_1, n_2) =_d \underline{\text{if}}\ n_2 = \mathbf{0}\ \underline{\text{then}}\ y \\ \underline{\text{else}}\ g(\hat{t}(x, y, n_1, p(n_2)), \hat{s}(x, n_1))\},$$
$$D(\hat{s}) = \{\hat{s}(x, n_1) =_d \underline{\text{if}}\ n_1 = \mathbf{0}\ \underline{\text{then}}\ x\ \underline{\text{else}}\ h(\hat{s}(x, p(n_1)))\}.$$

Intuitively we realize that $\hat{s}(x, \bar{2}) = h(h(x))$ and

$$\hat{t}(x, y, \bar{2}, \bar{1}) = g(\hat{t}(x, y, \bar{2}, \mathbf{0}), \hat{s}(x, \bar{2})) = g(y, h(h(x))).$$

The essence of term schemata is that they evaluate to terms under given parameter assignments. We restrict the evaluations to given $\omega\iota$-theories as only in theories equations are assigned to schematic term symbols. We first define the evaluations of terms in $\hat{\mathbf{T}}^\iota$ with respect to an $\omega\iota$-theory.

**Definition 4.1.22** Let $T^*\colon (T, T')$ be an $\omega\iota$-theory for

$$T = (\hat{t}, \mathcal{F}, D(\mathcal{F}), V, <) \text{ and } T' = (\dot{f}, \mathcal{F}', D(\mathcal{F}'), <),$$

where $V \subseteq \mathbf{V}_k^\iota$ and for $\hat{s} \in \mathcal{F}$, $\pi(\hat{s}) = (k, i, j)$ we have

$$D(\hat{s}) = \{\hat{s}(X_1, \ldots, X_i, n_1, \ldots, n_j) =_d \underline{\text{if}}\ n_j = \mathbf{0}\ \underline{\text{then}}\ \hat{s}_B\ \underline{\text{else}} \\ \hat{s}_S\{z \leftarrow \hat{s}(X_1, \ldots, X_i, n_1, \ldots, n_{j-1}, p(n_j))\}\}.$$

We are going to evaluate terms $t$ in $\hat{\mathbf{T}}^\iota(T, T')$. By $\downarrow$ we describe the evaluation of numeric ground terms as defined in Definition 4.1.9 and by $\downarrow_{T^*}$ we denote the evaluation of terms in $\hat{\mathbf{T}}^\iota$. Let $\sigma \in \mathcal{S}$ be a given parameter assigment; we define a two-place function $eval_{T^*}$ such that $eval_{T^*}(t, \sigma) = \sigma(t) \downarrow_{T^*}$ which denotes the evaluation of $t$ under $\sigma$.

- Let $t\colon X(t_1, \ldots, t_l)$ be a variable expression, where $X \in \mathbf{V}_l^\iota$, and $t_1, \ldots, t_l \in T_0$. Then
$$\sigma(t) \downarrow_{T^*} = X(\sigma(t_1) \downarrow, \ldots, \sigma(t_l) \downarrow).$$
- Let $f \in \mathbf{F}_j^\iota$, $t_1, \ldots, t_j \in \hat{\mathbf{T}}^\iota(\mathcal{F}, \mathcal{F}')$ and $t = f(t_1, \ldots, t_j)$. Then
$$\sigma(t) \downarrow_{T^*} = f(\sigma(t_1), \downarrow_{T^*}, \ldots, \sigma(t_j) \downarrow_{T^*}).$$

- Now let $\hat{s} \in \mathcal{F}$ and $\pi(\hat{s}) = (k, i, j)$ and $t = \hat{s}(X_1, \ldots, X_i, t_1, \ldots, t_j)$ for $X_1, \ldots, X_i \in V, t_1, \ldots, t_j \in \mathbf{T}^\omega$. Let $r_1 = \sigma(t_1) \downarrow, \ldots, r_j = \sigma(t_j) \downarrow$ and

$$\vartheta = \{n_1 \leftarrow r_1, \ldots, n_j \leftarrow r_j\}.$$

Note that the $r_\alpha$ are in *Num*. Then

$$\begin{aligned}\sigma(t) \downarrow_{T^*} = \hat{s}(X_1, \ldots, X_i, r_1, \ldots, r_j) \downarrow_{T^*} &= \hat{s}_S\vartheta\{z \leftarrow t'\} \downarrow_{T^*} \text{ for} \\ t' = \hat{s}(X_1, \ldots, X_i, r_1, \ldots, r_{j-1}, p(r_j) \downarrow) \downarrow_{T^*} &\text{ and } r_j > \mathbf{0}, \\ = \hat{s}_B\vartheta \downarrow_{T^*} &\text{ for } r_j = \mathbf{0}.\end{aligned}$$

**Example 4.1.12** We consider the theory from Example 4.1.10. We have

$$\begin{aligned}\hat{t}_S = g(X_1(n_1, n_2), z, \hat{s}(X_1, n_1)) &\text{ and } \hat{t}_B = g(X_1(n_1, \mathbf{0}), X_2(n_1, \mathbf{0}), c), \\ \hat{s}_S = h(X_1(\mathbf{0}, n_1), z) &\text{ and } \hat{s}_B = X_1(\mathbf{0}, \mathbf{0}).\end{aligned}$$

$$\begin{aligned}D(\hat{t}) = \{&\hat{t}(X_1, X_2, n_1, n_2) =_d \underline{\text{if}}\ n_2 = \mathbf{0}\ \underline{\text{then}}\ g(X_1(n_1, \mathbf{0}), X_2(n_1, \mathbf{0}), c) \\ &\underline{\text{else}}\ g(X_1(n_1, \mathbf{0}), \hat{t}(X_1, X_2, n_1, p(n_2)), \hat{s}(X_1, n_1))\}, \\ D(\hat{s}) = \{&\hat{s}(X_1, n_1) =_d \underline{\text{if}}\ n_1 = \mathbf{0}\ \underline{\text{then}}\ X_1(\mathbf{0}, \mathbf{0}) \\ &\underline{\text{else}}\ h(X_1(\mathbf{0}, n_1), \hat{s}(X_1, p(n_1)))\}.\end{aligned}$$

We evaluate the term $t = \hat{t}(X_1, X_2, \dot{g}(n_1, n_2), n_2)$, i.e. we determine $\sigma(t) \downarrow_{T^*}$. Let us assume $\sigma(n_1) = \sigma(n_2) = \bar{1}$. Then we obtain

$$r_1 = \sigma(\dot{g}(n_1, n_2)) \downarrow = \dot{g}(\bar{1}, \bar{1}) \downarrow = \bar{2}, r_2 = \sigma(n_2) \downarrow_{T^*} = \sigma(n_2) \downarrow = \bar{1}.$$

Therefore we obtain $\theta = \{n_1 \leftarrow \bar{2}, n_2 \leftarrow \bar{1}\}$. As $r_2 > \mathbf{0}$ we have

$$\begin{aligned}\sigma(t) \downarrow_{T^*} &= \hat{t}(X_1, X_2, \bar{2}, \bar{1}) \downarrow_{T^*} = \hat{t}_S\theta\{z \leftarrow t'\} \downarrow_{T^*} \text{ where} \\ \hat{t}_S\theta &= g(X_1(\bar{2}, \bar{1}), z, \hat{s}(X_1, \bar{2})), \\ t' &= \hat{t}(X_1; X_2, \bar{2}, \mathbf{0}) \downarrow_{T^*} = \hat{t}_B\theta \downarrow_{T^*} \text{ where} \\ \hat{t}_B\theta \downarrow_{T^*} &= g(X_1(\bar{2}, \mathbf{0}), X_2(\bar{2}, \mathbf{0}), c) \downarrow_{T^*} = g(X_1(\bar{2}, \mathbf{0}), X_2(\bar{2}, \mathbf{0}), c).\end{aligned}$$

Therefore

$$\hat{t}_S\theta\{z \leftarrow t'\} \downarrow_{T^*} = g(X_1(\bar{2}, \bar{1}), g(X_1(\bar{2}, \mathbf{0}), X_2(\bar{2}, \mathbf{0}), c), \hat{s}(X_1, \bar{2})) \downarrow_{T^*}).$$

It remains to evaluate $\hat{s}(X_1, \bar{2})$:

$$\begin{aligned}\hat{s}(X_1, \bar{2}) \downarrow_{T^*} &= h(X_1(\mathbf{0}, \bar{2}), \hat{s}(X_1, \bar{1})) \downarrow_{T^*} \\ &= h(X_1(\mathbf{0}, \bar{2}), h(X_1(\mathbf{0}, \bar{1}), \hat{s}(X_1, \mathbf{0}))) \downarrow_{T^*}\end{aligned}$$

$$= h(X_1(\mathbf{0}, \bar{2}), h(X_1(\mathbf{0}, \bar{1}), X_1(\mathbf{0}, \mathbf{0}))) \downarrow_{T^*}$$
$$= h(X_1(\mathbf{0}, \bar{2}), h(X_1(\mathbf{0}, \bar{1}), X_1(\mathbf{0}, \mathbf{0}))).$$

Putting things together we eventually obtain

$$\sigma(t) \downarrow_{T^*} =$$
$$g(X_1(\bar{2}, \bar{1}), g(X_1(\bar{2}, \mathbf{0}), X_2(\bar{2}, \mathbf{0}), c), h(X_1(\mathbf{0}, \bar{2}), h(X_1(\mathbf{0}, \bar{1}), X_1(\mathbf{0}, \mathbf{0}))).$$

Now we define the evaluation for simple $\omega\iota$-theories:

**Definition 4.1.23** Let $T^*\colon (T, T')$ be a simple $\omega\iota$-theory for $T = (\hat{t}, \mathcal{F}, D(\mathcal{F}), V, <)$, $T' = (\dot{f}, \mathcal{F}', D(\mathcal{F}'), <)$, where $V \subseteq \mathbf{V}_0^\iota$ and for $\hat{s} \in \mathcal{F}$, $\pi(\hat{s}) = (0, i, j)$ we have

$$D(\hat{s}) = \{\hat{s}(x_1, \ldots, x_i, n_1, \ldots, n_j) =_d \underline{\text{if}}\ n_j = \mathbf{0}\ \underline{\text{then}}\ \hat{s}_B\ \underline{\text{else}}$$
$$\hat{s}_S\{z \leftarrow \hat{s}(x_1, \ldots, x_i, n_1, \ldots, n_{j-1}, p(n_j))\}\}.$$

Now we are evaluating terms $t$ in $\hat{\mathbf{T}}^\iota_s(T^*)$. As in Definition 4.1.22 we write $\downarrow$ for the evaluation of numeric ground terms; moreover we add an operator $\downarrow_{T^*}$ which evaluates parameter-free terms in $\hat{\mathbf{T}}^\iota_s(T^*)$. Let $\sigma \in \mathcal{S}$ be a given parameter assignment; we define a two-place function $eval_s$ such that $eval_s(t, \sigma) = \sigma(t) \downarrow_{T^*}$ denotes the term obtained from $t$ under $\sigma$. If $t$ is parameter-free then $\sigma(t) \downarrow_{T^*} = t \downarrow_{T^*}$.

- Let $x$ be a variable in $V$ then $\sigma(x) \downarrow_{T^*} = x \downarrow_{T^*} = x$ for all $\sigma \in \mathcal{S}$.
- Let $f \in \mathbf{F}^\iota_j$, $t_1, \ldots, t_j \in \hat{\mathbf{T}}^\iota_s(\mathcal{F}, \mathcal{F}')$, $t = f(t_1, \ldots, t_j)$. We define

$$\sigma(t) \downarrow_{T^*} = f(\sigma(t_1) \downarrow_{T^*}, \ldots, \sigma(t_j) \downarrow_{T^*}).$$

- Now let $\hat{s} \in \mathcal{F}$ and $\pi(\hat{s}) = (0, i, j)$ and $t = \hat{s}(s_1, \ldots, s_i, t_1, \ldots, t_j)$ for $s_1, \ldots, s_i \in \hat{\mathbf{T}}^\iota_s(T^*)$, $t_1, \ldots, t_j \in \mathbf{T}^\omega(T)$. Let $s'_\alpha = \sigma(s_\alpha) \downarrow_{T^*}$ for $\alpha = 1, \ldots, i$ and $r_\beta = \sigma(t_\beta) \downarrow$ for $\beta = 1, \ldots, j$. Then

$$\sigma(t) \downarrow_{T^*} = \hat{s}(s'_1, \ldots, s'_i, r_1, \ldots, r_j) \downarrow_{T^*}.$$

Let $\theta = \{x_1 \leftarrow s'_1, \ldots, x_i \leftarrow s'_i, n_1 \leftarrow r_1, \ldots, n_j \leftarrow r_j\}$.
Then $\hat{s}(s'_1, \ldots, s'_i, r_1, \ldots, r_j) \downarrow_{T^*}$ is defined as follows

$$\hat{s}(s'_1, \ldots, s'_i, r_1, \ldots, r_j) \downarrow_{T^*} = \hat{s}_S\theta\{z \leftarrow t'\} \downarrow_{T^*}$$
$$\text{for } t' = \hat{s}(s'_1, \ldots, s'_i, r_1, \ldots, r_{j-1}, p(r_j) \downarrow) \downarrow_{T^*} \text{ and } r_j > \mathbf{0},$$
$$= \hat{s}_B\theta \downarrow_{T^*} \text{ for } r_j = \mathbf{0}.$$

**Example 4.1.13** We take the theory from Example 4.1.11 where

$$\hat{t}_S = g(z, \hat{s}(x_1, n_1)),\ \hat{t}_B = x_2,\ \hat{s}_S = h(z),\ \hat{s}_B = x_1 \text{ and}$$

$$D(\hat{t}) = \{\hat{t}(x_1, x_2, n_1, n_2) =_d \underline{\text{if}}\ n_2 = \mathbf{0}\ \underline{\text{then}}\ x_2$$
$$\underline{\text{else}}\ g(\hat{t}(x_1, x_2, n_1, p(n_2)), \hat{s}(x_1, n_1))\},$$
$$D(\hat{s}) = \{\hat{s}(x_1, n_1) =_d \underline{\text{if}}\ n_1 = \mathbf{0}\ \underline{\text{then}}\ x_1\ \underline{\text{else}}\ h(\hat{s}(x_1, p(n_1)))\}.$$

We evaluate the term $t = \hat{t}(\hat{t}(x, y, n, m), x, \dot{g}(n, m), n)$ under $\sigma(n) = \sigma(m) = \bar{1}$. By definition we have

$$\begin{aligned}\sigma(t) \downarrow_{T^*} &= \hat{t}(\sigma(\hat{t}(x, y, n, m)) \downarrow_{T^*}, \sigma(x) \downarrow_{T^*}, \sigma(\dot{g}(n, m)) \downarrow, \sigma(n) \downarrow) \downarrow_{T^*} \\ &= \hat{t}(\sigma(\hat{t}(x, y, n, m)) \downarrow_{T^*}, x, \bar{2}, \bar{1}) \downarrow_{T^*}.\end{aligned}$$

We need to compute $\sigma(\hat{t}(x, y, n, m)) \downarrow_{T^*}$ first. By definition we have

$$\begin{aligned}\sigma(\hat{t}(x, y, n, m)) \downarrow_{T^*} &= \hat{t}(x, y, \bar{1}, \bar{1}) \downarrow_{T^*}, \\ \theta &= \{x_1 \leftarrow x, x_2 \leftarrow y,\ n_1 \leftarrow \bar{1}, n_2 \leftarrow \bar{1}\} \text{ and} \\ \hat{t}(x, y, \bar{1}, \bar{1}) \downarrow_{T^*} &= \hat{t}_S\theta\{z \leftarrow t'\} \downarrow_{T^*} \\ &= g(z, \hat{s}(x, \bar{1}))\{z \leftarrow t'\} \downarrow_{T^*} \text{ where} \\ t' &= g(\hat{t}(x, y, \bar{1}, \mathbf{0}), \hat{s}(x, \bar{1})) \downarrow_{T^*}.\end{aligned}$$

Next we have to compute the term $t'$: $g(\hat{t}(x, y, \bar{1}, \mathbf{0}), \hat{s}(x, \bar{1})) \downarrow_{T^*}$:

$$\begin{aligned}t' &= g(\hat{t}(x, y, \bar{1}, \mathbf{0}) \downarrow_{T^*}, \hat{s}(x, \bar{1})) \downarrow_{T^*}, \\ \hat{t}(x, y, \bar{1}, \mathbf{0}) \downarrow_{T^*} &= y, \\ \hat{s}(x, \bar{1}) &= h(\hat{s}(x, \mathbf{0})) = h(x).\end{aligned}$$

$$\text{Therefore } t' = g(y, h(x)).$$

By going back we obtain

$$\begin{aligned}\hat{t}(x, y, \bar{1}, \bar{1}) &= g(z, h(x))\{z \leftarrow g(y, h(x))\} \\ &= g(g(y, h(x)), h(x)).\end{aligned}$$

Now we can compute $\hat{t}(\sigma(\hat{t}(x, y, n, m)) \downarrow_{T^*}, x, \bar{2}, \bar{1}) \downarrow_{T^*}$. By

$$\sigma(\hat{t}(x, y, n, m)) \downarrow_{T^*}= g(g(y, h(x)), h(x))$$

we can reduce the problem to the computation of

- $\hat{t}(g(g(y, h(x)), h(x)), x, \bar{2}, \bar{1}) \downarrow_{T^*}$.

Let $\theta' = \{x_1 \leftarrow g(g(y, h(x)), h(x)), x_2 \leftarrow x, n_1 \leftarrow \bar{2}, n_2 \leftarrow \bar{1}\}$. Then

$$\begin{aligned}\sigma(t) \downarrow_{T^*} &= \hat{t}(g(g(y, h(x)), h(x)), x, \bar{2}, \bar{1}) \downarrow_{T^*} \\ &= \hat{t}_S\theta'\{z \leftarrow t''\},\end{aligned}$$

$$\begin{aligned}t'' &= \hat{t}(g(g(y,h(x)),h(x)),x,\bar{2},\mathbf{0})\downarrow_{T^*}= x,\\ \hat{t}_S\theta'\{z \leftarrow t''\} &= g(z,\hat{s}(x_1,n_1))\theta'\{z \leftarrow x\}\downarrow_{T^*}\\ &= g(x,\hat{s}(g(g(y,h(x)),h(x)),\bar{2}))\downarrow_{T^*}\\ &= g(x,\hat{s}(g(g(y,h(x)),h(x)),\bar{2})\downarrow_{T^*})\end{aligned}$$

It remains to compute $\hat{s}(g(g(y,h(x)),h(x)),\bar{2})\downarrow_{T^*}$.

$$\hat{s}(g(g(y,h(x)),h(x)),\bar{2})\downarrow_{T^*}= h(\hat{s}(g(g(y,h(x)),h(x)),\bar{1})\downarrow_{T^*}) =\\ h(h(\hat{s}(g(g(y,h(x)),h(x)),\mathbf{0})\downarrow_{T^*})) = h(h(g(g(y,h(x)),h(x)))).$$

Finally we obtain

$$\sigma(t)\downarrow_{T^*}= g(x,h(h(g(g(y,h(x)),h(x))))).$$

We will first prove that, for any $\omega\iota$-theory $(T,T')$, for all terms $t \in \hat{\mathbf{T}}^\iota(T,T')$ and for all $\sigma \in \mathcal{S}$ $eval(t,\sigma) \in \mathbf{T}_0^\iota$, i.e. $eval(t,\sigma)$ is a first-order term. In proving this result we need three lemmas.

**Lemma 4.1.9** *Let $t$ be a term in $\mathbf{T}^\iota$ and $\sigma \in \mathcal{S}$; then $eval(t,\sigma) \in \mathbf{T}_0^\iota$.*

***Proof*** By induction on $\tau(t)$, the term depth of $t$.

- $\tau(t) = 0$. Then either $t \in \mathbf{F}_0^\iota$ ($t$ is a constant symbol) and thus $eval(t,\sigma) = t$, or $t = X(s_1,\ldots,s_k)$ for $X \in \mathbf{V}_k^\iota$ and $s_1,\ldots,s_k \in T_0$; in the latter case $eval(t,\sigma) = X(\sigma(s_1)\downarrow,\ldots,\sigma(s_k)\downarrow)$ by Definition 4.1.22. But $\sigma(s_i)\downarrow = r_i$ for $r_i \in Num$ and $i = 1,\ldots,k$ . So $eval(t,\sigma) = X(r_1,\ldots,r_k)$. By Definition 4.1.16, $X(r_1,\ldots,r_k)$ is a variable and thus in $\mathbf{T}_0^\iota$.
- $t = f(t_1,\ldots,t_k)$, $t_1,\ldots,t_k \in \mathbf{T}^\iota$ and $f \in \mathbf{F}_k^\iota$. By Definition 4.1.22

$$eval(t,\sigma) = f(eval(t_1,\sigma),\ldots,eval(t_k,\sigma))$$

As $\tau(t_p) < \tau(t)$ for $p = 1,\ldots,k$, the induction hypothesis yields $eval(t_p,\sigma) \in \mathbf{T}_0^\iota$. By definition of $\mathbf{T}_0^\iota$ also

$$f(eval(t_1,\sigma),\ldots,eval(t_k,\sigma)) \in \mathbf{T}_0^\iota.$$

**Definition 4.1.24** We define the set of subterms $Subt(t)$ and a measure $subt(t)$ on terms $t$ in an $\omega\iota$-theory $(T,T')$ counting the number of subterms in $t$.

- If $t$ is a variable expression or a schematic head term then $Subt(t) = \{t\}$ and $subt(t) = 1$.
- If $t = f(t_1,\ldots,t_k)$ for $f \in \mathbf{F}_k^\iota$ then

$$Subt(t) = \{t\} \cup \bigcup_{i=1}^{k} Subt(t_i) \text{ and } subt(t) = 1 + \sum_{i=1}^{k} subt(t_i).$$

**Lemma 4.1.10** *Let $(T, T')$ be an $\omega\iota$-theory, $t \in \hat{\mathbf{T}}^{\iota}(T, T')$ and $\sigma \in \mathcal{S}$. Let us assume that for all $t' \in Subt(t)$ which are schematic head terms $eval(t', \sigma) \in \mathbf{T}_0^{\iota}$. Then $eval(t, \sigma) \in \mathbf{T}_0^{\iota}$.*

***Proof*** By induction on $subt(t)$.

- If $subt(t) = 1$ then either $t$ is a constant symbol, a variable expression or a schematic head term. If $t$ is a constant symbol or a variable expression then $eval(t, \sigma) \in \mathbf{T}_0^{\iota}$ by Lemma 4.1.9; if $t$ is a schematic head term then $eval(t, \sigma) \in \mathbf{T}_0^{\iota}$ by assumption.
- Let us assume that the property holds for all terms $t$ with $subt(t) \leq k$ and assume that $subt(t) = k + 1$. Then there exists an $f \in \mathbf{F}_l^{\iota}$ for some $l \in \mathbb{N}$ and terms $t_1, \ldots, t_l \in \hat{\mathbf{T}}^{\iota}(T, T')$ such that $t = f(t_1, \ldots, t_l)$. By

$$subt(t) = 1 + \sum_{i=1}^{l} subt(t_i) \text{ we get that } subt(t_i) \leq k \text{ for } i = 1, \ldots, l.$$

By (IH) $eval(t_i, \sigma) \in \mathbf{T}_0^{\iota}$ for $i = 1, \ldots, l$. But

$$eval(t, \sigma) = f(eval(t_1, \sigma), \ldots, eval(t_l, \sigma)) \text{ and } f \in \mathbf{F}_l^{\iota} \text{ and so}$$

$eval(t, \sigma) \in \mathbf{T}_0^{\iota}$.

**Lemma 4.1.11** *Let $T^*\colon (T, T')$ be an $\omega\iota$-theory for $T = (\hat{t}, \mathcal{F}, D(\mathcal{F}), V, <)$ and $T' = (f, \mathcal{F}', D(\mathcal{F}'), <')$. Let $\hat{s} \in \mathcal{F}$ and $\pi(\hat{s}) = (k, i, j)$. Then, for all $t_1, \ldots, t_j \in \mathbf{T}^{\omega}(\mathcal{F}')$ and for $t = \hat{s}(X_1, \ldots, X_i, t_1, \ldots, t_j)$, $eval(t, \sigma) \in \mathbf{T}_0^{\iota}$.*

***Proof*** We proceed by induction on the ordering $<$ on $\mathcal{F}$.

- $\hat{s}$ is minimal in $\mathcal{F}$ and $t = \hat{s}(X_1, \ldots, X_i, t_1, \ldots, t_j)$ where

$$D(\hat{s}) = \{\hat{s}(X_1, \ldots, X_i, n_1, \ldots, n_j) =_d \underline{\text{if}}\ n_j = \mathbf{0}\ \underline{\text{then}}\ \hat{s}_B$$
$$\underline{\text{else}}\ \hat{s}_S\{z \leftarrow \hat{s}(X_1, \ldots, X_i, n_1, \ldots, p(n_j))\}\}.$$

By Definition 4.1.22 we have for

$$r_1 = \sigma(t_1)\downarrow, \ldots, r_j = \sigma(t_j)\downarrow,\ \theta = \{n_1 \leftarrow r_1, \ldots, n_j \leftarrow r_j\},$$
$$eval(t, \sigma) = \sigma(t)\downarrow_{T^*} = \hat{s}(X_1, \ldots, X_i, r_1, \ldots, r_j)\downarrow_{T^*} =$$
$$\hat{s}_S\theta\{z \leftarrow \hat{s}(X_1, \ldots, X_i, r_1, \ldots, r_{j-1}, p(r_j)\downarrow)\downarrow_{T^*} \text{ for } r_j > \mathbf{0},$$
$$= \hat{s}_B\theta\downarrow_{T^*} \text{ for } r_j = \mathbf{0}.$$

As $\hat{s}$ is minimal $\hat{s}_S, \hat{s}_B \in \mathbf{T}^{\iota}$. To show $eval(t, \sigma) \in \mathbf{T}_0^{\iota}$ we proceed by induction on $r_j$.
$r_j = \mathbf{0}$: Then $eval(t, \sigma) = \hat{s}_B\theta$. As $\hat{s}_B \in \mathbf{T}^{\iota}$ we get $\hat{s}_B\theta \in \mathbf{T}_0^{\iota}$.
(IH) Let us assume that $t\downarrow_{T^*} \in \mathbf{T}_0^{\iota}$ for $t\colon \hat{s}(X_1, \ldots, X_i, r_1, \ldots, r_j)$ for $r_1, \ldots, r_j \in Num$ and $r_j \leq q$ for $q \in Num$. Now let

$$t' \colon \hat{s}(X_1, \ldots, X_i, r_1, \ldots, r_j) \text{ and } r_j = s(q)$$

Then $r_j > \mathbf{0}$ and we obtain

$$t' \downarrow_{T^*} = \hat{s}_S\theta\{z \leftarrow \hat{s}(X_1, \ldots, X_i, r_1, \ldots, r_{j-1}, p(s(q)) \downarrow)\} \downarrow_{T^*}.$$

But $p(s(q)) \downarrow = q$ by definition of $T_0$ and, by (IH),

$$\hat{s}(X_1, \ldots, X_i, r_1, \ldots, r_{j-1}, q)) \downarrow_{T^*} = t_0$$

for some $t_0 \in \mathbf{T}_0^\iota$.
But $\hat{s}_S \in \mathbf{T}^\iota$ and, as $\theta$ replaces parameters by numerals, $\hat{s}_S\theta \in \mathbf{T}_0^\iota$ and therefore $\hat{s}_S\theta\{z \leftarrow t_0\} \in \mathbf{T}_0^\iota$, hence

$$t' \downarrow_{T^*} = \hat{s}_S\theta\{z \leftarrow t_0\} \in \mathbf{T}_0^\iota.$$

- (IH-main) Assume that for all $\hat{r} \in \mathcal{F}$ with $\hat{r} < \hat{s}$ and $t_1, \ldots, t_m \in \mathbf{T}^\omega$ and $t = \hat{r}(Y_1, \ldots, Y_l, t_1, \ldots, t_m)$, (for $\{Y_1, \ldots, Y_l\} \subseteq V$) $eval(t, \sigma) \in \mathbf{T}_0^\iota$.
We consider $t \colon \hat{s}(X_1, \ldots, X_i, s_1, \ldots, s_j)$ for $s_1, \ldots, s_j \in \mathbf{T}^\omega$. By Definition 4.1.22 we have (as above)

$$\begin{aligned}
&r_1 = \sigma(s_1) \downarrow, \ldots, r_j = \sigma(s_j) \downarrow,\\
&\theta = \{n_1 \leftarrow r_1, \ldots, n_j \leftarrow r_j\},\\
&eval(t, \sigma) = \sigma(t) \downarrow_{T^*} = \hat{s}(X_1, \ldots, X_i, r_1, \ldots, r_j) \downarrow_{T^*} =\\
&\hat{s}_S\theta\{z \leftarrow \hat{s}(X_1, \ldots, X_i, r_1, \ldots, r_{j-1}, p(r_j)) \downarrow_{T^*} \text{ for } r_j > \mathbf{0},\\
&\quad = \hat{s}_B\theta \downarrow_{T^*} \text{ for } r_j = \mathbf{0}.
\end{aligned}$$

Again, we proceed by inner induction on $r_j$. If $r_j = \mathbf{0}$ we obtain

$$eval(t, \sigma) = \hat{s}_B \downarrow_{T^*}.$$

By definition $\hat{s}_B$ contains only $\hat{r} \in \mathcal{F}$ with $\hat{r} < \hat{s}$. By definition of a theory the terms with head symbol $\hat{r}$ can only be of the form

$$t' \colon \hat{r}(Y_1, \ldots, Y_k, s_1, \ldots, s_m) \text{ for } s_p \in \mathbf{T}^\omega$$

$$\text{and } \{Y_1, \ldots, Y_l\} \subseteq V.$$

By (IH-main) $eval(t', \sigma) \in \mathbf{T}_0^\iota$. Therefore, for all $t' \in Subt(\hat{s}_B)$ which are schematic individual terms we have $eval(t', \sigma) \in \mathbf{T}_0^\iota$.
By Lemma 4.1.10 $eval(\hat{s}_B, \sigma) \in \mathbf{T}_0^\iota$.
If $r_j > \mathbf{0}$ the argument goes like in the step case for minimal $\hat{s}$.

By Definition of the ordering $<$ in $T$ $\hat{t}$ is the maximal element and is reachable by any increasing chain starting from minimal elements. So the induction yields the property for $\hat{t}$ as well.

Now we are ready to prove the main result of this section.

**Theorem 4.1.2** *Let $T^*\colon (T, T')$ be an $\omega\iota$-theory and $t \in \hat{\mathbf{T}}^\iota(T, T')$. Then, for all $\sigma \in \mathcal{S}$, $eval(t, \sigma) \in \mathbf{T}_0^\iota$.*

***Proof*** By induction on $subt(t)$.

- $subt(t) = 1$. If $t$ is a variable expression or a constant symbol then $eval(t, \sigma) \in \mathbf{T}_0^\iota$ by Lemma 4.1.9. If $t$ is a schematic individual term then $eval(t, \sigma) \in \mathbf{T}_0^\iota$ by Lemma 4.1.11.
- Let us assume that $eval(t', \sigma) \in \mathbf{T}^\iota$ for all $t' \in \hat{\mathbf{T}}^\iota$ with $subt(t) \leq k$ and let $subt(t) = k + 1$. Then there exists an $f \in \mathbf{F}_l^\iota$ for some $l \in \mathbb{N}$ and terms $t_1, \ldots, t_l \in \hat{\mathbf{T}}^\iota(T, T')$ such that $t = f(t_1, \ldots, t_l)$. By

$$subt(t) = 1 + \sum_{i=1}^{l} subt(t_i) \text{ we get that } subt(t_i) \leq k \text{ for } i = 1, \ldots, l.$$

Like in the proof of Lemma 4.1.10 we apply the induction hypothesis and conclude that $eval(t, \sigma) \in \mathbf{T}_0^\iota$.

We show now that also the evaluation of schematic simple terms is well-defined.

**Lemma 4.1.12** *Let $T^*\colon (T, T')$ be a simple $\omega\iota$-theory and $T = (\hat{t}, \mathcal{F}, D(\mathcal{F}), V, <)$. Let $\mathcal{F}' \subseteq \mathcal{F}$ and let us us assume that for all $\hat{s} \in \mathcal{F}'$ with $\pi(\hat{s}) = (0, i, j)$, $s_1, \ldots, s_i \in \mathbf{T}_0^\iota$, $r_1, \ldots, r_j \in Num$, we have $\hat{s}(s_1, \ldots, s_i, r_1, \ldots, r_j) \downarrow_{T^*} \in \mathbf{T}_0^\iota$. Then, for all terms $t \in \hat{\mathbf{T}}_s^\iota(\mathcal{F}')$ and for all $\sigma \in \mathcal{S}$, $eval_s(t, \sigma) \in \mathbf{T}_0^\iota$.*

***Proof*** The argumentation below is analogous to that in Lemma 4.1.4. Let $t$ be a term in $\hat{\mathbf{T}}_s^\iota(\mathcal{F}')$ for some $\mathcal{F}' \subseteq \mathcal{F}$. We write $occ_{\mathcal{F}'}(t)$ for the number of occurrences of symbols in $\mathcal{F}'$ in $t$. We proceed by induction on $occ_{\mathcal{F}'}(t)$.
(IB) $occ_{\mathcal{F}'}(t) = 0$. Then $t \in \mathbf{T}_s^\iota(\emptyset) = \mathbf{T}^\iota$ and, by Lemma 4.1.9 $eval(t, \sigma) \in \mathbf{T}_0^\iota$. But on terms in $\mathbf{T}^\iota$ the evaluation functions $eval$ and $eval_s$ do not differ and so $eval_s(t, \sigma) \in \mathbf{T}_0^\iota$.
(IH) Assume the lemma holds for all $t$ such that $occ_{\mathcal{F}'}(t) \leq n$.
Let $t \in \hat{\mathbf{T}}_s^\iota(\mathcal{F}')$ and $occ_{\mathcal{F}'}(t) = n + 1$. Then $t$ must contain a subterm $s$ of the form

$$s = \hat{s}(s_1, \ldots, s_i, t_1, \ldots, t_j) \text{ for } \pi(\hat{s}) = (0, i, j), occ_{\mathcal{F}'}(s_\alpha) = 0$$
$$\text{for } \alpha \in \{1, \ldots, i\} \text{ and } t_1, \ldots, t_j \in \mathbf{T}^\omega.$$

By Definition 4.1.23 $eval_s(s, \sigma) = \hat{s}(s_1', \ldots, s_i', r_1, \ldots, r_j) \downarrow_{T^*}$ where $s_\alpha' = eval_s(s_\alpha, \sigma)$ for $\alpha = 1, \ldots, i$ and $r_\beta = \sigma(t_\beta) \downarrow$ for $\beta = 1, \ldots, j$. As the $s_\alpha$ are in $\mathbf{T}^\iota$ $eval_s(s_\alpha, \sigma) \in \mathbf{T}_0^\iota$ by Lemma 4.1.9 (see also (IB)). By

Corollary 4.1.1 $r_\beta$ is well-defined and therefore the computation of $eval_s(s, \sigma)$ yields $\hat{s}(s'_1, \ldots, s'_i, r_1, \ldots, r_j) \downarrow_{T^*}$.
By assumption $\hat{s}(s_1, \ldots, s_i, r_1, \ldots, r_j) \downarrow_{T^*} = r_0$ for some $r_0 \in \mathbf{T}_0^\iota$. By replacing the occurrence of $\sigma(s)$ by $r_0$ in $t$ we obtain a term $t'$ with $occ_{\mathcal{F}'}(t') = n$ and by (IH) $eval_s(t', \sigma) \in \mathbf{T}_0^\iota$. But then

$$eval_s(t, \sigma) = eval_s(t', \sigma) \in \mathbf{T}_0^\iota.$$

**Lemma 4.1.13** *Let $T^* \colon (T, T')$ be a simple $\omega\iota$-theory and $T = (\hat{t}, \mathcal{F}, D(\mathcal{F}), V, <)$. Then for all $\hat{s} \in \mathcal{F}$ with $\pi(\hat{s}) = (0, i, j)$, $s_1, \ldots, s_i \in \mathbf{T}_0^\iota$, $r_1, \ldots, r_j \in Num$, we have $\hat{s}(s_1, \ldots, s_i, r_1, \ldots, r_j) \downarrow_{T^*} \in \mathbf{T}_0^\iota$.*

***Proof*** We proceed by induction on the ordering $<$ in $T$. Assume that $\hat{s} \in \mathcal{F}$, $\pi(\hat{s}) = (0, k, l)$ and $\mathcal{F}' = \{\hat{u} \mid \hat{u} \in \mathcal{F}, \hat{u} < \hat{s}\}$. Note that, if $\hat{s}$ is minimal in $\mathcal{F}$ then $\mathcal{F}' = \emptyset$ and the lemma trivially holds for $\mathcal{F}'$. Now we assume

(IH) $\hat{u}(s_1, \ldots, s_i, r_1, \ldots, r_j) \downarrow_{T^*} \in \mathbf{T}_0^\iota$ for all $\hat{u} \in \mathcal{F}'$ with $\pi(\hat{u}) = (0, i, j)$,

for all $s_1, \ldots, s_i \in \mathbf{T}_0^\iota$ and for all $r_1, \ldots, r_j \in Num$.
We have to show

$+$ $\hat{s}(s_1, \ldots, s_k, r_1, \ldots, r_l) \downarrow_{T^*} \in \mathbf{T}_0^\iota$ for $s_\alpha \in \mathbf{T}_0^\iota$ and $r_\beta \in Num$.

By definition of $\downarrow_{T^*}$ we have for

$$\begin{aligned}
&\theta = \{x_1 \leftarrow s_1, \ldots, x_k \leftarrow s_k, n_1 \leftarrow r_1, \ldots, n_{l-1} \leftarrow r_{l-1}\},\\
&\hat{s}(s_1, \ldots, s_k, r_1, \ldots, r_{l-1}, r_l) \downarrow_{T^*} = t_S^{\hat{s}}\theta\{z \leftarrow t'\} \downarrow_{T^*} \text{ for } r_l > \mathbf{0} \text{ and}\\
&t' = \hat{s}(s_1, \ldots, s_k, r_1, \ldots, r_{l-1}, p(r_l)\downarrow) \downarrow_{T^*},\\
&\ t_B^{\hat{s}}\theta \downarrow_{T^*} \text{ for } r_j = \mathbf{0}.
\end{aligned}$$

To prove + we proceed by induction on $r_l$.
For $r_l = \mathbf{0}$ we obtain

$$\hat{s}(s_1, \ldots, s_k, r_1, \ldots, r_{l-1}, \mathbf{0}) \downarrow_{T^*} = t_B^{\hat{s}}\theta \downarrow_{T^*}$$

for $\theta = \{x_1 \leftarrow s_1, \ldots, x_k \leftarrow s_k, n_1 \leftarrow r_1, \ldots, n_{l-1} \leftarrow r_{l-1}\}$. By definition of $D(\hat{s})$ we have $\hat{F}(t_B^{\hat{s}}) \subseteq \mathcal{F}'$ and so $t_B^{\hat{s}} \in \hat{\mathbf{T}}^\iota(\mathcal{F}')$ and also $t_B^{\hat{s}}\theta \in \hat{\mathbf{T}}^\iota(\mathcal{F}')$ (note that $\theta$ just substitutes variables by elements in $\mathbf{T}_0^\iota$ which is a subset of $\hat{\mathbf{T}}^\iota(\mathcal{F}')$). By (IH) we have $\hat{u}(s_1, \ldots, s_i, r_1, \ldots, r_j) \downarrow_{T^*} \in \mathbf{T}_0^\iota$ for all $\hat{u} \in \mathcal{F}'$ and all $s_\alpha, r_\beta$. By Lemma 4.1.12 we conclude that $eval_s(s, \sigma) \in \mathbf{T}_0^\iota$ for all $s \in \hat{\mathbf{T}}^\iota(\mathcal{F}')$ and $\sigma \in \mathcal{S}$.

Therefore, for all $\sigma \in \mathcal{S}$ we get $eval_s(t_B^{\hat{s}}\theta, \sigma) \in \mathbf{T}_0^\iota$ and, as $t_B^{\hat{s}}\theta$ contains no parameters, $eval_s(t_B^{\hat{s}}\theta, \sigma) = t_B^{\hat{s}}\theta \downarrow_{T^*} \in \mathbf{T}_0^\iota$. Putting things together we obtain

$$\hat{s}(s_1, \ldots, s_k, r_1, \ldots, r_{l-1}, \mathbf{0}) \downarrow_{T^*} = t_B^{\hat{s}}\theta \downarrow_{T^*} \in \mathbf{T}_0^\iota.$$

Now assume that

$$(*)\ \hat{s}(s_1,\ldots,s_k,r_1,\ldots,r_{l-1},r_l)\downarrow_{T^*}\in \mathbf{T}_0^\iota.$$

By definition of $\downarrow_{T^*}$ we obtain

$$\begin{aligned}&\text{for } \theta = \{x_1 \leftarrow s_1,\ldots,x_k \leftarrow s_k, n_1 \leftarrow r_1,\ldots,n_{l-1} \leftarrow r_{l-1}\},\\ &\hat{s}(s_1,\ldots,s_k,r_1,\ldots,r_{l-1},s(r_l))\downarrow_{T^*} = t_S^{\hat{s}}\theta\{z \leftarrow t'\}\downarrow_{T^*} \text{ and}\\ &t' = \hat{s}(s_1,\ldots,s_k,r_1,\ldots,r_{l-1},p(s(r_l))\downarrow)\downarrow_{T^*}.\end{aligned}$$

By $p(s(r_l))\downarrow = r_l$ and by $(*)$ there exists an $r_0 \in Num$ such that $t' = r_0$. Therefore

$$\hat{s}(s_1,\ldots,s_k,r_1,\ldots,r_{l-1},s(r_l))\downarrow_{T^*} = t_S^{\hat{s}}\theta\{z \leftarrow r_0\}\downarrow_{T^*}.$$

By definition of $D(\hat{s})$ we have $t_S^{\hat{s}} \in \hat{\mathbf{T}}^\iota(\mathcal{F}')$. As $\theta$ replaces all $x_i$ by elements in $\mathbf{T}_0^\iota$ and all $n_\beta$ by numerals we also have $t_S^{\hat{s}}\theta\{z \leftarrow r_0\} \in \hat{\mathbf{T}}^\iota(\mathcal{F}')$. By (IH) we have $\hat{u}(s_1,\ldots,s_i,r_1,\ldots,r_j)\downarrow_{T^*}\in \mathbf{T}_0^\iota$ for all $\hat{u} \in \mathcal{F}'$ and all $s_\alpha, r_\beta$. By Lemma 4.1.12 we conclude that $eval_s(s,\sigma) \in \mathbf{T}_0^\iota$ for all $s \in \hat{\mathbf{T}}^\iota(\mathcal{F}')$ and $\sigma \in \mathcal{S}$.

In particular $t_S^{\hat{s}}\theta\{z \leftarrow r_0\} \in \hat{\mathbf{T}}^\iota(\mathcal{F}')$ and therefore $eval_s(t_S^{\hat{s}}\theta\{z \leftarrow r_0\},\sigma) \in \mathbf{T}_0^\iota$ for all $\sigma$. As $t_S^{\hat{s}}\theta\{z \leftarrow r_0\}$ contains no variables we also have

- $eval_s(t_S^{\hat{s}}\theta\{z \leftarrow r_0\},\sigma) = t_S^{\hat{s}}\theta\{z \leftarrow r_0\}\downarrow_{T^*}$.

Finally,

$$\hat{s}(s_1,\ldots,s_k,r_1,\ldots,r_{l-1},s(r_l))\downarrow_{T^*} = t_S^{\hat{s}}\theta\{z \leftarrow r_0\}\downarrow_{T^*}\in \mathbf{T}_0^\iota.$$

This concludes the inner induction and we obtain

$$\hat{u}(s_1,\ldots,s_i,r_1,\ldots,r_j)\downarrow_{T^*}\in \mathbf{T}_0^\iota \text{ for all } \hat{u} \in \mathcal{F}' \cup \{\hat{s}\} \text{ with } \pi(\hat{u}) = (0,i,j)$$

and for all $s_1,\ldots,s_i \in \mathbf{T}_0^\iota$ and for all $r_1,\ldots,r_j \in Num$.

For $\hat{s} = \hat{t}$ we obtain the result.

**Lemma 4.1.14** *Let $T^*$ be a simple $\omega\iota$ and $t$ be a term in $\hat{\mathbf{T}}^\iota(T^*)$ such that $t = t[t']_\Lambda$ for a term $t' \in \hat{\mathbf{T}}^\iota(T^*)$ and an occurrence $\Lambda$ in $t$. Then $\sigma(t)\downarrow_{T^*} = \sigma(t[\sigma(t')\downarrow_{T^*}]_\Lambda)\downarrow_{T^*}$.*

***Proof*** By induction on the depth of the occurrence $\Lambda$ as in the proof of Lemma 4.1.3.

**Theorem 4.1.3** *Let $T^*\colon (T,T')$ be a simple $\omega\iota$-theory and $t \in \hat{\mathbf{T}}_s^\iota(T^*)$. Then, for all $\sigma \in \mathcal{S}$, $eval_s(t,\sigma) \in \mathbf{T}_0^\iota$.*

***Proof*** By induction on $occ(t,\hat{F})$, the number of symbol occurrences of elements in $\hat{F}$ in $t$.

(IB) $occ(t,\hat{F}) = 0$. Then $t \in \mathbf{T}^\iota$ and $\sigma(t)\downarrow_{T^*}\in \mathbf{T}_0^\iota$ by Lemma 4.1.9.

(IH) Assume that for $occ(t,\hat{F}) = n$ the lemma is valid.

Let $t$ be a term with $occ(t,\hat{F}) = n+1$. Then there exists a $\hat{t} \in \hat{F}$ such that a term of the form

$$\hat{t}(s_1, \ldots, s_k, t_1, \ldots, t_m) \text{ for } s_i \in \mathbf{T}^\iota \text{ and } t_i \in \mathbf{T}^\omega.$$

occurs in $t$ at some position $\Lambda$ in $t$ (note that all for all schematic term symbols $\hat{t}$ we have a profile of the form $(0, p_1, p_2)$ with $p_1, p_2 > 0$, so there must be a minimal schematic term). By Lemma 4.1.14 we obtain

$$\sigma(t) \downarrow_{T^*} = \sigma(t[\sigma(\hat{t}(s_1, \ldots, s_k, t_1, \ldots, t_m)) \downarrow_{T^*}]_\Lambda) \downarrow_{T^*}.$$

But $\sigma(\hat{t}(s_1, \ldots, s_k, t_1, \ldots, t_m)) \downarrow_{T^*} =$

$$\hat{t}(\sigma(s_1) \downarrow_{T^*}, \ldots, \sigma(s_k) \downarrow_{T^*}, \sigma(t_1) \downarrow, \ldots, \sigma(t_m) \downarrow) \downarrow_{T^*}.$$

By Lemma 4.1.9 the $\sigma(s_i) \downarrow$ are in $\mathbf{T}^\iota_0$ and the $\sigma(t_i) \downarrow \in \mathit{Num}$ by Corollary 4.1.1. So we may apply Lemma 4.1.13 and get

$$\hat{t}(\sigma(s_1) \downarrow_{T^*}, \ldots, \sigma(s_k) \downarrow_{T^*}, \sigma(t_1) \downarrow, \ldots, \sigma(t_m) \downarrow) \downarrow_{T^*} = t_0 \in \mathbf{T}^\iota_0.$$

Therefore

$$\sigma(t) \downarrow_{T^*} = \sigma(t[t_0]_\Lambda]_\Lambda) \downarrow_{T^*}.$$

But now we have $occ(t[t_0]_\Lambda) \leq n$ and by (IH) we get $\sigma(t[t_0]_\Lambda]) \downarrow_{T^*} \in \mathbf{T}^\iota_0$ and therefore also $\sigma(t) \downarrow_{T^*} \in \mathbf{T}^\iota_0$.

### *4.1.3 A Higher-Order Extension of* $\hat{\mathbf{T}}^\iota$

We have introduced two schematic term languages, $\hat{\mathbf{T}}^\iota$ (the set of general schematic terms) and $\hat{\mathbf{T}}^\iota_s$ (the set of simple schematic terms). Though $\hat{\mathbf{T}}^\iota$ is more general than $\hat{\mathbf{T}}^\iota_s$ the sets are defined in a different way and so are the semantics. The reason is that we did not want to introduce second-order substitutions.

For illustration let us consider the schematic term definition from Example 4.1.4:

$$\begin{aligned} \hat{m}(X, n) &= X(\mathbf{0}) \text{ for } n = \mathbf{0}, \\ &= \max(\hat{m}(X, p(n)), X(n)) \text{ for } n > \mathbf{0}. \end{aligned}$$

or in the if-then-else formalism developed in this chapter:

$$\hat{m}(X, n) =_d \underline{\text{if}}\ n = \mathbf{0}\ \underline{\text{then}}\ X(\mathbf{0})\ \underline{\text{else}}\ \max(\hat{m}(X, p(n)), X(n)).$$

$\hat{m}(X, n)$ is a term in $\hat{\mathbf{T}}^\iota$ which is not in $\hat{\mathbf{T}}^\iota_s$. While we could nest arbitrary simple schematic terms we did not allow such a nesting for $\hat{\mathbf{T}}^\iota$. Indeed the expression

$$\hat{m}(\hat{m}(X, n), n)$$

is syntactically incorrect as the types of $X$ (the type is $\omega \to \iota$) and of $\hat{m}(X, n)$ (the type is $\iota$) are different. However we could,using typed $\lambda$-calculus (see e.g. [38]), consider the expression $\lambda k.\hat{m}(X, k)$ (for $k \in \mathcal{N}$) instead (which has the same type as $X$, namely $\omega \to \iota$). Indeed,

$$\hat{m}(\lambda k.\hat{m}(X, k), n) \text{ is well-typed.}$$

Replacing $\hat{m}(X, n)$ in the recursive definition of $\hat{m}(X, n)$ abbove, i.e. defining

$$\hat{m}(X, n)\{X \leftarrow \lambda k.\hat{m}(X, k)\}$$

we obtain for $\sigma(n) = \bar{0}$ (using application and $\beta$-reduction in $\lambda$-calculus)

$$\begin{array}{l}\sigma(\hat{m}(X, n)\{X \leftarrow \lambda k.\hat{m}(X, k)\}) \downarrow = \sigma(\hat{m}(\lambda k.\hat{m}(X, k), n)) \downarrow = \\ (\lambda k.\hat{m}(X, k))\mathbf{0} \to_\beta \hat{m}(X, \mathbf{0}),\end{array}$$

and for $\sigma(n) = \alpha > \mathbf{0}$

$$\begin{array}{l}\sigma(\hat{m}(X, n)\{X \leftarrow \lambda k.\hat{m}(X, k)\}) \downarrow = \\ \sigma(\max(\hat{m}(\lambda k.\hat{m}(X, k), p(n)), (\lambda k.\hat{m}(X, k))n)) \downarrow = \\ \max(\hat{m}(\lambda k.\hat{m}(X, k), p(\alpha)) \downarrow, (\lambda k.\hat{m}(X, k))\alpha) \downarrow) = \\ \max(\hat{m}(\lambda k.\hat{m}(X, k), p(\alpha)) \downarrow, \hat{m}(X, \alpha) \downarrow).\end{array}$$

An easy induction argument shows that for $\sigma(n) = \alpha > \bar{0}$ $\hat{m}(\lambda k.\hat{m}(X, k), \alpha)$ evaluates to

$$\max(\hat{m}(X, \bar{0}) \downarrow, \ldots, \max(\hat{m}(X, p(\alpha)), \hat{m}(X, \alpha)) \ldots).$$

This result is nothing else than

$$\hat{m}(X, \alpha) \downarrow \{X(i) \leftarrow \hat{m}(X, i) \downarrow \mid i = \mathbf{0}, \ldots, \alpha\}.$$

We see that extension of the formalism to second-order substitutions extends the expressivity of the formalism—even beyond our needs in the applications. Let $\Theta(\alpha) = \{X(i) \leftarrow \hat{m}(X, i) \downarrow \mid i = \mathbf{0}, \ldots, \alpha\}$. In order to compute the substitutions $\Theta(\alpha)$ for all $\alpha$ we obtain a schema of substitutions with increasing and unbounded domains, a feature we want to avoid. Note that $s$-substitutions defined in Chap. 6 have schematic domains of bounded size only. In principle the extension of schemata via higher-order substitutions is an interesting problem which has not been studied so far and is beyond the scope of this book.

## 4.2 Formula Schemata

Like for term schemata we will distinguish basic (theory-free symbols) and theory symbols with recursive definitions. Formula schemata will be formalized by schematic atoms which evaluate to formulas under parameter assignments. Before we give the formal definitions we present two motivating examples.

**Example 4.2.1** Let $(T_1, T_2)$ be the simple $\omega\iota$-theory which was defined in Example 4.1.11. On the basis of the given term schemata we can define formula schemata based on a two-place predicate symbol $P$ and on schematic predicate symbols $\hat{P}$, $\hat{Q}$. Let us consider the following schema ($x, y, z, z_1, z_2 \in \mathbf{V}_0^\iota$):

$$
\begin{aligned}
&P(z, \hat{t}(x, y, \bar{1}, \mathbf{0})) \wedge \neg P(\hat{s}(z_1, \mathbf{0}), z_2),\\
&P(z, \hat{t}(x, y, \bar{1}, \bar{1})) \wedge (\neg P(\hat{s}(z_1, \mathbf{0}), z_2) \vee \neg P(\hat{s}(z_1, \bar{1}), z_2)),\\
&P(z, \hat{t}(x, y, \bar{1}, \bar{2})) \wedge (\neg P(\hat{s}(z_1, \mathbf{0}), z_2) \vee \neg P(\hat{s}(z_1, \bar{1}), z_2) \vee\\
&\ \neg P(\hat{s}(z_1, \bar{2}), z_2)),\\
&etc. \ldots
\end{aligned}
$$

This sequence of formulas is characterized by conjunctions of one atom with a disjunctions of other atoms. We use $\hat{p}$ to define the whole formulas and $\hat{q}$ to define the disjuncts:

$$
\begin{aligned}
\hat{p}(x, y, z, z_1, z_2, n) &= P(z, \hat{t}(x, y, \bar{1}, n)) \wedge \hat{q}(z_1, z_2, n)\\
\hat{q}(z_1, z_2, n) &= \hat{q}(z_1, z_2, p(n)) \vee \neg P(\hat{s}(z_1, n), z_2) \text{ for } n > \mathbf{0},\\
&= \neg P(\hat{s}(z_1, \mathbf{0}), z_2) \text{ for } n = \mathbf{0}.
\end{aligned}
$$

It is easy to see that $\hat{p}$ is the "main" schematic symbol and generates the sequence above. We know that the schematic terms $\hat{t}(x, y, \bar{1}, n)$ and $\hat{s}(z_1, n)$ evaluate to terms in $\mathbf{T}_0^\iota$ under all parameter assignments.

Therefore $\hat{p}(x, y, z, z_1, z_2, n)$ generates a sequence of formulas.

As in the case of simple schematic terms the formula schema in Example 4.2.1 is also simple in the sense that the variables in the formulas remain the same. The next example shows the phenomenon of increasing variable sets which we can decribe via second-order variables like for schematic individual terms.

**Example 4.2.2** Let $(T, \emptyset)$ be the $\omega\iota$-theory in which the numeric terms are built only by $s, p, \mathbf{0}$ and parameters and $T = (\hat{t}, \{\hat{t}\}, D(\{\hat{t}\}), \{x\}, \emptyset)$ for $x \in \mathbf{V}_0^\iota$, $\pi(\hat{t}) = (0, 1, 1)$ and

$$D(\hat{t}) = \{\hat{t}(x, n_1) =_d \underline{\text{if}}\ n_1 = \mathbf{0}\ \underline{\text{then}}\ x\ \underline{\text{else}}\ h(\hat{t}(x, p(n_1)))\}.$$

We are going to define the following sequence of formulas ($Q$ is a one-place predicate symbol, $h \in \mathbf{F}_1^\iota$):

$$Q(x_0),$$
$$Q(x_0) \lor Q(h(x_1)),$$
$$Q(x_0) \lor Q(h(x_1)) \lor Q(h(h(x_2))),$$
$$etc. \ldots$$

We use a variable $X \in \mathbf{V}^{\iota}_1$ and define the sequence via a schematic predicate symbol $\hat{q}$:

$$\hat{q}(X, n) = \hat{q}(X, p(n)) \lor Q(\hat{t}(X(n), n)) \text{ for } n > \mathbf{0},$$
$$= Q(X(\mathbf{0})) \text{ for } n = \mathbf{0}.$$

We see that the defining mechanism is similar to that in $\hat{\mathbf{T}}^{\iota}$.

Let $\mathbf{P}^o_k$ be the set of of predicate symbols of type $\iota^k \to o$ for every $k \geq 0$ (where $o$ stands for the type Boolean), where $\mathbf{P}^o_0 = \{\top, \bot\}$ and $\mathbf{P}^o_k$ is infinite for every $k \geq 1$. In case of formula schemata we have to consider two types: 1. the input type, i.e. formula schemata appearing in **LK**-proof schemata which form the input for the cut-elimination procedure and 2. formula schemata corresponding to the refutational phase of the CERES method (here to the refutation of a characteristic formula schema). Like for term schemata we distinguish simple schemata (based on variables in $\mathbf{V}^{\iota}_0$) and schemata based on second-order variables. We introduce schematic predicate symbols via profiles as in Definition 4.1.17.

**Definition 4.2.1** (*schematic predicate symbols*) To each profile $\pi$ we assign an infinite set of symbols $\hat{P}(\pi)$ which we call *schematic predicate symbols*. The set obtained by taking a union over all $\hat{P}(\pi)$ for profiles $\pi$ is denoted by $\hat{P}$.

For defining recursions for schematic predicate symbols we need formula variables; the set of all formula variables is defined as $\mathbf{V}^o$ and as a name for elements in $\mathbf{V}^o$ we use $\xi, \xi_1, \ldots$.

**Definition 4.2.2** (*basic formulas*) Let $T^*\colon (T, T')$ be an $\omega\iota$-theory. The set $\mathbf{F}^o_b(T^*)$ of basic formulas in $T^*$ is defined inductively as follows.

- $\mathbf{V}^o \subseteq \mathbf{F}^o_b(T^*)$.
- If $P \in \mathbf{P}^o_k$ and $t_1, \ldots, t_k \in \hat{\mathbf{T}}^{\iota}(T^*)$ then $P(t_1, \ldots, t_k) \in \mathbf{F}^o_b(T^*)$.
- If $F \in \mathbf{F}^o_b(T^*)$ then $\neg F \in \mathbf{F}^o_b(T^*)$.
- If $F_1, F_2 \in \mathbf{F}^o_b(T^*)$ then $F_1 \land F_2, F_1 \lor F_2 \in \mathbf{F}^o_b(T^*)$.

We denote the subset of $\mathbf{F}^o_b$ containg only terms in $\mathbf{T}^{\iota}$ (restriction of the second condition) by $\mathbf{F}^o_v$. The subset of $\mathbf{F}^o_v$ containing no formula variables (restriction of the first condition) is denoted by $\mathbf{F}^o_{-}$. The subset of $\mathbf{F}^o_{-}$ containing only terms in $\mathbf{T}^{\iota}_0$ is denoted by $\mathbf{F}^o_0$.

Below we define the class of quantifier-free schematic formulas which—for simplicity—we call *schematic formulas*.

**Definition 4.2.3** (*schematic formulas*) We define the set $\hat{F}$ (the set of *schematic formulas*) inductively as follows:

- $\mathbf{V}^o \subseteq \hat{F}$.
- If $P \in \mathbf{P}^o_k$ and $t_1, \ldots, t_k \in \hat{\mathbf{T}}^\iota$ then $P(t_1, \ldots, t_k) \in \hat{F}$.
- if $\pi = (i, j, k)$ is a profile and $\hat{p} \in \hat{P}(\pi)$, $X_l \in \mathbf{V}^\iota_i$ for $l = 1, \ldots, j, t_1, \ldots, t_k \in \mathbf{T}^\omega$ then $\hat{p}(X_1, \ldots, X_j, t_1, \ldots, t_k) \in \hat{F}$.
- If $F \in \hat{F}$ then $\neg F \in \hat{F}$.
- If $F_1, F_2 \in \hat{F}$ then $F_1 \wedge F_2, F_1 \vee F_2 \in \hat{F}$.

Let $\mathcal{F} \in \hat{F}$; with $\hat{P}(F)$ we denote the set of elements in $\hat{P}$ occurring in $F$.

The following concept of an $\omega\iota o$-theory is based on an extension of an $\omega\iota$-theory.

**Definition 4.2.4** *($\omega\iota o$-theory)* A tupel $T^+ \colon (T, T', T'')$ is called an $\omega\iota o$-theory of $\hat{p}$ if

- $T = (\hat{p}, \mathcal{P}, D(\mathcal{P}), V, <)$,
- $(T', T'')$ is an $\omega\iota$-theory,
- $T' = (\hat{t}, \mathcal{F}, D(\mathcal{F}), V', <)$,
- $T'' \colon (\hat{f}, \mathcal{F}', D(\mathcal{F}'), <')$,
- $\hat{p} \in \hat{P}$,
- $\mathcal{P}$ is a finite subset of $\hat{P}$,
- $V$ is a finite subset of $\mathbf{V}^\iota_k$ for some $k \in \mathbb{N}$,
- $<$ is an irreflexive, transitive relation on $\mathcal{P}$ such that for all $\hat{q} \in \mathcal{P}\backslash\{\hat{p}\} \colon \hat{q} < \hat{p}$ ($\hat{p}$ is maximal in $\mathcal{P}$). We extend $<$ to schematic formulas as follows: if $\hat{q} < \hat{r}$ then (for $\hat{q} \in \hat{P}(i, j, k)$ and $\hat{r} \in \hat{P}(i', j', k')$) then
$$\hat{q}(X_1, \ldots, X_j, t_1, \ldots, t_k) < \hat{r}(X'_1, \ldots, X'_{j'}, t'_1, \ldots, t'_{k'})$$
for all $t_\alpha, t'_\beta$; and for all schematic formulas $F'$ containing a symbol in $\mathcal{P}$ and for all $F \in \mathbf{F}^o_b$ we have $F < F'$, i.e. the formulas in $\mathbf{F}^o_b$ are smaller than all formulas containing schematic predicate symbols.
- $D(\mathcal{P}) = \bigcup_{\hat{q} \in \mathcal{P}} D(\hat{q})$ where $D(\hat{q})$ is a set of defining equations. We assume that $\pi(\hat{q}) = (k, i, j)$, $X_l \in V$ for $l = 1, \ldots, i$ and $n_1, \ldots, n_j \in \mathcal{N}$ and
$$\begin{aligned} D(\hat{q}) = \{\hat{q}(X_1, \ldots, X_i, n_1, \ldots, n_j) =_d \underline{\text{if}}\ n_j = \mathbf{0}\ \underline{\text{then}}\ \hat{q}_B \\ \underline{\text{else}}\ \hat{q}_S\{\xi \leftarrow \hat{q}(X_1, \ldots, X_i, n_1, \ldots, n_{j-1}, p(n_j))\}\}. \end{aligned}$$

Additionally we require
- $\hat{q}_S, \hat{q}_B \in \hat{F}(T^+)$, where $\hat{F}(T^+)$ is the subset of all formula schemata $F$ in $\hat{F}$ such that $\hat{P}(F) \in \mathcal{P}$, $\hat{F}(F) \subseteq \mathcal{F}$ and $\mathbf{F}^\omega(F) \subseteq \mathcal{F}'$,
- $\mathbf{V}^\iota(\hat{q}_S) \subseteq \{X_1, \ldots, X_i\}$, $\mathbf{V}^o(\hat{q}_S) \subseteq \{\xi\}$, $\mathcal{N}(\hat{q}_S) = \{n_1, \ldots, n_j\}$; $\mathbf{V}^\iota(\hat{q}_B) \subseteq \{X_1, \ldots, X_i\}$, $\mathbf{V}^o(\hat{q}_B) = \emptyset$, $\mathcal{N}(\hat{q}_B) \subseteq \{n_1, \ldots, n_{j-1}\}$.
- Let $\mathcal{G} = \hat{P}(\hat{q}_S) \cup \hat{P}(\hat{q}_B)$; then, for all $r \in \mathcal{G} : r < \hat{q}$.

If $V'$ is a finite subset of $\mathbf{V}^\iota_0$ then $T^+$ is called a *simple* $\omega\iota o$-theory. For simple theories we require that all schematic terms occurring in $\hat{q}_S, \hat{q}_B$ are in $\hat{\mathbf{T}}^\iota_s(T, T')$. A theory is called *supersimple* if $V \cup V' \subseteq \mathbf{V}^\iota_0$.

We apply the formal definition above to Example 4.2.1. In case of a symbol $\hat{q} \in \hat{P}$ for which $\hat{q}_S = \hat{q}_B$ we write

$$D(\hat{q}) \text{ as } \{\hat{q}(X_1, \ldots, X_i, n_1, \ldots, n_i) =_d \hat{q}_B\}.$$

**Example 4.2.3** we consider the $\omega\iota o$-theory $T^+ : (T, T', T'')$ where

- $T = (\hat{p}, \{\hat{p}, \hat{q}\}, D(\{\hat{p}, \hat{q}\}), V, <)$ where $V \subseteq \mathbf{V}_0^\iota$, $V = \{x_1, \ldots, x_5\}$, $\hat{q} < \hat{p}$, $\pi(\hat{p}) = (0, 5, 1)$, $\pi(\hat{q}) = (0, 2, 1)$ and

$$D(\hat{p}) = \{\hat{p}(x_1, \ldots, x_5, n_1) =_d P(x_3, \hat{t}(x_1, x_2, \bar{1}, n_1)) \wedge \hat{q}(x_4, x_5, n_1)\},$$
$$D(\hat{q}) = \{\hat{q}(x_1, x_2, n_1) =_d \underline{\text{if}}\ n_1 = \mathbf{0}\ \underline{\text{then}}\ \neg P(\hat{s}(x_1, \mathbf{0}), x_2)$$
$$\underline{\text{else}}\ \hat{q}(x_1, x_2, p(n_1)) \vee \neg P(\hat{s}(x_1, n_1), x_2)\}.$$

Here we have $\hat{p}_S = \hat{p}_B$.
$\hat{q}_S = \xi \vee \neg P(\hat{s}(x_1, n_1), x_2)$ for $\xi \in \mathbf{V}^o$, $\hat{q}_B = \neg P(\hat{s}(x_1, \mathbf{0}), x_2)$.

- $T' = (\hat{t}, \{\hat{t}, \hat{s}\}, D(\{\hat{t}, \hat{s}\}), \{x_1, x_2\}, <')$ where $\pi(\hat{t}) = (0, 2, 2)$, $\pi(\hat{s}) = (0, 1, 1)$, $x_1, x_2 \in \mathbf{V}_0^\iota$, $\hat{s} <' \hat{t}$, $g \in \mathbf{F}_2^\iota$, $h \in \mathbf{F}_1^\iota$, and

$$D(\hat{t}) = \{\hat{t}(x_1, x_2, n_1, n_2) =_d \underline{\text{if}}\ n_2 = \mathbf{0}\ \underline{\text{then}}\ x_2$$
$$\underline{\text{else}}\ g(\hat{t}(x_1, x_2, n_1, p(n_2)), \hat{s}(x_1, n_1))\},$$
$$D(\hat{s}) = \{\hat{s}(x_1, n_1) =_d \underline{\text{if}}\ n_1 = \mathbf{0}\ \underline{\text{then}}\ x_1\ \underline{\text{else}}\ h(\hat{s}(x_1, p(n_1)))\}.$$

Here we have $\hat{t}_S = g(z, \hat{s}(x_1, n_1))$, $\hat{t}_B = x_2$, $\hat{s}_S = h(z)$, $\hat{s}_B = x_1$.

- $T'' = (\dot{f}, \{\dot{f}, \dot{g}\}, D(\{\dot{f}, \dot{g}\}), <'')$ where $\dot{f}, \dot{g} \in \mathbf{F}_2^\omega$, $\dot{g} <'' \dot{f}$ and

$$D(\dot{f}) = \{\dot{f}(n_1, n_2) =_d \underline{\text{if}}\ n_2 = \mathbf{0}\ \underline{\text{then}}\ \mathbf{0}\ \underline{\text{else}}\ \dot{g}(n_1, \dot{f}(n_1, p(n_2)))\},$$
$$D(\dot{g}) = \{\dot{g}(n_1, n_2) =_d \underline{\text{if}}\ n_2 = \mathbf{0}\ \underline{\text{then}}\ n_1\ \underline{\text{else}}\ s(\dot{g}(n_1, p(n_2)))\}.$$

$\dot{f}_S = \dot{g}(n_1, z)$ for $z \in \mathcal{N}$, $\dot{f}_B = \mathbf{0}$, $\dot{g}_S = s(z)$, $\dot{g}_B = n_1$.
It is easy to see that $T^+$ is a supersimple $\omega\iota o$-theory.

Example 4.2.2 yields the $\omega\iota o$-theory given in the example below.

**Example 4.2.4** Consider $(T, T', \emptyset)$ where

- $T = (\hat{q}, \{\hat{q}\}, D(\{\hat{q}\}), \{X\}, \emptyset)$ where $X \in \mathbf{V}_1^{\iota}$, $\pi(\hat{q}) = (1, 1, 1)$ and

$$D(\hat{q}) = \{\hat{q}(X, n_1) =_d \underline{\text{if}} n_1 = \mathbf{0} \ \underline{\text{then}} \ Q(X(\mathbf{0})) \ \underline{\text{else}} \\ \hat{q}(X, p(n_1)) \vee Q(\hat{t}(X(n_1), n_1))\}.$$

Here we have $\hat{q}_S = \xi \vee Q(\hat{t}(X(n_1), n_1))$ for $\xi \in \mathbf{V}^o$, $\hat{q}_B = Q(X(\mathbf{0}))$.

- $T' = (\hat{t}, \{\hat{t}\}, D(\{\hat{t}\}), \{x\}, \emptyset)$ for $x \in \mathbf{V}_0^{\iota}$, $\pi(\hat{t}) = (0, 1, 1)$ and

$$D(\hat{t}) = \{\hat{t}(x, n_1) =_d \ \underline{\text{if}} \ n_1 = \mathbf{0} \ \underline{\text{then}} \ x \ \underline{\text{else}} \ \{h(\hat{t}(x, p(n_1)))\}.$$

$\hat{t}_S = h(z), \hat{t}_B = x$.

Still we have to define the semantics of formulas in $\hat{F}$, i.e. we need a function $eval_{T^+}$ such that, given an $\omega\iota o$-theory $T^+$, for every $\sigma \in \mathcal{S}$ and for every $F \in \hat{F}(T^+)$ , $eval_{T^+}(F, \sigma) \in \mathbf{F}_0^o$.

**Definition 4.2.5** (*evaluation of formula schemata*) Let us consider the evaluation of formula schemata $T^+ : (T, T', T'')$, where $T = (\hat{p}, \mathcal{P}, D(\mathcal{P}), V, <)$ and $T^*$: $(T', T'')$ the corresponding $\omega\iota$-theory. For any $F \in \hat{F}(T^+)$ which does not contain formula variables we set $eval_{T^+}(F, \sigma) = \sigma(F){\downarrow_{T^+}}$; if $F$ does not contain parameters we define $\sigma(F){\downarrow_{T^+}} = F{\downarrow_{T^+}}$. $\sigma(F){\downarrow_{T^+}}$ is defined inductively below.

According to Definition 4.2.3 we define:

- $\sigma(P(t_1, \ldots, t_k)){\downarrow_{T^+}} = P(\sigma(t_1) \downarrow_{T^*}, \ldots, \sigma(t_k) \downarrow_{T^*})$ for $P \in \mathbf{P}_k^o$ and $t_1, \ldots, t_k \in \mathbf{T}^{\iota}$.
- $\sigma(\neg F){\downarrow_{T^+}} = \neg(\sigma(F){\downarrow_{T^+}})$.
- $\sigma(F \wedge G){\downarrow_{T^+}} = \sigma(F){\downarrow_{T^+}} \wedge \sigma(G){\downarrow_{T^+}}$.
- $\sigma(F \vee G){\downarrow_{T^+}} = \sigma(F){\downarrow_{T^+}} \vee \sigma(G){\downarrow_{T^+}}$.

now let $\hat{q} \in \mathcal{P}$ with $\pi(\hat{q}) = (k, i, j)$ and

$$D(\hat{q}) = \{\hat{q}(X_1, \ldots, X_i, n_1, \ldots, n_j) =_d \ \underline{\text{if}} \ n_j = \mathbf{0} \ \underline{\text{then}} \ \hat{q}_B \\ \underline{\text{else}} \ \hat{q}_S\{\xi \leftarrow \hat{q}(X_1, \ldots, X_i, n_1, \ldots, n_{j-1}, p(n_j))\}\}.$$

We define $\sigma(\hat{q}(Y_1, \ldots, Y_i, t_1, \ldots, t_j)){\downarrow_{T^+}}$ for $t_1, \ldots, t_j \in \mathbf{T}^{\omega}(T'')$ and $Y_\alpha \in \mathbf{V}_k^{\iota}$.

Let $r_1 = \sigma(t_1) \downarrow, \ldots r_j = \sigma(t_j) \downarrow$ and

$$\theta = \{X_1 \leftarrow Y_1, \ldots, X_i \leftarrow Y_i, n_1 \leftarrow r_1, \ldots, n_j \leftarrow r_j\}.$$

$$\begin{aligned}
&\sigma(\hat{q}(Y_1, \ldots, Y_i, t_1, \ldots, t_j){\downarrow_{T^+}} = \\
&\hat{q}_S\theta\{\xi \leftarrow \hat{q}_S(Y_1, \ldots, Y_i, r_1, \ldots, r_{j-1}, p(r_j) \downarrow){\downarrow_{T^+}}\}{\downarrow_{T^+}}, \ \text{for } r_j > \mathbf{0}, \\
&\sigma(\hat{q}(Y_1, \ldots, Y_i, t_1, \ldots, t_j)){\downarrow_{T^+}} = \hat{q}_B\theta{\downarrow_{T^+}} \ \text{for } r_j = \mathbf{0}.
\end{aligned}$$

The definition of evaluation of simple term schemata is the same (only that the function $\downarrow_{T^*}$ is defined differently).

**Example 4.2.5** Let $T^+ = (T, T', T'')$ as in Example 4.2.3, $T^* = (T', T'')$ and $Q \in \mathbf{P}^o_1$. Let $\sigma(n_1) = \sigma(n_2) = \bar{1}$ and

$$F = Q(\hat{t}(x_1, x_2, n_1, n_2)) \wedge \hat{p}(x_1, \ldots, x_5, n_1).$$

We compute $\sigma(F)\downarrow_{T^+}$:

$$\sigma(F)\downarrow_{T^+} = \sigma(Q(\hat{t}(x_1, x_2, n_1, n_2)))\downarrow_{T^+} \wedge \sigma(\hat{p}(x_1, \ldots, x_5, n_1))\downarrow_{T^+}.$$

We first compute $\sigma(Q(\hat{t}(x_1, x_2, n_1, n_2)))\downarrow_{T^+}$:

$$\begin{aligned} \sigma(Q(\hat{t}(x_1, x_2, n_1, n_2)))\downarrow_{T^+} &= Q(\sigma(\hat{t}(x_1, x_2, n_1, n_2)) \downarrow_{T^*}) \\ &= Q(\hat{t}(x_1, x_2, \bar{1}, \bar{1}) \downarrow_{T^*}) \text{ via} \\ &\quad\ \theta = \{n_1 \leftarrow \bar{1}, n_2 \leftarrow \bar{1}\}. \end{aligned}$$

Now we get

$$\begin{aligned} \hat{t}(x_1, x_2, \bar{1}, \bar{1}) \downarrow_{T^*} &= \hat{t}_S\theta\{z \leftarrow \hat{t}(x_1, x_2, \bar{1}, \mathbf{0}) \downarrow_{T^*}\} \downarrow_{T^*} \\ &= g(\hat{t}(x_1, x_2, \bar{1}, \mathbf{0}) \downarrow_{T^*}, \hat{s}(x_1, \bar{1}) \downarrow_{T^*}). \end{aligned}$$

$$\begin{aligned} \hat{t}(x_1, x_2, \bar{1}, \mathbf{0}) \downarrow_{T^*} &= x_2, \\ \hat{s}(x_1, \bar{1}) \downarrow_{T^*} &= h(\hat{s}(x_1, \mathbf{0})) = h(x_1). \end{aligned}$$

Hence $\hat{t}(x_1, x_2, \bar{1}, \bar{1}) \downarrow_{T^*} = g(x_2, h(x_1))$ and

$$\sigma(Q(\hat{t}(x_1, x_2, n_1, n_2)))\downarrow_{T^+} = Q(g(x_2, h(x_1))).$$

Next we compute $\sigma(\hat{p}(x_1, \ldots, x_5, n_1))\downarrow_{T^+}$:

$$\begin{aligned} \sigma(\hat{p}(x_1, \ldots, x_5, n_1))\downarrow_{T^+} &= P(x_3, \hat{t}(x_1, x_2, \bar{1}, \bar{1}))\downarrow_{T^+} \wedge \hat{q}(x_4, x_5, \bar{1})\downarrow_{T^+}. \\ P(x_3, \hat{t}(x_1, x_2, \bar{1}, \bar{1}))\downarrow_{T^+} &= P(x_3, \hat{t}(x_1, x_2, \bar{1}, \bar{1}) \downarrow_{T^*}) \\ &= P(x_3, g(x_2, h(x_1))). \end{aligned}$$

$$\begin{aligned} \hat{q}(x_4, x_5, \bar{1})\downarrow_{T^+} &= \hat{q}(x_4, x_5, \mathbf{0})\downarrow_{T^+} \vee \neg P(\hat{s}(x_4, \bar{1}), x_5)\downarrow_{T^+}. \\ \hat{q}(x_4, x_5, \mathbf{0})\downarrow_{T^+} &= \neg P(\hat{s}(x_4, 0) \downarrow_{T^*}, x_5) = \neg P(x_4, x_5). \\ \neg P(\hat{s}(x_4, \bar{1}), x_5)\downarrow_{T^+} &= \neg P(h(x_4), x_5) \text{ and therefore} \end{aligned}$$

$$\hat{q}(x_4, x_5, \bar{1})\downarrow_{T^+} = \neg P(x_4, x_5) \vee \neg P(h(x_4), x_5),$$
$$\sigma(\hat{p}(x_1, \ldots, x_5, n_1))\downarrow_{T^+} = P(x_3, g(x_2, h(x_1))) \wedge (\neg P(x_4, x_5) \vee \neg P(h(x_4), x_5)).$$

Finally we obtain $\sigma(F)\downarrow_{T^+}=$

$$Q(g(x_2, h(x_1))) \wedge P(x_3, g(x_2, h(x_1))) \wedge (\neg P(x_4, x_5) \vee \neg P(h(x_4), x_5)).$$

We show now that, given a parameter assignment $\sigma$ and an $\omega\iota o$-theory $T^+$, $\sigma(F)\downarrow_{T^+}\in \mathbf{F}^o_0$ for all $F \in \hat{F}(T^+)$.

**Lemma 4.2.1** *Let $T^+$: $(\emptyset, T, T')$ be an $\omega\iota o$-theory (that is, a theory without schematic predicate symbols) and $F \in \mathbf{F}^o_b(T^+)$ such that $F$ contains no formula variables. Then, for all $\sigma \in \mathcal{S}$, $\sigma(F)\downarrow_{T^+}\in \mathbf{F}^o_0$.*

***Proof*** Let $T^* = (T, T')$. According to Definition 4.2.2 we have to distinguish the cases

1. If $P \in \mathbf{P}^o_k$ and $t_1, \ldots, t_k \in \hat{\mathbf{T}}^\iota(T^+)$ then $P(t_1, \ldots, t_k) \in \mathbf{F}^o_-(T^+)$.
2. If $F \in \mathbf{F}^o_-(T^+)$ then $\neg F \in \mathbf{F}^o_-(T^+)$.
3. If $F_1, F_2 \in \mathbf{F}^o_-(T^+)$ then $F_1 \wedge F_2, F_1 \vee F_2 \in \mathbf{F}^o_-(T^+)$.

By Definition 4.2.5 we have

$$\sigma(P(t_1, \ldots, t_k))\downarrow_{T^+}= P(\sigma(t_1) \downarrow_{T^*}, \ldots, \sigma(t_k) \downarrow_{T^*}).$$

By $\hat{\mathbf{T}}^\iota(T^+) = \hat{\mathbf{T}}^\iota(T^*)$ and by Theorem 4.1.2 $\sigma(t_\alpha) \downarrow_{T^*}\in \mathbf{T}^\iota_0$ for all $\alpha \in \{1, \ldots, k\}$; therefore $\sigma(P(t_1, \ldots, t_k))\downarrow_{T^+}\in \mathbf{F}^o_0$.

In cases 2 and 3 above we have by Definition 4.2.5

$$\sigma(\neg F)\downarrow_{T^+} = \neg(\sigma(F)\downarrow_{T^+}),$$
$$\sigma(F \wedge G)\downarrow_{T^+} = \sigma(F)\downarrow_{T^+} \wedge \sigma(G)\downarrow_{T^+},$$
$$\sigma(F \vee G)\downarrow_{T^+} = \sigma(F)\downarrow_{T^+} \vee \sigma(G)\downarrow_{T^+}.$$

So by assuming $\sigma(F)\downarrow_{T^+}, \sigma(G)\downarrow_{T^+}\in \mathbf{F}^o_0$ we immediately obtain

$$\sigma(\neg F)\downarrow_{T^+},\ \sigma(F \wedge G)\downarrow_{T^+},\ \sigma(F \vee G)\downarrow_{T^+}\in \mathbf{F}^o_0.$$

In proving $\sigma(F)\downarrow_{T^+}\in \mathbf{F}^o_0$ for all schematic formulas without formula variables we proceed like in the proof of Theorem 4.1.2 for schematic terms.

**Definition 4.2.6** Let $T^+$ be an $\omega\iota o$-theory, $P \in \mathbf{P}^o_k(T^+)$ and $t_1, \ldots, t_k \in \hat{\mathbf{T}}^\iota(T^+)$; then $P(t_1, \ldots, t_k)$ is called an *atom* in $T^+$. Let $\pi = (k, i, j)$ be the profile of $\hat{p} \in \hat{P}(T^+)$, $X_1, \ldots, X_i \in \mathbf{V}^\iota_i$ and $t_1, \ldots, t_j \in \mathbf{T}^\omega(T^+)$. Then $\hat{p}(X_1, \ldots, X_i, t_1, \ldots, t_j)$ is called a *schematic atom* (or simply $s$-atom) in $T^+$.

The following definition is analogous to Definition 4.1.24.

**Definition 4.2.7** We define the set of subformulas $\mathit{Subf}(F)$ of a schematic formula $F$ (not containing formula variables) in an $\omega\iota o$-theory $T^+$ and a measure $\mathit{subf}$ counting the number of subformulas in $F$.

- If $F$ is an atom or an $s$-atom then $\mathit{Subf}(F) = \{F\}$ and $\mathit{subf}(F) = 1$.
- If $F = \neg G$ then $\mathit{Subf}(F) = \mathit{Subf}(G) \cup \{F\}$ and $\mathit{subf}(F) = \mathit{subf}(G) + 1$.
- If $F = G_1 \circ G_2$ for $\circ \in \{\wedge, \vee\}$ then

$$\mathit{Subf}(F) = \{F\} \cup \mathit{Subf}(G_1) \cup \mathit{Subf}(G_2),$$

$$\mathit{subf}(F) = \mathit{subf}(G_1) + \mathit{subf}(G_2) + 1.$$

**Lemma 4.2.2** *Let $T^+$ be an $\omega\iota o$-theory, $F \in \hat{F}(T^+)$ be a formula without formula variables and $\sigma \in \mathcal{S}$. Assume that for all $s$-atoms $A$ in $\mathit{Subf}(F)$ we have $\sigma(A){\downarrow_{T^+}} \in \mathbf{F}_0^o$. Then $\sigma(F){\downarrow_{T^+}} \in \mathbf{F}_0^o$.*

***Proof*** By induction on $\mathit{subf}(F)$.

If $\mathit{subf}(F) = 1$ then $F$ is either an atom or an $s$-atom. If $F$ is an atom then $\sigma(F){\downarrow_{T^+}} \in \mathbf{F}_0^o$ by Lemma 4.2.1. If $F$ is an $s$-atom then $\sigma(F){\downarrow_{T^+}} \in \mathbf{F}_0^o$ by assumption. Now we assume $\mathit{subf}(F) > 1$. Then $F$ is either of the form

(a) $\neg G$ or
(b) $G_1 \circ G_2$ for $\circ \in \{\wedge, \vee\}$.

In case (a) the set of all $s$-atoms in $\neg G$ is equal to the set of all $s$-atoms in $G$. By assumption we have $\sigma(A){\downarrow_{T^+}} \in \mathbf{F}_0^o$ for all $s$-atoms $A$ in $\mathit{Subf}(G)$. So, by induction hyypothesis, $\sigma(G){\downarrow_{T^+}} \in \mathbf{F}_0^o$. But

$$\sigma(\neg G){\downarrow_{T^+}} = \neg(\sigma(G){\downarrow_{T^+}}) \text{ and thus } \sigma(F){\downarrow_{T^+}} \in \mathbf{F}_0^o.$$

In case (b) we observe that the set of all $s$-atoms in $G_1 \circ G_2$ is the set of $s$-atoms in $G_1$ union the set of $s$-atoms in $G_2$. Now assume that for all $s$-atoms $A$ in $\mathit{Subf}(G_1) \cup \mathit{Subf}(G_2)$ we have $\sigma(A){\downarrow_{T^+}} \in \mathbf{F}_0^o$. By induction hypothesis $\sigma(G_1){\downarrow_{T^+}}, \sigma(G_2){\downarrow_{T^+}} \in \mathbf{F}_0^o$. But

$$\sigma(G_1 \circ G_2){\downarrow_{T^+}} = \sigma(G_1){\downarrow_{T^+}} \circ \sigma(G_2){\downarrow_{T^+}} \text{ and so } \sigma(F){\downarrow_{T^+}} \in \mathbf{F}_0^o.$$

**Lemma 4.2.3** *Let $T^+ = (T, T', T'')$ be an $\omega\iota o$-theory and $T = (\hat{p}, \mathcal{F}, D(\mathcal{F}), V, <)$. Assume that $\hat{q} \in \mathcal{F}$ and $\pi(\hat{q}) = (k, i, j)$. Then, for all $\sigma \in \mathcal{S}$ and for all $t_1, \ldots, t_j \in \mathbf{T}^\omega(T^+)$, $Y_1, \ldots, Y_i \in \mathbf{V}_k^\iota$ we have $\sigma(\hat{q}(Y_1, \ldots, Y_i, t_1, \ldots, t_j)){\downarrow_{T^+}} \in \mathbf{F}_0^o$.*

***Proof*** By induction on the ordering $<$ on $\mathcal{F}$.
We assume that $\hat{q}$ is minimal in $\mathcal{F}$ and $\pi(\hat{q}) = (k, i, j)$. Then

$$D(\hat{q}) = \{\hat{q}(X_1, \ldots, X_i, n_1, \ldots, n_j) =_d \underline{\text{if}}\ n_j = \mathbf{0}\ \underline{\text{then}}\ \hat{q}_B$$
$$\underline{\text{else}}\ \hat{q}_S\{\xi \leftarrow \hat{q}(X_1, \ldots, X_i, n_1, \ldots, n_{j-1}, p(n_j))\}\}.$$

for some $\hat{q}_S, \hat{q}_B$ with $\hat{q}_B \in \mathbf{F}^o_-$ and $\hat{q}_S \in \mathbf{F}^o_\upsilon$, and $V^o(\hat{q}_S) \subseteq \{\xi\}$.
Let $\sigma \in \mathcal{S}$ and $\hat{q}(Y_1, \ldots, Y_i, t_1, \ldots, t_j)$ be an $s$-atom with schematic symbol $\hat{q}$. Then, by Definition 4.2.5 $\sigma(\hat{q}(Y_1, \ldots, Y_i, t_1, \ldots, t_j))\downarrow_{T^+}$ is evaluated as follows:

Let $r_1 = \sigma(t_1) \downarrow, \ldots, r_j = \sigma(t_j) \downarrow$ and
$\theta = \{X_1 \leftarrow Y_1, \ldots, X_i \leftarrow Y_i, n_1 \leftarrow r_1, \ldots, n_j \leftarrow r_j\}$. Then
$\sigma(\hat{q}(Y_1, \ldots, Y_i, t_1, \ldots, t_j))\downarrow_{T^+}=$
$\hat{q}_S\theta\{\xi \leftarrow \hat{q}(Y_1, \ldots, Y_i, r_1, \ldots, r_{j-1}, p(r_j) \downarrow)\downarrow_{T^+}\}\downarrow_{T^+}$ for $r_j > \mathbf{0}$,
$\hat{q}_B\downarrow_{T^+}$ for $r_j = \mathbf{0}$.

As $\hat{q}_B \in \mathbf{F}^o_0$, $\mathcal{N}(\hat{q}_B\theta) = \emptyset$ (note that we have $\mathcal{N}(\hat{q}_B) \subseteq \{n_1, \ldots, n_{j-1}\}$) we have $\sigma(\hat{q}_B\theta)\downarrow_{T^+}= \hat{q}_B\theta\downarrow_{T^+}$. By Lemma 4.2.1 we obtain $\hat{q}_B\theta\downarrow_{T^+}\in \mathbf{F}^o_0$. Therefore

$$\text{for } t_j \downarrow= r_j = \mathbf{0} \text{ we have } \sigma(\hat{q}(Y_1, \ldots, Y_i, t_1, \ldots, t_j))\downarrow_{T^+}\in \mathbf{F}^o_0$$

which is the proof of the induction basis.
(IH) Assume that for $t_j \downarrow= r$ (where $r \in Num$) we have

$$\sigma(\hat{q}(Y_1, \ldots, Y_i, t_1, \ldots, t_j))\downarrow_{T^+}\in \mathbf{F}^o_0.$$

We compute $\sigma(\hat{q}(Y_1, \ldots, Y_i, t_1, \ldots, t_j))\downarrow_{T^+}$ for $t_j \downarrow= s(r)$. Then

$$\theta = \{X_1 \leftarrow Y_1, \ldots, X_i \leftarrow Y_i, n_1 \leftarrow r_1, \ldots, n_{j-1} \leftarrow r_{j-1}, n_j \leftarrow s(r)\}.$$

as $s(r) > \mathbf{0}$ we obtain

$\sigma(\hat{q}(Y_1, \ldots, Y_i, t_1, \ldots, t_j))\downarrow_{T^+}=$
$\hat{q}_S\theta\{\xi \leftarrow \hat{q}(Y_1, \ldots, Y_i, r_1, \ldots, r_{j-1}, p(s(r)) \downarrow)\downarrow_{T^+}\}\downarrow_{T^+}$.
by (IH) $\hat{q}(Y_1, \ldots, Y_i, r_1, \ldots, r_{j-1}, r)\downarrow_{T^+}\in \mathbf{F}^o_0$.
As $p(s(r)) \downarrow= r$ we obtain $\hat{q}(Y_1, \ldots, Y_i, r_1, \ldots, r_{j-1}, p(s(r)) \downarrow)\downarrow_{T^+}\in \mathbf{F}^o_0$.

As $\hat{q}_S \in \mathbf{F}^o_\upsilon$ also $\hat{q}_S\theta \in \mathbf{F}^o_\upsilon$. But we also have $V^o(\hat{q}_S\theta) \subseteq \{\xi\}$ which implies that

$$\hat{q}_S\theta\{\xi \leftarrow \hat{q}(Y_1, \ldots, Y_i, r_1, \ldots, r_{j-1}, p(s(r)) \downarrow)\downarrow_{T^+}\}\downarrow_{T^+}\in \mathbf{F}^o_0.$$

This concludes the proof for minimal $\hat{q}$ (and of the main induction basis).

Induction hypothesis (main): Assume that $\hat{q} \in \mathcal{F}$ and for all $\hat{r} < \hat{q}$, where $\pi(\hat{r}) = (k, i', j')$, and for $Y_1, \ldots, Y_{i'} \in \mathbf{V}^\iota_k, t_1, \ldots, t_{j'} \in \mathbf{T}^\omega$ we have

$$\sigma(\hat{r}(Y_1, \ldots, Y_{i'}, t_1, \ldots, t_{j'}))\downarrow_{T^+}\in \mathbf{F}^o_0.$$

We will prove that for $Y_1, \ldots, Y_i \in \mathbf{V}^\iota_k$ and $t_1, \ldots, t_j \in \mathbf{T}^\omega$ we get

$$\sigma(\hat{q}(Y_1, \ldots, Y_i, t_1, \ldots, t_j))\downarrow_{T^+}\in \mathbf{F}^o_0.$$

Again we consider $D(\hat{q})(X_1, \ldots, X_i, n_1, \ldots, n_j)$ and

$$\theta = \{X_1 \leftarrow Y_1, \ldots, X_i \leftarrow Y_i, n_1 \leftarrow r_1, \ldots, n_j \leftarrow r_j\} \text{ for } r_\alpha = \sigma(t_\alpha)\downarrow.$$

(IB) If $r_j = \mathbf{0}$ then

$$\sigma(\hat{q}(Y_1, \ldots, Y_i, t_1, \ldots, t_j))\downarrow_{T^+} = \hat{q}_B\theta\downarrow_{T^+}.$$

By definition of $D(\hat{q})$ $\hat{r} < \hat{q}$ for all schematic predicate symbols $\hat{r}$ occuring in $\hat{q}_B$—and thus also for all $\hat{r}$ occurring in $\hat{q}_B\theta$. Therefore, by the main induction hypothesis, all $s$-atoms in $\hat{q}_B\theta$ evaluate to $\mathbf{F}_0^o$. By Lemma 4.2.2 $\sigma(\hat{q}_B\theta)\downarrow_{T^+} = \hat{q}_B\theta\downarrow_{T^+} \in \mathbf{F}_0^o$. This proves (IB).
(IH) Assume that

$$\sigma(\hat{q}(Y_1, \ldots, Y_i, t_1, \ldots, t_j))\downarrow_{T^+} \in \mathbf{F}_0^o \text{ for } \sigma(t_j)\downarrow = r.$$

Then, for $\sigma(t_j)\downarrow = s(r)$ we obtain

$\theta = \{X_1 \leftarrow Y_1, \ldots, X_i \leftarrow Y_i, n_1 \leftarrow r_1, \ldots, n_{j-1} \leftarrow r_{j-1}, n_j \leftarrow s(r)\}$ and
$\sigma(\hat{q}(Y_1, \ldots, Y_i, t_1, \ldots, t_j))\downarrow_{T^+} =$
$\hat{q}_S\theta\{\xi \leftarrow \hat{q}(Y_1, \ldots, Y_i, r_1, \ldots, r_{j-1}, p(s(r))\downarrow)\downarrow_{T^+}\}\downarrow_{T^+} =$
$\hat{q}_S\theta\{\xi \leftarrow \hat{q}(Y_1, \ldots, Y_i, r_1, \ldots, r_{j-1}, r)\downarrow_{T^+}\}\downarrow_{T^+}$.
By (IH) $\hat{q}(Y_1, \ldots, Y_i, r_1, \ldots, r_{j-1}, r)\downarrow_{T^+} \in \mathbf{F}_0^o$.

Now let $\hat{r}(...)$ be an $s$-atom in $Subf(\hat{q}_S)$. Then, by definition of $D(\hat{q})$ we must have $\hat{r} < \hat{q}$. Obviously this holds also for all $s$-atoms in $Subf(\hat{q}_S\theta)$. Now let $\hat{q}(Y_1, \ldots, Y_i, r_1, \ldots, r_{j-1}, r)\downarrow_{T^+} = r'$ for a numeral $r'$. Then $\hat{q}_S\theta\{\xi \leftarrow r'\}$ does not contain formula variables and has the same $s$-atoms as $\hat{q}_S\theta$ (where all of them are smaller that $\hat{q}$). By (IH) all these $s$-atoms evaluate to $\mathbf{F}_0^o$ under any $\sigma$. Using Lemma 4.2.2 again we conclude that $\sigma(\hat{q}_S\theta\{\xi \leftarrow r'\})\downarrow_{T^+} \in \mathbf{F}_0^o$. But $\sigma(\hat{q}_S\theta\{x_i \leftarrow r'\})\downarrow_{T^+}$ is just $\sigma(\hat{q}(Y_1, \ldots, Y_i, t_1, \ldots, t_j))\downarrow_{T^+}$, which concludes the proof.

**Lemma 4.2.4** *Let $T^+$ be an $\omega\iota o$-theory, $\sigma \in \mathcal{S}$ and $F \in \hat{F}(T^+)$ such that there exists a position $\Lambda$ in $F$ and an $F' \in \hat{F}(T^+)$ such that $F = F[F']_\Lambda$. Let $\sigma \in \mathcal{S}$, then*

- $\sigma(F)\downarrow_{T^+} = \sigma(F[F'']_\Lambda)\downarrow_{T^+}$ *for* $F'' = \sigma(F')\downarrow_{T^+}$.

***Proof*** By induction on the depth of $\Lambda$ in $F$.

**Theorem 4.2.1** *Let $T^+$ be an $\omega\iota o$-theory, $\sigma \in \mathcal{S}$ and $F \in \hat{F}(T^+)$. Then $\sigma(F)\downarrow_{T^+} \in \mathbf{F}_0^o$.*

***Proof*** By Lemmas 4.2.3 and 4.2.4 by using induction on the number of occurrences of schematic atoms in $F$.

**Theorem 4.2.2** *Let $T^+$ be a simple $\omega\iota o$-theory, $\sigma \in \mathcal{S}$ and $F \in \hat{F}(T^+)$. Then $\sigma(F)\downarrow_{T^+} \in \mathbf{F}_0^o$.*

***Proof*** The proof is completely analogous to that of Theorem 4.2.1. We have only to change the evaluation function for schematic terms in $\hat{\mathbf{T}}^\iota$ to that for $\hat{\mathbf{T}}^\iota_s$ defined in Definition 4.1.23. All other steps in the proof remain unchanged.

We distinguish schemata of **LK**-proofs and refutation schemata in refutational calculi for quantifier-free logic. We call the formula schemata for **LK**-proofs input-schemata as the schematic **LK**-proofs form the input of our proof analysis method.

**Definition 4.2.8** (*input schemata*) Consider a simple $\omega\iota o$-theory $(T, T', T'')$, where $T = (\hat{p}, \mathcal{P}, D(\mathcal{P}), V, <)$.

1. Let $P \in \mathbf{P}^o_k$ and $t_1, \ldots, t_k \in \hat{\mathbf{T}}^\iota(T', T'')$ then $P(t_1, \ldots, t_k) \in \mathbf{F}_{inp}$.
2. Let $\hat{q} \in \mathcal{P}$, $\pi(\hat{q}) = (0, i, j)$, $x_1, \ldots, x_i \in \mathbf{V}^\iota_0$ and $s_1, \ldots, s_j \in \mathbf{T}^\omega$ then $\hat{q}(x_1, \ldots, x_i, s_1, \ldots, s_j) \in \mathbf{F}_{inp}$
3. If $F \in \mathbf{F}_{inp}$ then $\neg F \in \mathbf{F}_{inp}$.
4. If $F_1, F_2 \in \mathbf{F}_{inp}$ then $F_1 \wedge F_2, F_1 \vee F_2, F_1 \rightarrow F_2 \in \mathbf{F}_{inp}$.
5. If $F \in \mathbf{F}_{inp}$ and $v \in \mathbf{V}^\iota_0$ then $\forall v.F, \exists v.F \in \mathbf{F}_{inp}$.

For proof analysis via CERES the proof schemata have to be regularized, i.e. different introductions of strong quantifiers require different eigenvariables. To realize regularization in schemata we have to take into account the parameters. In Chap. 7 regularization of proofs will be achieved by using global variables and variable expressions of the form $X(n_1, \ldots, n_k)$ in replacing original eigenvariables. Hence the regularization may change the syntax defined in Definition 4.2.8.

**Definition 4.2.9** Let $X \in \mathbf{V}^\iota_k$ and $n_1, \ldots, n_k \in \mathcal{N}$; then the variable expression $X(n_1, \ldots, n_k)$ is called a *general variable expression*. The set of all general variable expressions is denoted by $\mathrm{var}^g$.

**Definition 4.2.10** (*q-schemata*) Let $F$ be an input-schema, and $\{x_1, \ldots, x_k\}$ a set of free variables in $F$. Let $\theta = \{x_1 \leftarrow t_1, \ldots, x_k \leftarrow t_k\}$ where $t_1, \ldots, t_k \in \mathrm{var}^g$. Then $F\theta$ is called a q-schema.

**Definition 4.2.11** (*semantics of input schemata*) Let $(T, T', T'')$ be a simple $\omega\iota o$-theory where $T = (\hat{p}, \mathcal{P}, D(\mathcal{P}), V, <)$and $\sigma \in \mathcal{S}$. We extend the semantics$\downarrow_{T^+}$ for simple formula schemata as defined in Definition 4.2.3 to input formulas schemata. For every $F \in \mathbf{F}_{inp}$ we define a function$\downarrow_{T_i^+}$ such that $\sigma(F)\downarrow_{T_i^+}$ is a formula in first-order predicate logic. We proceed according to Definition 4.2.8:

1. If $P \in \mathbf{P}^o_k$ and $t_1, \ldots, t_k \in \hat{\mathbf{T}}^\iota(T', T'')$ then $\sigma(P(t_1, \ldots, t_k))\downarrow_{T_i^+} = \sigma(P(t_1, \ldots, t_k))\downarrow_{T^+}$.
2. Let $\hat{q} \in \mathcal{P}$, $\pi(\hat{q}) = (0, i, j)$, $x_1, \ldots, x_i \in \mathbf{V}^\iota_0$ and $s_1, \ldots, s_j \in \mathbf{T}^\omega$ then $\sigma(\hat{q}(x_1, \ldots, x_i, s_1, \ldots, s_j))\downarrow_{T_i^+} = \sigma(\hat{q}(x_1, \ldots, x_i, s_1, \ldots, s_j))\downarrow_{T^+}$.
3. $\sigma(\neg F)\downarrow_{T_i^+} = \neg\sigma(F)\downarrow_{T_i^+}$.
4. If $\circ \in \{\wedge, \vee, \rightarrow\}$ then $\sigma(F_1 \circ F_2)\downarrow_{T_i^+} = \sigma(F_1)\downarrow_{T_i^+} \circ \sigma(F_2)\downarrow_{T_i^+}$.
5. Let $q \in \{\forall, \exists\}$. Then $\sigma(qv.F)\downarrow_{T_i^+} = qv.\sigma(F)\downarrow_{T_i^+}$.

**Theorem 4.2.3** *Let $F$ be a schematic formula in $\mathbf{F}_{inp}$ w.r.t. a simple theory $T^+$ and $\sigma \in \mathcal{S}$. Then $\sigma(F)\downarrow_{T_i^+}$ is a formula in predicate logic.*

***Proof*** By induction on the definition of a formula $F$ in $\mathbf{F}_{inp}$:

- If $F$ is of the form $P(t_1, \ldots, t_k)$ or $\hat{q}(x_1, \ldots, x_i, s_1, \ldots, s_j)$ then, by Definition 4.2.11

$$\sigma(F)\downarrow_{T_i^+} = \sigma(F)\downarrow_{T^+}$$

  and the result follows from Theorem 4.2.2.
- Assume that $F = \neg G$. Then, by Definition 4.2.11, $\sigma(\neg G)\downarrow_{T_i^+} = \neg\sigma(G)\downarrow_{T_i^+}$. As, by induction hypothesis, $\sigma(G)\downarrow_{T_i^+} \in \mathbf{F}_0^o$ so $\neg\sigma(G)\downarrow_{T_i^+} \in \mathbf{F}_0^o$.
- $F = F_1 \circ F_2$ for $\circ \in \{\wedge, \vee, \rightarrow\}$. Then $\sigma(F_1 \circ F_2)\downarrow_{T_i^+} \in \mathbf{F}_0^o$ by the induction hypothesis and $\sigma(F_i)\downarrow_{T_i^+} \in \mathbf{F}_0^o$ for $i = 1, 2$.
- $F = qv.G$ for $q \in \{\forall, \exists\}$. Then, by definition,

$$\sigma(qv.G)\downarrow_{T_i^+} = qv.\sigma(G)\downarrow_{T_i^+}$$

  and we just apply the induction hypothesis.

# Chapter 5
# Point Transition Systems

In Chap. 4 we have defined the class $\mathbf{T}^\omega$ of numeric terms. The definition mechanism we used was that of primitive recursion; in this type of recursion an ordering of the schematic function symbols is required. However, in practice of recursive definitions, cyclic calls among different defined functions cannot be avoided (we will see examples in Chap. 7).

**Example 5.0.1** To give just a flavor of the problem we define functions $f, g$ and assume that $p, +, *$ are already defined:

$$
\begin{aligned}
f(x, y) &= f(p(x), y) + g(x, p(y)) \text{ for } x > 0 \wedge y > 0,\\
f(x, y) &= x \text{ for } x > 0 \wedge y = 0,\\
f(x, y) &= y \text{ for } x = 0 \wedge y > 0,\\
f(x, y) &= 0 \text{ for } x = 0 \wedge y = 0.
\end{aligned}
$$

$$
\begin{aligned}
g(x, y) &= f(x, p(y)) + f(x, p(y)) * x \text{ for } y > 0,\\
g(x, y) &= x * x \text{ for } y = 0.
\end{aligned}
$$

When we are primarily interested in showing that $f$ and $g$ are well-defined—and not in their values—we may abstract from the function symbols $p, +, *$ from which we know that they are well-defined and specify total functions. The structure we are using here will be defined formally as *point transition system* in this chapter. The points will be the arguments of the functions, the labels the names of the functions. We use the label "end" for the cases where at most the functions $s, p, +, *$ are used in the definitions. We give the point transition system for the definitions above (the formal definitions then follow below).



A. Leitsch et al., *First-Order Schemata and Inductive Proof Analysis*, Computer Science Foundations and Applied Logic, https://doi.org/10.1007/978-3-032-05741-9_5

$$
\begin{aligned}
(f, (x, y)) &\to \{(f, (p(x), y)), (g, (x, p(y))\}: x > 0 \wedge y > 0,\\
(f, (x, y)) &\to \{(\text{end}, (x, y))\}: x > 0 \wedge y = 0,\\
(f, (x, y)) &\to \{(\text{end}, (x, y))\}: x = 0 \wedge y > 0,\\
(f, (x, y)) &\to \{(\text{end}, (x, y))\}: x = 0 \wedge y = 0.
\end{aligned}
$$

$$
\begin{aligned}
(g, (x, y)) &\to \{(f, (x, p(y))\}: y > 0,\\
(g, (x, y)) &\to \{(\text{end}, (x, y))\}: y = 0.
\end{aligned}
$$

By using an adequate tupel ordering for the arguments of the functions $f$ and $g$ we can show that $f$ and $g$ are indeed well-defined. Such orderings and the concept of canonic point transition systems will be defined in this chapter.

We are now developing the formal framework of point transition systems.

**Definition 5.0.1** (*Point*) A *point* is an element of $\mathcal{P}^\alpha$ for $\alpha \geq 1$ and $\mathcal{P} \subseteq T_0$ for $T_0$ as in Definition 4.1.1.

**Definition 5.0.2** (*Labeled point*) Let $\Delta$ be a countably infinite set of symbols called *labels*. A pair $(\delta, p)$, where $p$ is a point and $\delta$ is a label is called a *labeled point*.

**Definition 5.0.3** (*Condition*) An *atomic condition* is either $\top, \bot$ or of the form $s < t, s > t, s = t, s \neq t$ for $s, t \in T_0$. Let ATC the set of atomic conditions; the set COND of all conditions is defined as follows:

- ATC $\subseteq$ COND,
- if $C \in$ COND then $\neg C \in$ COND,
- if $C_1, C_2 \in$ COND then $C_1 \wedge C_2 \in$ COND and $C_1 \vee C_2 \in$ COND.

**Example 5.0.2** $n = 0, m < k + 1$ are atomic conditions, $n = 0 \wedge \neg m < k + 1$ and $(n = 0 \wedge m < k + 1) \vee m - 1 > k$ are conditions.

The semantics of a condition $C$ is obtained by interpreting the variables by a parameter assigment $\sigma$ and by evaluating $C$ under $\sigma$.

**Definition 5.0.4** For every $C \in$ COND and $\sigma \in \mathcal{S}$ we define $\sigma[C] \in \{\top, \bot\}$, the truth value of $C$ under $\sigma$:

- $\sigma[\top] = \top,\ \sigma(\bot) = \bot$,
- $\sigma[s = t] = \top$ iff $\sigma(s)\downarrow = \sigma(t)\downarrow$,
- $\sigma[s < t] = \top$ iff $\sigma(s)\downarrow < \sigma(t)\downarrow$,
- $\sigma[s > t] = \top$ iff $\sigma(t)\downarrow < \sigma(s)\downarrow$,
- $\sigma[s \neq t] = \top$ iff $\sigma[s = t] = \bot$,
- $\sigma[\neg C] = \top$ iff $\sigma[C] = \bot$,
- $\sigma[C_1 \wedge C_2] = \top$ iff $\sigma[C_1] = \top$ and $\sigma[C_2] = \top$,
- $\sigma[C_1 \vee C_2] = \top$ iff $\sigma[C_1] = \top$ or $\sigma[C_2] = \top$.

A condition $C$ is called *valid* if for all $\sigma \in \mathcal{S}$: $\sigma[C] = \top$, and *unsatisfiable* if for all $\sigma \in \mathcal{S}$: $\sigma[C] = \bot$.

**Definition 5.0.5** (*Point transition*) A *point transition* is an expression of the form

$$(\delta, p) \to \mathcal{L}: C$$

where $(\delta, p)$ is a labeled point, $\mathcal{L}$ is a nonempty set of labeled points, and $C$ is a condition. To extract the different parts of a point transition we define the functions $l, r, c$ as follows: if $t = (\delta, p) \to \mathcal{L}: C$ then $l(t) = (\delta, p), r(t) = \mathcal{L}$ and $c(t) = C$.

To every label $\delta$ we assign points of the same arity; the result is a so-called point transition cluster.

**Definition 5.0.6** (*Point transition cluster*) A finite set of point transitions $\mathbf{P}$ is called a *point transition cluster* if for all $\delta \in \Delta$ and points $p, q$ such that $(\delta, p)$ and $(\delta, q)$ occur in $\mathbf{P}$ (either as $l(t)$ or as $r(t)$ for a $t \in \mathbf{P}$) there exists an $\alpha \geq 1$ such that $p, q \in \mathcal{P}^\alpha$. We write $\Delta(\mathbf{P})$ for the set of all $\delta \in \Delta$ occurring in $\mathbf{P}$; with $A(\delta)$ we denote the arity of the points corresponding to $\delta$.

**Example 5.0.3** Let

$$\begin{aligned}\mathbf{P}_1 = \{&(\delta, (m+1, n)) \to \{\delta, (m-1, n+1), \\ &(\delta', (m, n, m)\}: m > 0 \wedge n < m, \\ &(\delta, (m, n)) \to \{(\delta', (n, n, n)\}: m < n\end{aligned}$$

$\mathbf{P}_1$ is a point transition cluster: we have $A(\delta) = 2$, $A(\delta') = 3$. The set $\mathbf{P}_2$ defined as

$$\begin{aligned}\mathbf{P}_2 = \{&(\delta, (m+1, n)) \to \{\delta, (m-1, n+1), \\ &(\delta', (m, n, m)\}: m > 0 \wedge n < m, \\ &(\delta, (m, n)) \to \{(\delta', (n, n)\}: m < n\end{aligned}$$

is not a point transition cluster as $A(\delta')$ cannot be defined consistently.

Consider the point transition cluster $\mathbf{P}_1$ in Example 5.0.3. The conditions $m > 0 \wedge n < m$ and $m < n$ exclude each other, but they are not exhaustive: we do not know what happens when $m = n$. In point transition systems (restricted forms of point transition clusters) the conditions must define a partition and the point transitions must be of a restricted form.

**Definition 5.0.7** (*Partition*) Let $\mathcal{C} = \{C_1, \ldots, C_\alpha\}$ be a set of conditions. $\mathcal{C}$ is called a *partition* if $C_1 \vee \cdots \vee C_\alpha$ is valid and for all $i, j \in \{1, \ldots, \alpha\}$ such that $i \neq j$ we have that $C_i \wedge C_j$ is unsatisfiable.

We do not consider arbitrary partitions but only those of "if-then-else type". Consider e.g. the following if-then-else commands:

(1) If $C_1$ <u>then</u> $x_1$ <u>else</u> $x_2$,
(2) If $C_1$ <u>then</u> $x_1$ <u>else</u> <u>if</u> $C_2$ <u>then</u> $x_2$ <u>else</u> $x_3$,
(3) If $C_1$ <u>then</u> $x_1$ <u>else</u> <u>if</u> $C_2$ <u>then</u> $x_2$ <u>else</u> <u>if</u> $C_3$ <u>then</u> $x_3$ <u>else</u> $x_4$.

The partitions induced by the commands (1)–(3) are the following ones:

- for (1) $\mathcal{C}_1 : \{C_1, \neg C_1\}$,
- for (2) $\mathcal{C}_2 : \{C_1, \neg C_1 \wedge C_2, \neg(C_1 \vee C_2)\}$,
- for (3) : $\mathcal{C}_3 : \{C_1, \neg C_1 \wedge C_2, \neg C_1 \wedge \neg C_2 \wedge C_3, \neg(C_1 \vee C_2 \vee C_3)\}$.

We see that any set of conditions $C_1, \ldots, C_n$ defines a partition

- $\{C_1, C_2', \ldots, C_n', D\}$

for appropriate sets $C_2', \ldots, C_d', D$.

**Definition 5.0.8** Let $\mathcal{L}: (C_1, \ldots, C_n)$ be a list of conditions. Then the *if-then-else partition induced* by $\mathcal{L}$ is of the form

$$\{D_1, \ldots, D_n, D_{n+1}\}$$

where

- $D_1 = C_1$,
- $D_i = \neg C_1 \wedge \cdots \wedge C_{i-1} \wedge C_i$ for $i = 2, \ldots, n$,
- $D_{n+1} = \neg(C_1 \vee \cdots \vee C_n)$.

It is obvious (use elementary propositional logic) that an if-then-else partition is indeed a partition. From now on we consider only if-then-else partitions which can be modified for practical purposes by replacing the $D_i$ by equivalent formulas $D_i'$.

**Example 5.0.4** Let $C_1 = m > 0 \wedge n > 0$, $C_2 = n > 0$, $C_3 = m > 0$ and $\mathcal{L} = (C_1, C_2, C_3)$ then $\{D_1, D_2, D_3, D_4\}$ is an if-then-else partition induced by $\mathcal{L}$ for

$$\begin{aligned} D_1 &= m > 0 \wedge n > 0, \\ D_2 &= \neg(m > 0 \wedge n > 0) \wedge n > 0, \\ D_3 &= \neg(m > 0 \wedge n > 0) \wedge \neg n > 0 \wedge m > 0, \\ D_4 &= \neg((m > 0 \wedge n > 0) \vee n > 0 \vee m > 0). \end{aligned}$$

It is easy yo see that $D_2$ is equivalent to $m = 0 \wedge n > 0$, $D_3$ is equivalent to $m > 0 \wedge n = 0$ and $D_4$ to $m = 0 \wedge n = 0$. We only have to use simple rules like the de Morgan rules, $\neg m > 0 \sim m = 0$, $\neg n > 0 \sim n = 0$, $n = 0 \wedge n > 0 \sim \bot$ and distributivity. So we obtain a modified partition of the form $\mathcal{C}: \{C_1', C_2', C_3', C_4'\}$ where

$$C_1' = m > 0 \wedge n > 0,\ C_2' = m = 0 \wedge n > 0,\ C_3' = m > 0 \wedge n = 0,$$
$$C_4' = m = 0 \wedge n = 0.$$

From now on we do not distinguish between if-then-else partitions and modified if-then-else partitions.

**Definition 5.0.9** (*Regular point transition*) Let $t\colon (\delta, p) \to \mathcal{L}\colon C$ be a point transition in a point transition cluster and $A(\delta) = \alpha$. $t$ is called *regular* if

- $p = (n_1, \ldots, n_\alpha)$ for distinct parameters $n_1, \ldots, n_\alpha$,
- for all $(\delta', q) \in \mathcal{L}$ $V(q) \subseteq \{n_1, \ldots, n_\alpha\}$,
- $V(C) \subseteq \{n_1, \ldots, n_\alpha\}$.

Note that the first point transition in the point transition cluster $\mathbf{P}_1$ in Example 5.0.3 is not regular as $p = (m + 1, n)$, but it should be of the form $(m, n)$.

**Definition 5.0.10** (*Point transition system*) Let $\mathbf{P}$ be a point transition cluster and $\delta_0 \in \Delta(\mathbf{P})$. Then the tuple $\mathbf{P}^*\colon (\delta_0, \Delta^*, \Delta_e, \mathbf{P})$ is called a *point transition system* if the following conditions are fulfilled:

- $\delta_0 \in \Delta^*$.
- $\Delta^* = \Delta(\mathbf{P})$.
- All point transitions in $\mathbf{P}$ are regular.
- Let $t_1, t_2 \in \mathbf{P}$ such that $l(t_1) = (\delta, p)$ and $l(t_2) = (\delta, q)$; then $p = q$—i.e. every left-hand side of a point transition corresponding to $\delta$ has the same point. We call $p$ the *source* of $\delta$.
- Let $\mathbf{P}(\delta) = \{(\delta, p) \to \mathcal{L}_1\colon C_1, \ldots, (\delta, p) \to \mathcal{L}_\alpha\colon C_\alpha\}$ be the set of all point transitions $t$ in $\mathbf{P}$ with $l(t) = (\delta, p)$. Then, for $\mathbf{P}(\delta) \neq \emptyset$, $\{C_1, \ldots, C_\alpha\}$ is an if-then-else partition.
- $\mathbf{P}(\delta_0) \neq \emptyset$.
- $\mathbf{P}(\delta) = \emptyset$ for $\delta \in \Delta_e$.

The label $\delta_0$ is called the *start label* of $\mathbf{P}^*$, any $\delta \in \Delta_e$ is called an *end label* of $\mathbf{P}$.

**Proposition 5.0.1** *It is decidable whether a point transition cluster is a point transition system.*

***Proof*** All conditions are purely syntactic and trivially decidable. Also if-then-else conditions are purely syntactic (they are induced by a list of conditions), modified if-then-else conditions can be obtained by normalization.

**Example 5.0.5** Consider Example 5.0.1. We reformulate the point transition system given there such that it is closer to the formal definition above. For $f$ we choose $\delta_0$, $\delta_1$ for $g$ and $\delta_e$ for "end". Then $\Delta^* = \{\delta_0, \delta_1, \delta_e\}$, $\Delta_e = \{\delta_e\}$ and

$$\begin{aligned}
\mathbf{P}(\delta_0) = \{&(\delta_0, (x, y)) \to \{(\delta_0, (p(x), y)), (\delta_1, (x, p(y))\}\colon x > 0 \land y > 0,\\
&(\delta_0, (x, y)) \to \{(\delta_e, (x, y))\}\colon x > 0 \land y = 0,\\
&(\delta_0, (x, y)) \to \{(\delta_e, (x, y))\}\colon x = 0 \land y > 0,\\
&(\delta_0, (x, y)) \to \{(\delta_e, (x, y))\}\colon x = 0 \land y = 0\}.\\
\mathbf{P}(\delta_1) = \{&(\delta_1, (x, y)) \to \{(\delta_0, (x, p(y))\}\colon y > 0,\\
&(\delta_1, (x, y)) \to \{(\delta_e, (x, y))\}\colon y = 0\}.
\end{aligned}$$

$$\mathbf{P}(\delta_e) = \emptyset.$$

Indeed, all point transitions are regular and the conditions belonging to $\mathbf{P}(\delta_0)$ and $\mathbf{P}(\delta_1)$ define an if-then-else partition.

Every point transition system defines a computation with respect to parameter assignments. We represent such a computation by a tree.

**Definition 5.0.11** (*σ-tree*) Let $\mathbf{P}^*\colon (\delta_0, \Delta^*, \Delta_e, \mathbf{P})$ be a point transition system, $\sigma \in \mathcal{S}$, and $(\delta, p)$ a labeled point in $\mathcal{P}$. We define the *σ-tree* $T(\mathbf{P}, (\delta, p), \sigma)$ inductively:

We define trees $T(\mathbf{P}, (\delta, p), \sigma, \alpha)$ for all $\alpha \in \mathbb{N}$.

$T(\mathbf{P}, (\delta, p), \sigma, 0)$ is the tree consisting just of the root node $\nu$ labeled with $(\delta, \sigma(p)\downarrow)$. Let us assume that we have already defined the tree $T_\alpha\colon T(\mathcal{P}, (\delta, p), \sigma, \alpha)$ and let $\nu$ be a leaf in $T_\alpha$ and $\nu$ is labelled by $(\delta', q)$ where the arity of $q$ is that of the lefthandside of $\mathcal{P}(\delta')$ and $q$ is a tuple of numerals. If $\mathcal{P}(\delta') = \emptyset$ then $\nu$ is also a leaf of $T(\mathcal{P}, (\delta, p), \sigma, \alpha + 1)$ with the same label.
If $\mathcal{P}(\delta') \neq \emptyset$ then

$$\mathcal{P}(\delta') = \{(\delta', p') \to \mathcal{P}_1\colon C_1, \ldots, (\delta', p') \to \mathcal{P}_\beta\colon C_\beta\}$$

As all point transitions are regular and $q$ has the right arity there exists a $\sigma' \in \mathcal{S}$ such that $\sigma'(p') = q$. By Definition 5.0.10 the $C_i s$ define a partition and therefore there exists exactly one $C_i$ such that $\sigma'(C_i) = \top$. Now let

$$\mathcal{P}_i = \{(\delta^i_1, q^i_1), \ldots, (\delta^i_{r(i)}, q^i_{r(i)})\}.$$

Now we define nodes $\mu_1, \ldots, \mu_{r(i)}$ labeled with

$$(\delta^i_1, \sigma'(q^i_1)\downarrow), \ldots, (\delta^i_{r(i)}, \sigma'(q^i_{r(i)})\downarrow).$$

As all productions in $\mathbf{P}$ are regular the $\sigma'(q^i_j)$ are ground and $\sigma'(q^i_j)\downarrow$ are numerals of the right arity.

Now let $T(\mathbf{P}, (\delta, p), \sigma, \alpha) = (V_\alpha, E_\alpha)$ for all $\alpha \in \mathbb{N}$. Note that, for all $\alpha$, we have $V_\alpha \subseteq V_{\alpha+1}$ and $E_\alpha \subseteq E_{\alpha+1}$. When we define $V_\infty = \bigcup_{\alpha\in\mathbb{N}} V_\alpha$ and $E_\infty = \bigcup_{\alpha\in\mathbb{N}} E_\alpha$ then $(V_\infty, E_\infty)$ is a tree; this tree is called a *σ-tree on* $(\delta, p)$ and is denoted by $T(\mathbf{P}, (\delta, p), \sigma)$. We define the tree $T_{\mathbf{P}^*}[\sigma]\colon T(\mathbf{P}, (\delta_0, p_0), \sigma)$ as the *computation-tree* of $\mathbf{P}^*$ with respect to $\sigma$.

The tree $\mathbf{P}^*[\sigma]$ defines the computation of the point transition system on input $\sigma$. $\mathbf{P}^*[\sigma]$ may be finite or infinite; in the latter case the computation is nonterminating. We have seen that $\mathbf{P}^*[\sigma]$ may be finite, but for another $\sigma'$ infinite. The desirable property is the termination for all $\sigma \in \mathcal{S}$.

**Definition 5.0.12** (*Terminating point transition system*) Let $\mathbf{P}^*$ be a point transition system. $\mathbf{P}^*$is called *terminating* if, for all $\sigma \in \mathcal{S}$, $\mathbf{P}^*[\sigma]$ is finite.

In order to ensure the termination of a point transition system some syntactic criteria are necessary. To this aim we will define first ordering on points, then ordering on labels and finally combine these orderings to a well-founded ordering on labeled points.

**Definition 5.0.13** (*Point ordering*) Let $\mathcal{P}_0^\alpha$ be the subclass of $\mathcal{P}^\alpha$ consisting of $\alpha$-tuples of numerals and let $<_\alpha$ be a well-founded ordering on $\mathcal{P}_0^\alpha$ (e.g. the lexicographic ordering).

**Definition 5.0.14** (*Label ordering*) Let $\Delta^*$ be a nonempty finite subset of $\Delta$. We define a tree inductively which consists of nodes which are disjoint subsets of $\Delta^*$. We define a tree $T_0$ consisting of the root node $\Delta^0$ where $\Delta^0$ is a nonempty subset of $\Delta^*$ and contains the label $\delta_0$.
Suppose we have defined a tree $T_r$ with nodes $\Delta^0, \ldots, \Delta^k$ such that the $\Delta_j$ are nonempty subsets of $\Delta^*$ and are pairwise disjoint. If the union of the $\Delta^i$ is equal to $\Delta^*$ we are done. So let us assume that

$$\Delta' = \Delta^* \backslash (\Delta^0 \cup \ldots \cup \Delta^k) \text{ and } \Delta' \neq \emptyset.$$

Then we select a subset $\Delta''$ of $\Delta'$ and partition $\Delta''$ in nonempty disjoint subsets $\Delta_1^+, \ldots, \Delta_m^+$ which are becoming new nodes in the tree. Furthermore we select a $\Delta^j$ and append the edges $(\Delta^j, \Delta_1^+), \ldots, (\Delta^j, \Delta_m^+)$ to $\Delta^j$. The result is a new tree which we call $T_{r+1}$. Note that the construction terminates as $\Delta^*$ is finite and we add a nonempty set of labels in every step.
The final tree is called $T(\Delta^*)$. On $T(\Delta^*)$ we define the usual path ordering on trees: let $\mu, \nu$ be nodes in $T(\Delta_0)$ and $\mu \neq \nu$; we define $\nu <_T \mu$ if there exists a path from $\mu$ to $\nu$ in $T(\Delta_0)$. There exists a single maximal node, the root node $\Delta^0$ and (potentially many) minimal nodes, the leaves.

Note that, in general, there are various ways to define a tree $T(\delta_0)$ of the basis of $\Delta_0$.

**Example 5.0.6** Consider the point transition system of Example 5.0.5. We have $\Delta^* = \{\delta_0, \delta_1, \delta_e\}$. We first define $\Delta^0 = \{\delta_0, \delta_1\}$; then $\Delta^* \backslash \Delta_0 = \{\delta_e\}$. We obtain the new node $\Delta_1 = \{\delta_e\}$ and create the edge $(\Delta_0, \Delta_1)$. As we have exhausted the labels in $\Delta^*$ the construction stops and our tree consisting of the root $\Delta_0$ and the edge $(\Delta_0, \Delta_1)$ is our tree $T(\Delta^*)$. It is this tree which can be used to prove termination of the point transition system. However, there are several other possibilities to define trees over $\Delta^*$; even for defining the root we have three possibilities.

**Definition 5.0.15** (*Ordering on labeled points*) Let $\Delta^*$ be a finite subset of $\Delta$ and $\mathcal{P}_0 = \bigcup_{\alpha \in \mathbb{N}} \mathcal{P}_0^\alpha$. We define an ordering on labeled points $(\delta, p)$ for $\delta \in \Delta^*$ and $p \in \mathcal{P}_0$ based on the orderings $<_\alpha$ and $<_T$ (based on $T(\Delta^*)$):

$(\delta, p) <_l (\delta', q)$ if either

(p1) There exists a node $N$ in $T(\Delta^*)$ such that $\delta, \delta' \in N$, $A(\delta) = A(\delta') = \alpha$ and $p <_\alpha q$ or
(p2) $\delta \in M$ and $\delta' \in N$ for nodes $M, N \in T(\Delta^*)$ and $M <_T N$.

**Example 5.0.7** Let $\Delta^* = \{\delta_0, \delta_1, \delta_e\}$ and $T(\Delta^*)$ be the tree from Example 5.0.6 with $\Delta^0 = \{\delta_0, \delta_1\}$ and $\Delta_1 = \{\delta_e\}$. By construction we have $\Delta_1 <_T \Delta_0$ and therefore, for any $p, q \in \mathcal{P}_0^2$ we have $(\delta_e, p) <_l (\delta_0, q)$ and $(\delta_e, p) <_l (\delta_1, q)$.
So let us consider the node $\Delta^0$. We have $\delta_0, \delta_1 \in \Delta_0$ and $A(\delta_0) = A(\delta_1) = 2$. Let $<_2$ the standard lexicographic order on $\mathcal{P}_0^2$ which gives us $(2, 3) <_2 (3, 2)$. Then we have $(\delta_0, (2, 3)) <_l (\delta_1, (3, 2))$ and $(\delta_1, (2, 3)) <_l (\delta_0, (3, 2))$, but not $(\delta_0, (3, 2)) < (\delta_1, (2, 3))$ or $(\delta_1, (3, 2)) < (\delta_1, (2, 3))$.

The relation $<_l$ depends on the choice of $\Delta^*$ and $T(\Delta^*)$. For simplicity we do not indicate $T(\Delta^*)$ in the relation but assume that definitions of $\Delta^*$ and $T(\Delta^*)$ are given.

**Proposition 5.0.2** *The relation $<_l$ is well-founded.*

***Proof*** We show first that $<_l$ is irreflexive and transitive. Irreflexivity is trivial by (p1) and (p2) in Definition 5.0.15 and the fact that both $<_\alpha$ and $<_T$ are irreflexive. For showing transitivity we have to check some possible cases appearing in the tree $T(\Delta_0)$. So let us assume that $(\delta, p) <_l (\delta', q)$ and $(\delta', q) <_l (\delta'', r)$.

- case 1: $\delta, \delta', \delta'' \in N$ for some node $N$ in $T(\Delta_0)$.
  Then, by (p1), $p <_\alpha q$ and $q <_\alpha r$ for some $\alpha \in \mathbb{N}$. As $<_\alpha$ is transitive we obtain $(\delta, p) <_l (\delta'', r)$.
- case 2: $\delta, \delta' \in M$, $\delta'' \in N$ for two different nodes $M, N \in T(\Delta_0)$.
  Then, by $\delta' \in M$ and $\delta'' \in N$, $(\delta', q) <_l (\delta'', r)$ implies $M <_T N$. But then also $(\delta, p) <_l (\delta'', r)$ by $M <_T N$.
- case 3: $\delta \in M$ and $\delta', \delta'' \in N$ for two different nodes $M, N \in T(\Delta_0)$.
  By $\delta \in M$, $\delta' \in N$ and $(\delta, p) <_l (\delta', q)$ we must have $M <_T N$. But then also $(\delta, p) <_l (\delta'', r)$ by $M <_T N$.
- case 4: there are $M, N, L$ in the nodes of $T(\Delta_0)$ such that $M, N, L$ are pairwise different. Then $(\delta, p) <_l (\delta', q)$ and $(\delta', q) <_l (\delta'', q)$ implies $M <_T N$ and $N <_T L$. By transitivity of $<_T$ we get $M <_T L$ and thus $(\delta, p) <_l (\delta'', r)$.

This shows that $<_l$ is transitive.
Now assume that $<_l$ is not well founded. Then there exists an infinite sequence $\xi\colon (\delta_i, p_i)_{i \in \mathbb{N}}$ such that $(\delta_{i+1}, p_{i+1}) <_l (\delta_i, p_i)$ for all $i \in \mathbb{N}$. As $\xi$ is infinite, but $\Delta_0$ is finite, there exists an infinite subsequence $\zeta$ of $\xi$ such that all labeled points are labeled by one and the same label, let us say $\delta$. Then $\zeta$ is of the form

$$(\delta, q_0), (\delta, q_1), \ldots, (\delta, q_i), (\delta, q_{i+1}), \ldots,$$

and $(\delta, q_{i+1}) <_l^* (\delta, q_i)$ for the transitive closure $<_l^*$ of $<_l$. But we have shown that $<_l$ is transitive and so $(\delta, q_{i+1}) <_l (\delta, q_i)$ for all $i \in \mathbb{N}$. Let $A(\delta) = \alpha$. Then, by (p1) in Definition 5.0.15, we must have $q_{i+1} <_\alpha q_i$ for all $i$, which is impossible by the well-foundedness of $<_\alpha$. This shows that $<_l$ is well-founded.

For defining termination conditions for point transition systems we have to extend the ordering from points consiting of numerals to arbitrary points.

**Definition 5.0.16** Let $(\delta, p)$, $(\delta', p')$ two labeled points and $C$ be a condition. We define $(\delta, p) <_C (\delta', p')$ if for all $\sigma \in \mathcal{S}$ such that $\sigma[C] = \top$ we get $(\delta, \sigma(p)\downarrow) <_l (\delta', \sigma(p')\downarrow)$.

**Example 5.0.8** We illustrate that the conditions in the ordering are vital. Consider the standard ordering on the numerals, extend it lexicographically to 3-tuples and consider a correponding $<_l$-ordering; let $\delta$ be a label, then

$$(\delta, (n, p(m), s(k))) <_C (\delta, (n, m, k)) \text{ for } C\colon m > 0.$$

Indeed, for all $\sigma$ with $\sigma[C] = \top$ (i.e. $\sigma(m) \neq \bar{0}$), we have

$$(\sigma(n), \sigma(p(m))\downarrow, \sigma(s(k)) <_3 (\sigma(n), \sigma(m), \sigma(k))$$

as $\sigma(p(m))\downarrow$ is a proper subterm of $\sigma(m)\downarrow$ and thus $p(\bar{\alpha})\downarrow < \bar{\alpha}$ for $\bar{\alpha} = \sigma(m)\downarrow$.

However,

$$(\delta, (n, p(m), s(k))) <_C (\delta, (n, m, k)) \text{is wrong for } C\colon \top.$$

Choose $\sigma(n) = \sigma(m) = \sigma(k) = \bar{0}$. Then substituting $\sigma$ in the relation above (note that $\sigma[\top] = \top$) we obtain

$$(\delta, (\bar{0}, \bar{0}, \bar{1})) <_l (\delta, (\bar{0}, \bar{0}, \bar{0}))$$

which is obviously wrong.

**Proposition 5.0.3** *Let $(\delta, p)$, $(\delta', q)$ be labeled points and $C$ be a condition. Then $(\delta, p) <_C (\delta', q)$ is decidable.*

***Proof*** Let $T(\Delta^*)$ be the tree corresponding to the relation $<_C$. By Definition 5.0.16 $(\delta, p) <_C (\delta', q)$ iff

$$(*) \text{ for all } \sigma \in \mathcal{S} \text{ such that } \sigma[C] = \top \text{ we have} (\delta, \sigma(p)\downarrow) <_l (\delta', \sigma(q)\downarrow).$$

We distinguish two cases

(a) $\delta, \delta'$ are in different nodes $M, N$ of $T(\Delta^*)$, and $\delta \in M, \delta' \in N$.
(b) There exists a node $N$ in $T(\Delta^*)$ such that $\delta, \delta' \in N$.

In case (a) we check whether $M <_T N$. If this is the case then, by Definition 5.0.15 $(\delta, r) <_l (\delta', s)$ for all points $r, s \in \mathcal{P}_0$ of the right arity. In particular $(*)$ trivially holds. If $M \not<_T N$ then, for no $s, r$, we have $(\delta, r) <_l (\delta', s)$ and thus $(*)$ does not hold.

In case (b) we check whether $A(\delta) = A(\delta')$. If not then, by Definition 5.0.15, $(\delta, r) \not<_l (\delta', s)$ for no points $r, s \in \mathcal{P}_0$ and so (*) does not hold. If $A(\delta) = A(\delta')$ then we consider the condition $D\colon \neg C \vee p < q$. Then, by Definition 5.0.15, $(*)$ is valid iff $D$ is valid. The validity of $D$ can be decided by transforming $D$ into a formula of Preburger's arithmetic which is decidable (see e.g. [10]).

The transformation of the formula $D$ above to Presburger's arithmetic via universal quantification over all parameters in $D$ and by substituting expressions like $p(n) = m$ by $(n = 0 \rightarrow m = 0) \wedge (n > 0 \rightarrow \exists y < n.y + 1 = m)$ is clearly inefficient. More efficient algorithms are unknown to the authors. But better algorithms could be developed by transforming $\neg C \vee p < q$ into clause form (conjunctive normal form) and checking whether each clause is a tautology; hereby we have to use the semantic equivalences $\neg n = 0 \sim n > 0$, $\neg n > 0 \sim n = 0$, $\neg n = m \sim n > m \vee m > n$, $\neg n > m \sim n = m \vee m > n$ (and similar for $n < m$); finally we also need $n = m \vee n < m \vee m < n = \top$. However, the handling of expressions $p < q$ which contain function symbols remains nontrivial. A splitting in $n = 0$ and $n > 0$ in presence of a term $p(n)$ might be helpful. Consider e.g. Example 5.0.8 and the relation $(\delta, (n, p(m), s(k))) <_C (\delta, (n, m, k))$ for some condition $C$. Then $C \sim (C \wedge m = 0) \vee (C \wedge m > 0)$ and we split the relation into two parts where $C_1 = (C \wedge m = 0)$ and $C_2 = (C \wedge m > 0)$ and we test

$$(\delta, (n, 0, s(k))) <_{C_1} (\delta, (n, 0, k)) \text{ and } (\delta, (n, m - 1, s(k))) <_{C_2} (\delta, (n, m, k))$$

where $m - 1$ is the subtraction of 1 from $m$. It is easy to see that the first relation is false, while the second is true in Example 5.0.8, no matter what $C$ actually is. More research is necessary to find efficient algorithms and subclasses of problems admitting such algorithms.

**Definition 5.0.17** (*Canonic point transition systems*) Consider a point transition system $\mathbf{P}^*\colon (\delta_0, \Delta^*, \Delta_e, \mathbf{P})$. $\mathbf{P}^*$ is called *canonic* if there exists a tree $T(\Delta^*)$ such that

$$\text{for all } (\delta, p) \rightarrow \mathcal{P}\colon C \in \mathbf{P} \text{ we have } (\delta', p') <_C (\delta, p) \text{ for all } (\delta', p') \in \mathcal{P}.$$

**Proposition 5.0.4** *It is decidable whether a given point transition system is canonic.*

***Proof*** We assume a fixed well-founded ordering on the $\mathcal{P}_0^\alpha$, e.g. the lexicographic ordering. For defining the ordering $<_C$ we check all possible trees $T(\Delta^*)$—there are only finitely many ones. As for all $T(\Delta^*)$ the relation $<_C$ is decidable (by Proposition 5.0.3) canonicity is decidable as (for a given $T(\Delta^*)$) it requires only finitely many $<_C$-checks.

**Example 5.0.9** Consider the point transition system in Example 5.0.5, where $\Delta^* = \{\delta_0, \delta_1, \delta_e\}$, $\Delta_e = \{\delta_e\}$ and

$$\mathbf{P}(\delta_0) = \{(\delta_0, (x, y)) \rightarrow \{(\delta_0, (p(x), y)), (\delta_1, (x, p(y))\}\colon x > 0 \wedge y > 0,$$

$$\begin{aligned}
&(\delta_0, (x, y)) \to \{(\delta_e, (x, y))\}\colon x > 0 \wedge y = 0,\\
&(\delta_0, (x, y)) \to \{(\delta_e, (x, y))\}\colon x = 0 \wedge y > 0,\\
&(\delta_0, (x, y)) \to \{(\delta_e, (x, y))\}\colon x = 0 \wedge y = 0\}.\\
\mathbf{P}(\delta_1) = \{&(\delta_1, (x, y)) \to \{(\delta_0, (x, p(y)))\}\colon y > 0,\\
&(\delta_1, (x, y)) \to \{(\delta_e, (x, y))\}\colon y = 0\}.\\
\mathbf{P}(\delta_e) = \emptyset&.
\end{aligned}$$

Take the tree $T(\Delta^*)$ defined in Example 5.0.6. We have the nodes $\Delta_0\colon \{\delta_0, \delta_1\}$, $\Delta_1\colon \{\delta_e\}$ and the edge $(\Delta_0, \Delta_1)$. In particular $\Delta_1 <_T \Delta_0$. As $\delta_0, \delta_1 \in \Delta_0$ and $A(\delta_0) = A(\delta_1) = 2$ we have to check (for the first point transitions in $\mathbf{P}(\delta_0), \mathbf{P}(\delta_1)$) whether the points are decreasing for all $\sigma \in \mathcal{S}$ fulfilling the conditions. Indeed, for $\mathbf{P}(\delta_0)$ we have

$$(\sigma(p(x))\downarrow, \sigma(y)) <_2 (\sigma(x), \sigma(y)),\ (\sigma(x), \sigma(p(y))\downarrow) <_2 (\sigma(x), \sigma(y))$$

for all $\sigma$ with $\sigma(x) > 0$ and $\sigma(y) > 0$. Similarly, for $\mathbf{P}(\delta_1)$, we get

$$(\sigma(x), \sigma(p(y))\downarrow <_2 (\sigma(x), \sigma(y)) \text{ for all } \sigma \text{ with } \sigma(y) > 0.$$

For all other point transitions we have that $\delta_e \in \Delta_1$ and $\Delta_1 <_T \Delta_0$ and so the conditions for being canonic are fulfilled. Note that the tree $T(\Delta^*)$ definable over $\Delta^*$ is the only one yielding a relation $<_C$ verifying the canonicity of the point transition system.

**Theorem 5.0.1** *Canonic point transition systems are terminating.*

***Proof*** By contradiction. Assume that $\mathbf{P}^*$ is canonic and nonterminating. Then there exists a $\sigma \in \mathcal{S}$ such that $T_{\mathbf{P}^*}[\sigma]$ is infinite. So $T_{\mathbf{P}^*}[\sigma]$ is an infinite tree with finite node degree. By König's lemma $T_{\mathbf{P}^*}[\sigma]$ has an infinite path $\eta$ for

$$\eta = (\delta_0, p_0), \ldots, (\delta_i, p_i), (\delta_{i+1}, p_{i+1}) \ldots$$

By Definition of $T_{\mathbf{P}^*}[\sigma]$ , for each $(\delta_i, p_i)$ there exists a point transition

$$(\delta_i, p) \to \mathcal{P}\colon C \in \mathcal{P}$$

matching $(\delta_i, p_i)$. By Definition 5.0.11 there exists a parameter assignment $\sigma' \in \mathcal{S}$ such that $\sigma'(p)\downarrow = p_i$. As $\eta$ is infinite we have $\mathcal{P}(\delta_i) \neq \emptyset$ and

$$\mathcal{P}(\delta_i) = \{(\delta_i, p) \to \mathcal{P}_1\colon C_1, \ldots, (\delta_i, p) \to \mathcal{P}_\beta\colon C_\beta\}$$

By Definition 5.0.10 the $C_j - s$ define a partition and therefore there exists exactly one condition $C_k$ such that $\sigma'[C_k] = \top$. Now let

$$\mathcal{P}_k = \{(\delta_1', q_1), \ldots, (\delta_m', q_m)\}.$$

and $\mu_1, \ldots, \mu_m$ be the successor nodes of $(\delta_i, p_i)$. To these nodes corresponds the set of labels

$$L\colon \{(\delta'_1, \sigma'(q_1)\downarrow), \ldots, (\delta'_m, \sigma'(q_m)\downarrow)\}.$$

Now, by definitions of a $\sigma$-tree and of $\eta$ $(\delta_{i+1}, p_{i+1}) \in L$. Let us say that $(\delta_{i+1}, p_{i+1}) = (\delta'_r, \sigma'(q_r)\downarrow)$. As $\mathbf{P}^*$ is canonic and

$$(\delta_i, p) \to \mathcal{P}_k\colon C_k \text{ is a point transition occurring in } \mathbf{P}^*$$

we have $(\delta'_r, q_r) <_{C_k} (\delta_i, p)$, which means that for all $\sigma''$ with $\sigma''[C_k] = \top$ we have

$$(\delta'_r, \sigma''(q_r))\downarrow <_l (\delta_i, \sigma''(p)).$$

But $\sigma'$ is such a $\sigma''$ as $\sigma'[C_k] = \top$. Moreover, $(\delta'_r, \sigma'(q_r)\downarrow) = (\delta_{i+1}, p_{i+1})$ and $\sigma'(p)\downarrow = p_i$; therefore

$$(\delta_{i+1}, p_{i+1}) <_l (\delta_i, p_i).$$

That means $\eta$ represents an infinite sequence of points which is descending under $<_l$. But this is impossible as $<_l$ is well-founded by Proposition 5.0.2.

# Chapter 6
# Schematic Substitution and Unification

## 6.1 Schematic Substitution

Similar to terms and schematic terms *schematic substitutions* are expressions which evaluate to substitutions under parameter assignments. Below we define the syntactic expressions which will turn out useful in schematic refutational calculi . In this section we use only schematic $\omega\iota$-theories which are simple( see Definition 4.1.21).

**Definition 6.1.1** Let $T^*\colon (T_1, T_2)$ be a simple $\omega\iota$-theory. A mapping

$$\Theta\colon \mathbf{V}_0^\iota \to \hat{\mathbf{T}}^\iota(T^*)$$

is called an *s-substitution* over $T^*$ if $\Theta(x) \neq x$ only for finitely many $x \in \mathbf{V}_0^\iota$. We define the *domain* of $\Theta$ as

$$\begin{aligned} dom(\Theta) &= \{x \mid x \in \mathbf{V}_0^\iota, \Theta(x) \neq x\} \text{ and the range of } \Theta \text{ as} \\ rg(\Theta) &= \{\Theta(x) \mid x \in dom(\Theta)\}. \end{aligned}$$

If $dom(\Theta) = \{x_1, \ldots, x_\alpha\}$ and $\Theta(x_i) = t_i$ for $i = 1, \ldots, \alpha$ then we write $\Theta$ as $\{x_1 \leftarrow t_1, \ldots, x_\alpha \leftarrow t_\alpha\}$.

**Example 6.1.1** Let $T^*\colon (T_1, T_2)$ be a simple $\omega\iota$-theory where

$$T_1 = (\hat{t}, \{\hat{t}, \hat{s}\}, D(\{\hat{t}, \hat{s}\}), \{x, y\}, <),$$

$\hat{s} < \hat{t}$ and

$$\begin{aligned} D(\hat{t}) &= \{\hat{t}(x, y, n_1, n_2) =_d \underline{\text{if}}\ n_2 = \bar{0}\ \underline{\text{then}}\ y \\ &\qquad \underline{\text{else}}\ g(\hat{t}(x, y, n_1, p(n_2)), \hat{s}(x, n_1))\}, \\ D(\hat{s}) &= \{\hat{s}(x, n_1) =_d \underline{\text{if}}\ n_1 = \bar{0}\ \underline{\text{then}}\ x\ \underline{\text{else}}\ h(\hat{s}(x, p(n_1)))\}. \end{aligned}$$



A. Leitsch et al., *First-Order Schemata and Inductive Proof Analysis*, Computer Science Foundations and Applied Logic, https://doi.org/10.1007/978-3-032-05741-9_6

Let $u, v, x_1, x_2 \in \mathbf{V}_0^\iota$. Then we define

$$\Theta = \{u \leftarrow h(\hat{t}(x_1, x_2, n_1, n_2)),\ v \leftarrow \hat{s}(g(x_1, x_2), n_1)\}$$

$\Theta$ is an s-substitution over $T^*$ where

$$dom(\Theta) = \{u, v\},\ rg(\Theta) = \{h(\hat{t}(x_1, x_2, n_1, n_2)), \hat{s}(g(x_1, x_2), n_1)\}.$$

In the same way as schematic terms $t$ evaluate to first-order terms $\sigma(t){\downarrow_{T^*}}$ under a parameter assignment $\sigma$, $s$-substitutions evaluate to ordinary first-order substitutions.

**Definition 6.1.2** Let $T^*$: $(T_1, T_2)$ be a simple $\omega\iota$-theory and $\Theta$: $\{x_1 \leftarrow t_1, \ldots, x_\alpha \leftarrow t_\alpha\}$ be an $s$-substitution over $T^*$. Then, for $\sigma \in \mathcal{S}$,

$$\Theta[\sigma] = \{x_1 \leftarrow \sigma(t_1){\downarrow_{T^*}}, \ldots, x_\alpha \leftarrow \sigma(t_\alpha){\downarrow_{T^*}}\}$$

is called the $\sigma$-evaluation of $\Theta$.

**Proposition 6.1.1** *Let $T^*$ be a simple $\omega\iota$-theory, $\sigma \in \mathcal{S}$ and $\Theta$ be an $s$-substitution over $T^*$. Then $\Theta[\sigma]$ is a first-order substitution.*

***Proof*** Let $\Theta = \{x_1 \leftarrow t_1, \ldots, x_\alpha \leftarrow t_\alpha\}$ be an $s$-substitution over $T^*$ and $\sigma \in \mathcal{S}$. By 4.1.23 $t_i'$: $\sigma(t_i){\downarrow_{T^*}} \in \mathbf{T}_0^\iota$ for $i = 1, \ldots, \alpha$. Therefore $t_i'$ are first-order terms and

$\Theta[\sigma] = \{x_1 \leftarrow t_1', \ldots, x_\alpha \leftarrow t_\alpha'\}$ is a first-order substitution.

**Example 6.1.2** We take $T^*$ and $\Theta$ from Example 6.1.1 and define $\sigma$ by $\sigma(n_1) = \bar{2}$, $\sigma(n_2) = \bar{1}$. Then

$$\Theta[\sigma] = \{u \leftarrow h(\hat{t}(x_1, x_2, \bar{2}, \bar{1})){\downarrow_{T^*}},\ v \leftarrow \hat{s}(g(x_1, x_2), \bar{2}){\downarrow_{T^*}}\}.$$

$\hat{t}(x_1, x_2, \bar{2}, \bar{1}){\downarrow_{T^*}} = g(x_2, h(h(x_1)))$ and $\hat{s}(g(x_1, x_2), \bar{2}){\downarrow_{T^*}} = h(h(g(x_1, x_2)))$, therefore we obtain

$$\Theta[\sigma] = \{u \leftarrow h(g(x_2, h(h(x_1)))),\ v \leftarrow h(h(g(x_1, x_2)))\}.$$

Now we have to define formally how $s$-substitutions apply to schematic terms and formulas. In particular we define $s$-substitutions on terms in $\mathbf{T}^\iota$ and $\hat{\mathbf{T}}^\iota$ first. In our first step we avoid terms containing higher-order variables, so our variables are ordinary variables in $\mathbf{V}_0^\iota$.

**Definition 6.1.3** (*$s$-substitutions on individual term*) Let $\Theta$: $\{x_1 \leftarrow t_1, \ldots, x_\alpha \leftarrow t_\alpha\}$ be an $s$-substitution, then we define

- $c\Theta = c$ for $c \in \mathbf{F}_0^\iota$,
- if $x \in \mathbf{V}_0^\iota$ and $x \notin \{x_1, \ldots, x_\alpha\}$ then $x\Theta = x$,

- if $x = x_i \in \{x_1, \dots, x_\alpha\}$ then $x_i\Theta = t_i$,
- if $f \in \mathbf{F}^\iota_k$ then $f(t_1, \dots, t_k)\Theta = f(t_1\Theta, \dots, t_k\Theta)$.

In the next step we extend $s$-substitutions to term schemata.

**Definition 6.1.4** Let $\Theta$ be an $s$-substitution.

- If $t_1, \dots, t_k \in \hat{\mathbf{T}}^\iota$ and $f \in \mathbf{F}^\iota_k$ then $f(t_1, \dots, t_k)\Theta = f(t_1\Theta, \dots, t_k\Theta)$,
- if $\pi = (j, k)$ is the profile of $\hat{t} \in \hat{F}(\pi)$, $s_l \in \hat{\mathbf{T}}^\iota_s$ for $l = 1, \dots, j$, $t_1, \dots, t_k \in \mathbf{T}^\omega$ then
$$\hat{t}(s_1, \dots, s_j, t_1, \dots, t_k)\Theta = \hat{t}(s_1\Theta, \dots, s_j\Theta, t_1, \dots, t_k).$$

**Example 6.1.3** Let $T^*\colon (T_1, T_2)$ be the theory in Example 6.1.1. Then
$$t\colon h(\hat{t}(\hat{t}(x, y, n, m), \hat{s}(x, n), n, m)) \in \hat{\mathbf{T}}^\iota(T^*).$$
Let $\Theta = \{x \leftarrow \hat{s}(u, k), y \leftarrow g(u, v)\}$. Then
$$t\Theta = h(\hat{t}(\hat{t}(\hat{s}(u, k), g(u, v), n, m), \hat{s}(\hat{s}(u, k), n), n, m)).$$

Even when the theories do not contain $n$-ary variable expressions for $n > 0$ (which concerns the recursive definition of schematic terms) the use of nontrivial variable expressions cannot be avoided due to the definition of formula schemata. We give an example below.

**Example 6.1.4** Let $(T, T', \emptyset)$ the $\omega\iota o$-theory defined below.

- $T = (\hat{q}, \{\hat{q}\}, D(\{\hat{q}\}), \{X\}, \emptyset)$ where $X \in \mathbf{V}^\iota_1$, $\pi(\hat{q}) = (1, 1, 1)$ and
$$D(\hat{q}) = \{\hat{q}(X, n) =_d \underline{\text{if}}\ n = \bar{0}\ \underline{\text{then}}\ Q(X(\bar{0}))\ \underline{\text{else}}\ \hat{q}(X, p(n)) \vee Q(\hat{t}(X(n), n))\}.$$
Here we have $\hat{q}_S = \xi \vee Q(\hat{t}(X(n), n))$ for $\xi \in \mathbf{V}^o$, $\hat{q}_B = Q(X(\bar{0}))$.
- $T' = (\hat{t}, \{\hat{t}\}, D(\{\hat{t}\}), \{x\}, \emptyset)$ for $x \in \mathbf{V}^\iota_0$, $\pi(\hat{t}) = (0, 1, 1)$ and
$$D(\hat{t}) = \{\hat{t}(x, n) =_d \underline{\text{if}}\ n = \bar{0}\ \underline{\text{then}}\ x\ \underline{\text{else}}\ h(\hat{t}(x, p(n)))\}.$$

$\hat{t}_S = h(z)$, $\hat{t}_B = x$.
The theory $T^*\colon (T, T', \emptyset)$ contains $(T', \emptyset)$ which is an $\omega\iota$-theory. However, due to the definition of $\hat{q}$ we obtain schematic formulas of the form $\hat{q}(X, n) \vee Q(\hat{t}(X(s(n)), n))$ where the term $\hat{t}(X(s(n)), s(n))$ contains a unary variable expression—which is not a variable as $n$ is a parameter. We took care of this phenomenon as $\hat{t}(X(n), n)$ is a term in $\hat{\mathbf{T}}^\iota(T^*)$ (see Definition 4.1.20).

**Definition 6.1.5** (*extended s-substitution*) Let $T^*$ be a simple $\omega\iota o$-theory. An extended $s$-substitution over $T^*$ is an expression of the form

$$\{V_1 \leftarrow t_1, \ldots, V_\alpha \leftarrow t_\alpha\}$$

where, for $i = 1, \ldots, \alpha$, $V_i$ are variable expressions and $t_i$ are terms in $\hat{\mathbf{T}}^\iota(T^*)$, and for all $\sigma \in \mathcal{S}$ and all $i, j \in \{1, \ldots, \alpha\}$ such that $i \neq j$ we have $\sigma(V_i){\downarrow} \neq \sigma(V_j){\downarrow}$. If there exists a $\sigma \in \mathcal{S}$ such that $\sigma(V_i){\downarrow} = \sigma(V_j){\downarrow}$ then we say that $\{V_i, V_j\}$ is *parameter-unifiable*.

Below we define the semantics of extended $s$-substitutions and prove that they evaluate to first-order substitutions by normalization under parameter assignments.

**Definition 6.1.6** Let

$$\Theta\colon \{V_1 \leftarrow t_1, \ldots, V_\alpha \leftarrow t_\alpha\}$$

be an extended $s$-substitution over a theory $T^*$ and let $\sigma \in \mathcal{S}$. Then we define

$$\Theta[\sigma] = \{\sigma(V_1){\downarrow} \leftarrow \sigma(t_1){\downarrow}, \ldots, \sigma(V_\alpha) \leftarrow \sigma(t_\alpha){\downarrow}\}.$$

**Proposition 6.1.2** *Let $\Theta$ be an extended s-substitution and $\sigma \in \mathcal{S}$. Then $\Theta[\sigma]$ is a first-order substitution.*

***Proof*** Let $\Theta = \{V_1 \leftarrow t_1, \ldots, V_\alpha \leftarrow t_\alpha\}$. In $\Theta[\sigma]$, $\sigma(V_i){\downarrow}$ are first-order variables and $\sigma(t_i){\downarrow}$ are first-order terms. Moreover, by Definition 6.1.5, $\sigma(V_i){\downarrow} \neq \sigma(V_j){\downarrow}$ for $i \neq j$ and so all resulting first-order variables are pairwise different; hence $\Theta[\sigma]$ is a first-order substitution.

**Example 6.1.5** Let

$$\Theta = \{X(n, n) \leftarrow g(X(\bar{0}, n)),\ X(s(n), m) \leftarrow h(X(m, m)), Y(m) \leftarrow a\}$$

where $X, Y$ are variable classes and $g, h$ are function symbols.
$\Theta$ is an extended $s$-substitution: it is easy to see that for all $\sigma \in \mathcal{S}$ the sets $\{X(n, n), X(s(n), m)\}$, $\{X(n, n), Y(m)\}$ and $\{X(s(n), m), Y(m)\}$ are not parameter-unifiable. Let $\sigma \in \mathcal{S}$ such that $\sigma(n) = \bar{0}$, $\sigma(m) = \bar{1}$. Then

$$\Theta[\sigma] = \{X(\bar{0}, \bar{0}) \leftarrow g(X(\bar{0}, \bar{0})),\ X(\bar{1}, \bar{1}) \leftarrow h(X(\bar{1}, \bar{1})), Y(\bar{1}) \leftarrow a\}.$$

Clearly, $\Theta[\sigma]$ is a first-order substitution which, under variable renaming, is equivalent to

$$\{x \leftarrow g(x),\ y \leftarrow h(y), z \leftarrow a\}.$$

In extending schematic terms to extended schematic terms it makes sense to restrict the arguments of the variable expressions to simplify the handling of extended $s$-substitutions. Below we define a restricted class which is large enough for schematic `CERES`.

**Definition 6.1.7** A term $t$ in $\hat{\mathbf{T}}^{\iota}$ is called a *standard term* if all for variable expressions $X(\vec{s})$ occurring in $t$ for $X \in \mathbf{V}_k^{\iota}$ and $k \geq 1$ the following conditions are fulfilled:

- assigned to $X$ is a unique parameter list $(n_1, \ldots, n_k)$,
- the only variable expressions starting with $X$ are of the form $X(r_1, \ldots, r_k)$ where each $r_i$ is of one of the forms $n_i, \bar{0}, p(n_i)$ or $s(n_i)$. We call these variable expressions *standard*.

**Definition 6.1.8** A *standard $s$-substitution* is an expression of the form

$$\{X_1(r_1^1, \ldots, r_{\alpha_1}^1) \leftarrow t_1, \ldots, X_k(r_1^k, \ldots, r_{\alpha_k}^k) \leftarrow t_k\}$$

where

- $X_1, \ldots, X_k$ are variable classes,
- the parameter list assigned to $X_i$ is $(n_1^i, \ldots, n_{\alpha_i}^i)$ for $i \in \{1, \ldots, k\}$,
- the $X_i(r_1^i, \ldots, r_{\alpha_i}^i)$ are standard variable expressions,
- the $t_i$ are standard terms in $\hat{\mathbf{T}}^{\iota}$,
- the sets $\{X_i(r_1^i, \ldots, r_{\alpha_i}^i), X_j(r_1^j, \ldots, r_{\alpha_j}^j)\}$ for $i \neq j$ are not parameter-unifiable.

**Example 6.1.6** Let $X$ be a variable class with assigned parameter $n$, $Y$ be a variable class with assigned parameters $m, n$ and $t_1, t_2, t_3$ standard terms in $\hat{\mathbf{T}}^{\iota}$. The $s$-substitution

$$\{X(\bar{0}) \leftarrow t_1, X(s(n)) \leftarrow t_2, Y(n, m) \leftarrow t_3\}$$

is a standard $s$-substitution.

The parameter unifiability in Definition 6.1.8 is easy to check: either the second-order variables of the variable expressions are different, or if they are equal and the expressions are of the form $X(r_1, \ldots, r_k), X(s_1, \ldots, s_k)$ it suffices to find a pair $\{r_i, s_i\}$ which is not parameter unfiable. As $r_i$ and $s_i$ are of the form $n_i, p(n_i), s(n_i), \bar{0}$ this is trivial to check.
Below we analyze the application of standard $s$-substitutions to standard terms.

**Definition 6.1.9** (*state*) Let $X(r_1, \ldots, r_\alpha)$ be a standard variable expression. By definition we have $r_i \in \{n_i, p(n_i), s(n_i), \bar{0}\}$ for $i = 1, \ldots, \alpha$. For every $i$ we define three conditions $C_{i,1}, C_{i,2}, C_{i,3}$ as follows:

$$\begin{aligned} C_{i,1} &= n_i = \bar{0}, \\ C_{i,2} &= n_i \neq \bar{0} \wedge p(n_i) = \bar{0}, \\ C_{i,3} &= n_i \neq \bar{0} \wedge p(n_i) \neq \bar{0}. \end{aligned}$$

Note that $C_i \colon \{C_{i,1}, C_{i,2}, C_{i,3}\}$ is a partition.
A *state* for $X(r_1, \ldots, r_\alpha)$ is a conjunction of the form

$$D_1 \wedge \cdots \wedge D_\alpha$$

where for all $i = 1, \ldots, \alpha$ $D_i \in \mathcal{C}_i$. Note that there exist $3^\alpha$ different states (not all of them are satisfiable) and that the set of all states—written as $\mathcal{D}(X(r_1, \ldots, r_\alpha))$ also forms a partition. Let $p$ be a state for $X(r_1, \ldots, r_\alpha)$, $\beta \leq \alpha$ such that $Y(r'_1 \ldots, r'_\beta)$ is a standard variable expression; then $p$ is also a state for $Y(r'_1 \ldots, r'_\beta)$.

Application of standard $s$-substitutions to standard terms is not a trivial operation. E.g. just by writing $X(r, s)\Theta$ we cannot guarantee the uniqueness of application.

**Example 6.1.7** Let $\Theta = \{X(n_1, n_2) \leftarrow t\}$ for some term $t$ different from $X(n_1, n_2)$. Consider now the application

$$X(\bar{0}, \bar{0})\Theta$$

If $\sigma(n) = \bar{0}$ and $\sigma(m) = \bar{0}$ then $X(\bar{0}, \bar{0})\Theta[\sigma] = \sigma(t)\downarrow$, for all other parameter assignments we get $X_i(\bar{0}, \bar{0})\Theta[\sigma] = X_i(\bar{0}, \bar{0})$. That means for the state $n = \bar{0} \wedge m = \bar{0}$ and we get $t$, and for all other states we obtain $X_i(\bar{0}, \bar{0})$. We see that a unique result can only be obtained by taking into account the states.

We are now going to define the application of a standard $s$-substitution w.r.t. the set of states of a standard variable expression. To this aim we define, for any state $p$, a replacement operator $\psi_p$ which performs a partial evaluation of terms and substitutions when the value of some parameters is $\bar{0}$. We define $\psi_p$ for terms in $\hat{\mathbf{T}}^\iota$.

**Definition 6.1.10** Let $X(r_1, \ldots, r_\alpha)$ be a standard variable expression over the parameters $\{n_1, \ldots, n_\alpha\}$ and let

$$p\colon D_1 \wedge \cdots \wedge D_\alpha$$

be a state for $V\colon X(r_1, \ldots, r_\alpha)$. We define $\psi_p$ for $X(r_1, \ldots, r_\alpha)$.

Let $p(i)$ be the $i$th conjunction of $p$, i.e. $p(i) = D_i$. For every $p(i)$ we have to distinguish whether $r_i$ evaluates to $\bar{0}$ under a $\sigma \in \mathcal{S}$ such that $\sigma(p(i)) = \top$. For every $p$ we define $\psi_p(V)$ inductively. We define $\psi_p^0(V) = V$. Assume now that for all $j < i$ $\psi_p^j(V)$ are already defined. We distinguish the following cases below:

- If $p(i) = C_{i,1}$ (i.e. $p(i)\colon n_i = \bar{0}$) we define $\psi_p^i(V) = \psi_p^{i-1}(V)[r_i/\ \bar{0}]$ (i.e. we replace $r_i$ by $\bar{0}$) if $r_i \in \{n_i, p(n_i), \bar{0}\}$ and $\psi_p^i(V) = \psi_p^{i-1}(V)[r_i/\ \bar{1}]$ if $r_i = s(n_i)$.
- If $p(i) = C_{i,2}$ (we have $p(i)\colon n_i \neq \bar{0} \wedge p(n_i) = \bar{0}$) we define $\psi_p^i(V) = \psi_p^{i-1}(V)[r_i/\bar{0}]$ if $r_i = p(n_i)$, $\psi^i(V) = \psi_p^{i-1}(V)[r_i/\bar{1}]$ if $r_i = n_i$, and $\psi_p^i(V) = \psi_p^{i-1}(V)[r_i/\bar{2}]$ if $r_i = s(n_i)$. For $r_i = \bar{0}$ we get $\psi_p^i(V) = \psi_p^{i-1}(V)$.
- If $p(i) = C_{i,3}$ (i.e. $p(i) = n_i \neq \bar{0} \wedge p(n_i) \neq \bar{0}$) then we define $\psi_p^i(V) = \psi_p^{i-1}(V)$.

Finally we define $\psi_p(V) = \psi_p^\alpha(V)$.

We extend now $\psi_p$ to $\mathbf{T}^\iota$:

- If $t \in \mathbf{F}_0^\iota$ then $\psi_p(t) = t$.
- For a standard variable expressions $X(r_1, \ldots, r_\alpha)$ $\psi_p(X(r_1, \ldots, r_\beta))$ is defined above.
- If $t = f(t_1, \ldots, t_k)$ then $\psi_p(f(t_1, \ldots, t_k)) = f(\psi_p(t_1), \ldots, \psi_p(t_k))$.

Finally we extend $\psi_p$ to $\hat{\mathbf{T}}^\iota$:

- If $t \in \mathbf{T}^\iota$ then $\psi_p(t)$ is defined as above.
- If $t = f(t_1, \ldots, t_k)$ for $f \in \mathbf{F}_k^\iota$ then

$$\psi_p(f(t_1, \ldots, t_k)) = f(\psi_p(t_1), \ldots, \psi_p(t_k)).$$

- if $t = \hat{t}(s_1, \ldots, s_j, t_1, \ldots, t_k)$ for a profile $\pi = (j, k)$ and $\hat{t} \in \hat{F}(\pi)$, $s_l \in \hat{\mathbf{T}}^\iota$ for $l = 1, \ldots, j, t_1, \ldots, t_k \in \mathbf{T}^\omega$ then %

$$\psi_p(t) = \hat{t}(\psi_p(s_1), \ldots, \psi_p(s_j), t_1, \ldots, t_k).$$

**Example 6.1.8** Let

$$t = h(X(n_1, n_2), Y(\bar{0}, s(n_2)))$$

for $h \in \mathbf{F}_2^\iota$ and

$$p = n_1 \neq \bar{0} \wedge p(n_1) \neq \bar{0} \wedge n_2 = \bar{0}.$$

Then

$$\psi_p(t) = h(X(n_1, \bar{0}), Y(\bar{0}, \bar{1})).$$

The terms $t$ and $\psi_p(t)$ evaluate to the same first-order term if $\sigma(p){\downarrow} = \top$, a property we will need in soundness results in this section.

**Lemma 6.1.1** *Let $t$ be a standard term in $\hat{\mathbf{T}}^\iota$ such that the set of parameters in $t$ is contained in $\mathcal{P}$: $\{n_1, \ldots, n_\alpha\}$, let $\mathcal{S}(p)$ be the set of all parameter assigments such that $\sigma(p){\downarrow} = \top$. Then $\sigma(t){\downarrow} = \sigma(\psi_p(t)){\downarrow}$ for all states $p$ over $\mathcal{P}$ and $\sigma \in \mathcal{S}(p)$.*

***Proof*** By the homomorphic extension of $\psi_p$ from variable expressions to terms, it is enough to show that for all variable expressions $X(r_1, \ldots, r_\beta)$ occurring in $t$ (where $\beta \leq \alpha$) we obtain

$$(\star)\ \sigma(X(r_1, \ldots, r_\beta)){\downarrow} = \sigma(\psi_p(X(r_1, \ldots, r_\beta))){\downarrow} \text{ for } \sigma \in \mathcal{S}(p).$$

So let us assume we have the variable expression $X(r_1, \ldots, r_\beta)$ in $t$. In order to prove $(\star)$ we recall the definition of $\psi_p$ via $\psi_p^i$ as in Definition 6.1.10 and proceed by induction on $i$.

$i = 0$: by definition we have $\psi_p^0(X(r_1, \ldots, r_\beta)) = X(r_1, \ldots, r_\beta)$ and $(\star)$ trivially holds when we replace $\psi_p$ by $\psi_p^0$. Now let us assume that

$$(\text{IH})\sigma(X(r_1,\dots,r_\beta))\downarrow = \sigma(\psi_p^j(X(r_1,\dots,r_\beta)))\downarrow \text{ for } j < i.$$

By definition of the $\psi_p^j$ there exist terms $r'_1,\dots,r'_{i-1}$ such that

$$\psi_p^{i-1}(X(r_1,\dots,r_\beta)) = X(r'_1,\dots,r'_{i-1},r_i,\dots,r_\beta).$$

In order to define $\chi^i$ from $\chi^{i-1}$ we have to consider the condition $p(i)$. We distinguish three cases

- $p(i) = n_i = \bar{0}$: as $\sigma(p)\downarrow = \top$ we also have $\sigma(p(i))\downarrow = \top$ and therefore $\sigma(n_i) = \bar{0}$. By definition of $\chi_p^i$ we have

  $$\psi_p^i(X(r'_1,\dots,r'_{i-1},r_i,\dots,r_\beta) = X(r'_1,\dots,r'_{i-1},\bar{0},\dots,r_\beta)$$

  if $r_i \in \{n_i, p(n_i), \bar{0}\}$; in this case we obtain $\sigma(r_i)\downarrow = \sigma(\bar{0})\downarrow = \bar{0}$ and therefore

  $$\sigma(X(r_1,\dots,r_\beta))\downarrow = \sigma(\psi_p^i(X(r_1,\dots,r_\beta)))\downarrow.$$

  If $r_i = s(n_i)$ we obtain

  $$\psi_p^i(X(r'_1,\dots,r'_{i-1},r_i,\dots,r_\beta) = X(r'_1,\dots,r'_{i-1},\bar{1},\dots,r_\beta)$$

  But then $\sigma(r_i)\downarrow = \bar{1} = \sigma(\bar{1})\downarrow$, and again

  $$\sigma(X(r_1,\dots,r_\beta))\downarrow = \sigma(\psi_p^i(X(r_1,\dots,r_\beta)))\downarrow.$$

- $p(i) = n_i \neq \bar{0} \wedge p(n_i) = \bar{0}$: as $\sigma(p)\downarrow = \top$ we get $\sigma(n_i) = \bar{1}, \sigma(p(n_i)) = \bar{0}$. By definition of $\chi$ we obtain $\psi_p^i(X(r'_1,\dots,r'_{i-1},r_i,\dots,r_\beta)) =$

  $$\begin{aligned}
  &= X(r'_1,\dots,r'_{i-1},\bar{0},\dots,r_\beta) \text{ if } r_i = p(n_i),\\
  &= X(r'_1,\dots,r'_{i-1},\bar{1},\dots,r_\beta) \text{ if } r_i = n_i,\\
  &= X(r'_1,\dots,r'_{i-1},\bar{2},\dots,r_\beta) \text{ if } r_i = s(n_i),\\
  &= \psi^{i-1}p(X(r'_1,\dots,r'_{i-1},r_i,\dots,r_\beta)) \text{ if } r_i = \bar{0}.
  \end{aligned}$$

  It is easy to see, that by $\sigma(C_{i,2}) = \top$ we obtain

  $$\sigma(X(r_1,\dots,r_\beta))\downarrow = \sigma(\psi_p^i(X(r_1,\dots,r_\beta)))\downarrow.$$

- $p(i) = n_i \neq \bar{0} \wedge p(n_i) \neq \bar{0}$: by definition of $\psi_p$ we get $\psi_p^i = \psi_p^{i-1}$ and thus by (IH)

  $$\sigma(X(r_1,\dots,r_\beta))\downarrow = \sigma(\psi_p^i(X(r_1,\dots,r_\beta)))\downarrow.$$

We extend $\psi_p$ further to standard $s$-substitutions.

**Definition 6.1.11** Let

$$\Theta\colon \{A_1 \leftarrow t_1, \ldots, A_\alpha \leftarrow t_\alpha\}$$

be a standard $s$-substitution over the parameters $\{n_1, \ldots, n_\beta\}$ and $p$ be a state over these parameters. Then

$$\psi_p(\Theta) = \{\psi_p(A_1) \leftarrow \psi_p(t_1), \ldots, \psi_p(A_\alpha) \leftarrow \psi_p(t_\alpha)\}.$$

**Example 6.1.9** Let

$$\begin{aligned}\Theta = \{&X(\bar{0}) \leftarrow Z(n_1),\ X(s(n_1)) \leftarrow g(Z(n_1)),\\ &Y_1(n_1, n_2) \leftarrow h(Y_2(n_1, n_2))\},\\ p = n_1 &\neq \bar{0} \wedge p(n_1) = \bar{0} \wedge n_2 \neq \bar{0} \wedge p(n_2) \neq \bar{0}.\end{aligned}$$

Then

$$\psi_p(\Theta) = \{X(\bar{0}) \leftarrow Z(\bar{1}),\ X(\bar{2}) \leftarrow g(Z(\bar{1})),\ Y_1(\bar{1}, n_2) \leftarrow h(Y_2(\bar{1}, n_2))\}.$$

**Proposition 6.1.3** *Let $\Theta$ be a standard $s$-substitution, $p$ be a state over the parameters of $\Theta$. Then $\psi_p(\Theta)$ is an extended $s$-substitution and for all parameter assigments $\sigma \in \mathcal{S}(p)$ we have $\Theta[\sigma] = \psi_p(\Theta)[\sigma]$.*

***Proof*** Let $\Theta = \{A_1 \leftarrow t_1, \ldots, A_\alpha \leftarrow t_\alpha\}$. Then

$$\psi_p(\Theta) = \{\psi_p(A_1) \leftarrow \psi_p(t_1), \ldots, \psi_p(A_\alpha) \leftarrow \psi_p(t_\alpha)\}.$$

As $\Theta$ is a standard $s$-substitution we have for all $A_i$, $A_j$ such that $i \neq j$ and all $\sigma \in \mathcal{S}$ that

$$\sigma(A_i)\!\downarrow \neq \sigma(A_j)\!\downarrow.$$

Now consider $A_i$, $A_j$ for $i \neq j$ and $\psi_p(A_i)$, $\psi_p(A_j)$, and assume that

$$\sigma(\psi_p(A_i))\!\downarrow = \sigma(\psi_p(A_j))$$

for some $\sigma \in \mathcal{S}$. Let $A_i = X(r_1, \ldots, r_\alpha)$ and $A_j = X(s_1, \ldots, s_\alpha)$. Then

$$\psi_p(A_i) = X(r_1', \ldots, r_\alpha')$$

and

$$\psi_p(A_j) = X(s_1', \ldots, s_\alpha')$$

where, for all $i = 1, \ldots, \alpha$ either $s_i = s_i', r_i = r_i'$ or $r_i' \in \{\bar{0}, \bar{1}, \bar{2}\}$ and $s_i' \in \{\bar{0}, \bar{1}, \bar{2}\}$. As $\sigma(\psi_p(A_i))\!\downarrow = \sigma(\psi_p(A_j))$ we have if, for some $i$, $r_i' \in \{\bar{0}, \bar{1}, \bar{2}\}$ and $s_i' \in \{\bar{0}, \bar{1}, \bar{2}\}$ ($s_i', r_i'$ are thus numerals) then $r_i' = s_i'$. Now let $J \subseteq I$ (for $I = \{1, \ldots, \alpha\}$) be the subset of positions in $\{1, \ldots, \alpha\}$ where $r_i' = s_i'$ and $r_i', s_i' \in \{\bar{0}, \bar{1}, \bar{2}\}$ As $\sigma(\psi_p(A_i))\!\downarrow =$

$\sigma(\psi_p(A_j))\downarrow$ we must have $\sigma(r_j)\downarrow = \sigma(s_j)\downarrow$ for all $j \in I\backslash J$. We define now $\sigma'(n_j) = \sigma(n_j)$ for $j \in I\backslash J$ and $\sigma'(n_j)$ for $j \in J$ is defined in a way that $\sigma'(r_j)\downarrow = \bar{k}$ for $r'_j = \bar{k}$. Then

$$\sigma'(A_i)\downarrow = \sigma'(\psi_p(A_i))\downarrow = \sigma'(\psi_p(A_j))\downarrow = \sigma'(A_j)\downarrow$$

contradicting the assumption that $\Theta$ is an $s$-substitution.
That for $\sigma \in \mathcal{S}(p)$ we have $\Theta[\sigma] = \psi_p(\Theta)[\sigma]$ is trivial by

$$\sigma(A_i)\downarrow = \sigma(\psi_p(A_i)\downarrow \text{ and } \sigma(t_i)\downarrow = \sigma(\psi_p(t_i)\downarrow \text{ for } i = 1, \ldots, \alpha$$

which follows from Lemma 6.1.1.

We are now in the possession of the tools to define the application of a standard $s$-substitution to a standard term.

**Definition 6.1.12** Let $\Theta$ be a standard $s$-substitution and $t$ be a standard term and $\mathcal{D}$ the set of states over the parameters occurring in $\Theta$ and $s$. Then we define

$$Ap(\Theta, t) = \{p\colon \psi_p(t)\psi_p(\Theta) \mid p \in \mathcal{D}\}.$$

We can compactify the representation of $Ap(\Theta, t)$ in Definition 6.1.12. If there are just a few states $\{p_1, \ldots, p_\beta\}$ such that $\psi_p(t)\psi_p(\Theta) \neq t$ for $p \in \{p_1, \ldots, p_\beta\}$ and $\psi_p(t)\psi_p(\Theta) = t$ for $p \in \mathcal{D}' = \mathcal{D}\backslash\{p_1, \ldots, p_\beta\}$ we can write

$$Ap(\Theta, t) = \{p_1\colon \psi_{p_1}(t)\psi_{p_1}(\Theta), \ldots, p_\beta\colon \psi_{p_\beta}(t)\psi_{p_\beta}(\Theta)\} \cup \{\mathcal{D}'\colon t\}.$$

**Definition 6.1.13** (*semantics of Ap*) Let $Ap(\Theta, t)$ be defined as in Definition 6.1.12. Then

$$Ap(\Theta, t)|_p = \psi_p(t)\psi_p(\Theta),$$
$$Ap(\Theta, t)[\sigma] = \sigma(Ap(\Theta, t)|_p)\downarrow \text{ if } \sigma(p)\downarrow = \top.$$

Note that, as the set of all states form a partition, $Ap(\Theta, t)[\sigma]$ in Definition 6.1.13 is defined for all $\sigma \in \mathcal{S}$.

**Example 6.1.10** Let $\Theta = \{X(n_1, n_2) \leftarrow g(Y(n_1))\}$ like in Example 6.1.7. Consider now the syntactic application $X(\bar{0}, \bar{0})\Theta$ which yields $X(\bar{0}, \bar{0})$. Obviously $X(\bar{0}, \bar{0})\Theta$ is unsound for $\sigma(n_1) = \sigma(n_2) = \bar{0}$ as then

$$\sigma(X(\bar{0}, \bar{0})\Theta)\downarrow = X(\bar{0}, \bar{0}) \neq g(Y(\bar{0})) = X(\bar{0}, \bar{0})\Theta[\sigma].$$

The problem is resolved with the states $p \in \mathcal{D}(X(n_1, n_2))$.
Let $p_1 = n_1 = \bar{0} \wedge n_2 = \bar{0}$, then

$$\psi_p(X(\bar{0}, \bar{0}))\psi_p(\Theta) = X(\bar{0}, \bar{0})\{X(\bar{0}, \bar{0}) \leftarrow g(Y(\bar{0})) = g(Y(\bar{0})).$$

For the other 8 states $p' \in \{p_2, \ldots, p_9\}$ we get

$$\psi_{p'}(X(\bar{0},\bar{0}))\psi_{p'}(\Theta) = X(\bar{0},\bar{0}), \text{ and therefore}$$
$$Ap(\Theta,t) = \{p_1 : g(Y(\bar{0}))\} \cup \{\{p_2,\dots,p_9\} : X(\bar{0},\bar{0})\}.$$

We show now that the definition of application is sound. To this aim we need the following lemma which shows that *syntactic* application of substitutions in sound for terms $\psi_p(t)$ and substitutions $\psi_p(\Theta)$.

**Lemma 6.1.2** *Let $\Theta$ be a standard s-substitution and t be a standard term. Then for states all p over the parameters of $\Theta$ and t and for all $\sigma \in \mathcal{S}(p)$ we have $\sigma(\psi_p(t))\downarrow \psi_p(\Theta)[\sigma] = \sigma(\psi_p(t)\psi_p(\Theta))\downarrow$.*

***Proof*** It is enough to show that $\sigma(\psi_p(B))\downarrow \psi_p(\Theta)[\sigma] = \sigma(\psi_p(B)\psi_p(\Theta))\downarrow$ for variable expressions $B$ and $\sigma \in \mathcal{S}(p)$. So let $B = X(r_1,\dots,r_\beta)$ and

$$\Theta = \{A_1 \leftarrow t_1,\dots,A_\gamma \leftarrow t_\gamma\}$$

where $B$ and $\Theta$ range over the parameters $\{n_1,\dots,n_\alpha\}$. If $X$ does not occur in the domain of $\Theta$ then for all $\sigma$

$$\sigma(\psi_p(B))\downarrow \psi_p(\Theta)[\sigma] = \sigma(\psi_p(B))\downarrow = \sigma(\psi_p(B)\psi_p(\Theta))\downarrow$$

and the lemma trivially holds. Therefore it remains to assume

$$X(s_1,\dots,s_\beta) \leftarrow t \in \Theta$$

where $t$ is some of the $t_1,\dots,t_\gamma$. Let $r_i' = \psi_p(r_i)$ and $s_i' = \psi_p(s_i)$ for $i = 1,\dots,\beta$. We show that

- $X(r_1',\dots,r_\beta') = X(s_1',\dots,s_\beta')$ iff there exists a $\sigma \in \mathcal{S}(p)$ such that

$$(\star)\sigma(X(r_1',\dots,r_\beta'))\downarrow = \sigma(X(s_1',\dots,s_\beta'))\downarrow.$$

The direction from left to right is trivial. So it is enough to prove that the existence of a $\sigma$ which yields $(\star)$ implies that $X(r_1',\dots,r_\beta') = X(s_1',\dots,s_\beta')$. We assume $(\star)$ and $X(r_1',\dots,r_\beta') \neq X(s_1',\dots,s_\beta')$ and derive a contradiction. So let us assume that $r_i' \neq s_i'$ and consider the definition of $\psi_p^i(r_i')$, $\psi^i(s_i')$ in Definition 6.1.10:

- If $p(i) = C_{i,1}$ (i.e. $p(i)\colon n_i = \bar{0}$) we define $\psi_p^i(V) = \psi_p^{i-1}(V)[r_i/\bar{0}]$ (i.e. we replace $r_i$ by $\bar{0}$) if $r_i \in \{n_i, p(n_i), \bar{0}\}$ and $\psi_p^i(V) = \psi_p^{i-1}(V)[r_i/\bar{1}]$ if $r_i = s(n_i)$. That means that, in this case $r_i'$ and $s_i'$ are constant symbols and, as $\sigma(X(r_1',\dots,r_\beta'))\downarrow = \sigma(X(s_1',\dots,s_\beta'))\downarrow$ and thus also $\sigma(r_i')\downarrow = \sigma(s_i')\downarrow$)—which implies that $r_i' = s_i'$.
- If $p(i) = C_{i,2}$ (we have $p(i)\colon n_i \neq \bar{0} \wedge p(n_i) = \bar{0}$) we define $\psi_p^i(V) = \psi_p^{i-1}(V)[r_i/\bar{0}]$ if $r_i = p(n_i)$, $\psi^i(V) = \psi_p^{i-1}(V)[r_i/\bar{1}]$ if $r_i = n_i$, and $\psi_p^i(V) = \psi_p^{i-1}(V)[r_i/bar2]$ if $r_i = s(n_i)$. For $r_i = \bar{0}$ we get $\psi_p^i(V) = \psi_p^{i-1}(V)$. Here $r_i'$

and $s_i'$ are both constants and therefore, as in the case above, $r_i' = s_i'$ as, otherwise, $\sigma(r_i')\downarrow \neq \sigma(s_i')\downarrow$.

- If $p(i) = C_{i,3}$ (i.e. $p(i) = n_i \neq \bar{0} \wedge p(n_i) \neq \bar{0}$) then we define $\psi_p^i(V) = \psi_p^{i-1}(V)$. We consider all cases where $r_i' \neq s_i'$:

1. $r_i' = \bar{0}$, $s_i' = n_i$: as $\sigma(n_i) \neq \bar{0}$ we get $\sigma(r_i')\downarrow \neq \sigma(s_i')\downarrow$.
2. $r_i' = \bar{0}$, $s_i' = p(n_)i$: as $\sigma(p(n_i)) \neq \bar{0}$ we get $\sigma(r_i')\downarrow \neq \sigma(s_i')\downarrow$.
3. $r_i' = \bar{0}$, $s_i' = s(n_i)$: $\sigma(r_i')\downarrow \neq \sigma(s_i')\downarrow$.
4. $r_i' = n_i$, $s_i' = s(n_i)$: for all $\sigma$ $\sigma(r_i')\downarrow \neq \sigma(s_i')\downarrow$.
5. $r_i' = n_i$, $s_i' = p(n_i)$: for $\sigma \in \mathcal{S}(p)$ we have $\sigma(p(n_i))\downarrow \neq \bar{0}$ and thus $\sigma(p(n_i))\downarrow \neq \sigma(n_i)$ which yields $\sigma(r_i')\downarrow \neq \sigma(s_i')\downarrow$.
6. $r_i' = p(n_i)$, $s_i' = s(n_i)$: for all $\sigma$ $\sigma(r_i')\downarrow \neq \sigma(s_i')\downarrow$.

As all remaining cases are symmetric (it does not matter whether we start with $r_i'$ or $s_i'$) we obtain that for all $\sigma \in \mathcal{S}(p)$ $\sigma(r_i')\downarrow \neq \sigma(s_i')\downarrow$—which contradicts ($\star$). We can conclude that $\{X(r_1', \ldots, r_\beta'), X(s_1', \ldots, s_\beta')\}$ is parameter-unifiable iff $X(r_1', \ldots, r_\beta') = X(s_1', \ldots, s_\beta')$. Therefore

$$X(r_1', \ldots, r_\beta')\{X(s_1', \ldots, s_\beta') \leftarrow \psi(t)\} = \psi(t) \text{ iff for all } \sigma \in \mathcal{S}(p)$$
$$\sigma(X(r_1', \ldots, r_\beta'))\downarrow \{\sigma(X(s_1', \ldots, s_\beta'))\downarrow \leftarrow \sigma(\psi(t))\downarrow\} = \sigma(\psi(t))\downarrow$$

which in the end yields $\sigma(\psi_p(t))\downarrow \psi_p(\Theta)[\sigma] = \sigma(\psi_p(t)\psi_p(\Theta))\downarrow$ for $\sigma \in \mathcal{S}(p)$.

**Theorem 6.1.1** *Application is sound, i.e. for all* $\sigma \in \mathcal{S}(p)$ *we have*

$$\sigma(Ap(\Theta, t)|_p)\downarrow = \sigma(t)\downarrow \Theta[\sigma].$$

***Proof*** Let $Ap(\Theta, t)$ be defined as in Definition 6.1.12. Then

$$\begin{aligned} Ap(\Theta, t)|_p &= \psi_p(t)\psi_p(\Theta) \text{ and therefore} \\ \sigma(Ap(\Theta, t)|_p)\downarrow &= \sigma(\psi_p(t)\psi_p(\Theta))\downarrow \text{ and by Lemma 7.1.2} \\ \sigma(\psi_p(t))\downarrow \psi_p(\Theta)[\sigma] &= \sigma(\psi_p(t)\psi_p(\Theta))\downarrow. \end{aligned}$$

As $\sigma(\psi_p(t))\downarrow = \sigma(t)\downarrow$ and $\Theta[\sigma] = \psi_p(\Theta)[\sigma]$ for $\sigma \in \mathcal{S}(p)$ the theorem follows.

We illustrate now the application of a substitution $\Theta$ in case of variable expressions $X(r_1, r_2)$ for $X \in \mathbf{V}_2^\iota$.

**Example 6.1.11** Given a standard variable expression $X(r_1, r_2)$ and $\Theta = \Theta_1 \cup \cdots \cup \Theta_i$ we compute the term $X(r_1, r_2)\Theta$. We assume that the $\Theta_i$ are the substitutions for the variable class $X_i$. Therefore it is sufficient to determine the result of $X(s, t)\Theta_i$ for just one $i \in \{1, \ldots, \alpha\}$: if $X \neq X_i$ then we have $X(s, t)\Theta_i = \{\top\colon X(s, t)\}$.
To this aim we have to distinguish all different forms of $X(r_1, r_2)$ and all forms of replacements $X(r_1', r_2') \leftarrow t'$ in $\Theta_i$ and all states $p$ of $X_i(r_1, r_2)$. According to Definition 6.1.9 we obtain 9 states i.e. $\mathcal{D} = \{p_1, \ldots, p_9\}$ (for comfort we use the names $n, m$ for $n_1, n_2$).

$$
\begin{aligned}
p_1 &= n = \bar{0} \wedge m = \bar{0},\\
p_2 &= n = \bar{0} \wedge m \neq \bar{0} \wedge p(m) \neq \bar{0},\\
p_3 &= n = \bar{0} \wedge m \neq \bar{0} \wedge p(m) = \bar{0},\\
p_4 &= n \neq \bar{0} \wedge m = \bar{0} \wedge p(n) \neq \bar{0},\\
p_5 &= n \neq \bar{0} \wedge m = \bar{0} \wedge p(n) = \bar{0},\\
p_6 &= n \neq \bar{0} \wedge m \neq \bar{0} \wedge p(n) = \bar{0} \wedge p(m) = \bar{0},\\
p_7 &= n \neq \bar{0} \wedge m \neq \bar{0} \wedge p(n) = \bar{0} \wedge p(m) \neq \bar{0},\\
p_8 &= n \neq \bar{0} \wedge m \neq \bar{0} \wedge p(n) \neq \bar{0} \wedge p(m) = \bar{0},\\
p_9 &= n \neq \bar{0} \wedge m \neq \bar{0} \wedge p(n) \neq \bar{0} \wedge p(m) \neq \bar{0}.
\end{aligned}
$$

Let us assume that $X(r_1, r_2) = X(n, m)$ and

$$\Theta_i = \{X(\bar{0}, m) \leftarrow g(Y(n)), X(s(n), m) \leftarrow Y(n)\}.$$

According to Definition 6.1.12 we have to determine the set

$$S = \{p_1 \colon \psi_1(X(n, m))\psi_1(\Theta_i), \ldots, p_9 \colon \psi_9(X(n, m))\psi_9(\Theta_i)\}.$$

where we write $\psi_i$ for $\psi_{p_i}$.

$\psi_1(X(n, m)) = X(\bar{0}, \bar{0})$, $\psi_1(\Theta_i) = \{X(\bar{0}, \bar{0}) \leftarrow g(Y(\bar{0})), X(s(\bar{0}), \bar{0}) \leftarrow Y(\bar{0})\}$
therefore $p_1 \colon X(\bar{0}, \bar{0})\{X(\bar{0}, \bar{0}) \leftarrow g(Y(\bar{0})), X(s(\bar{0}), \bar{0}) \leftarrow Y(\bar{0})\} \in S$.
resulting in $p_1 \colon g(Y(\bar{0})) \in S$.
$\psi_2(X(n, m)) = X(\bar{0}, m)$, $\psi_2(\Theta_i) = \{X(\bar{0}, m) \leftarrow g(Y(\bar{0})), X(s(\bar{0}), m) \leftarrow Y(\bar{0})\}$
therefore $p_2 \colon g(Y(\bar{0})) \in S$.
$\psi_3(X(n, m)) = X(\bar{0}, \bar{1})$, $\psi_3(\Theta_i) = \{X(\bar{0}, \bar{1}) \leftarrow g(Y(\bar{0})), X(\bar{1}, \bar{1}) \leftarrow Y(\bar{0})\}$
therefore $p_3 \colon g(Y(\bar{0})) \in S$.
$\psi_4(X(n, m)) = X(n, \bar{0})$, $\psi_4(\Theta_i) = \{X(\bar{0}, \bar{0}) \leftarrow g(Y(n)), X(s(n), \bar{0}) \leftarrow Y(n)\}$
therefore $p_4 \colon X(n, \bar{0}) \in S$.
$\psi_5(X(n, m)) = X(\bar{1}, \bar{0})$, $\psi_5(\Theta_i) = \{X(\bar{0}, \bar{0}) \leftarrow g(Y(\bar{1})), X(\bar{2}, \bar{0}) \leftarrow Y(\bar{1})\}$
therefore $p_5 \colon X(\bar{1}, \bar{0}) \in S$.
$\psi_6(X(n, m)) = X(\bar{1}, \bar{1})$, $\psi_6(\Theta_i) = \{X(\bar{0}, \bar{1}) \leftarrow g(Y(\bar{1})), X(\bar{2}, \bar{1}) \leftarrow Y(\bar{1})\}$
therefore $p_6 \colon X(\bar{1}, \bar{1}) \in S$.
$\psi_7(X(n, m)) = X(\bar{1}, m)$, $\psi_7(\Theta_i) = \{X(\bar{0}, m) \leftarrow g(Y(\bar{1})), X(\bar{2}, m) \leftarrow Y(\bar{1})\}$
therefore $p_7 \colon X(\bar{1}, m) \in S$.
$\psi_8(X(n, m)) = X(n, \bar{1})$, $\psi_8(\Theta_i) = \{X(\bar{0}, \bar{1}) \leftarrow g(Y(\bar{n})), X(s(n), \bar{1}) \leftarrow Y(\bar{1})\}$
therefore $p_8 \colon X(n, \bar{1}) \in S$.
$\psi_9(X(n, m) = X(n, m)$, $\psi_9(\Theta_i) = \Theta_i$ and therefore $p_9 \colon X(n, m) \in S$.

Now we can determine $Ap(\Theta_i, X(n, m))$; we obtain

$$Ap(\Theta_i, X(n,m)) =$$
$$\{p_1\colon g(Y(\bar{0})),\ \{p_2, p_3\}\colon g(Y(\bar{0})),\ p_4\colon X(n,\bar{0}),\ p_5\colon X(\bar{1},\bar{0}),$$
$$p_6\colon X(\bar{1},\bar{1}),\ p_7\colon X(\bar{1},m),\ p_8\colon X(n,\bar{1}),\ p_9\colon X(n,m)\}.$$

It is easy to see that, e.g., $Ap(\Theta_i, X(s(n), s(m))) = \{\mathcal{D} : X(s(n), s(m))\}$.

**Example 6.1.12** Let $\Theta = \Theta_i$ be the standard $s$-substitution defined in Example 6.1.11. We consider the term

$$t = r(X(n,m), X(s(n),\bar{0}), Y(n)).$$

By Definition 6.1.3 we have, for every state $p$,

$$Ap(\Theta, t)_p = r(\psi_p(X(n,m))\chi_p(\Theta), \psi_p(X(s(n),\bar{0})\chi_p(\Theta), \psi_p(Y(n))\chi_p(\Theta).$$

Again we have only two parameters so $\mathcal{D}(X(n,m)) = \{p_1, \ldots, p_9\}$ as in Example 6.1.11. Therefore we obtain

$$Ap(\Theta, t) = \{p\colon Ap(\Theta, t)|_p \mid p \in \{p_1, \ldots, p_9\}$$

Below we compute the subset $S\colon \{Ap(\Theta, t)|_{p_1}, Ap(\Theta, t)|_{p_2}, Ap(\Theta, t)|_{p_9}\}$ of $Ap(\Theta, t)$. Recall that

$$\Theta = \{X(\bar{0}, m) \leftarrow g(Y(n)), X(s(n), m) \leftarrow Y(n)\} \text{ and}$$

$$\begin{aligned}
p_1 &= n = \bar{0} \wedge m = \bar{0},\\
p_2 &= n = \bar{0} \wedge m \neq \bar{0} \wedge p(m) \neq \bar{0},\\
p_9 &= n \neq \bar{0} \wedge m \neq \bar{0} \wedge p(n) \neq \bar{0} \wedge p(m) \neq \bar{0}.\\
Ap(\Theta, t)|_{p_1} &= r(g(Y(\bar{0}))Y(\bar{0}), Y(n))),\\
Ap(\Theta, t)|_{p_2} &= r(g(Y(\bar{0}), Y(n), Y(n))),\\
Ap(\Theta, t)|_{p_3} &= r(X(n,m), X(s(n),\bar{0}), Y(n)).
\end{aligned}$$

Therefore

$$S = \{p_1\colon r(g(Y(\bar{0}))Y(\bar{0}), Y(n))),\ p_2\colon r(g(Y(\bar{0}), Y(n), Y(n))),$$
$$p_9\colon r(X(n,m), X(s(n),\bar{0}), Y(n))\}.$$

Though the computation of application is still complicated—also for standard terms and standard $s$-substitutions—the complexity is fixed when the set of parameters if fixed. The number of different states for a standard substitution over two parameters is always 9—thus a fixed number. Fixed is also the number of possible arguments for a variable $X \in \mathbf{V}_2^{\iota}$ and $X(n,m)$, namely $(r,s)\colon r \in \{\bar{0}, n, s(n), p(n)\}, s \in \{\bar{0}, m, s(m), p(m)\}$—i.e. 16. In fact the substitutions can be computed via predefined tables! Fixed numbers (though larger ones) we get also for more than 2 parameters. Thus the complexity of the representation only depends on the number of parameters and not on the complexity of terms. This holds also for the concatenation of substitutions (under a given state concatenation of substitutions are uniquely defined) We are now able to define the composition of standard $s$-substitutions.

**Definition 6.1.14** Let $\Theta_1$ and $\Theta_2$ be standard $s$-substitutions such that the set of parameters in the domains and range of $\Theta_1, \Theta_2$ is contained in $\{n_1, \ldots, n_\gamma\}$. Then for any $p \in \mathcal{D}(X(n_1, \ldots, n_\gamma))$ we define

$$(\Theta_1 \circ \Theta_2)|_p = \psi_p(\Theta_1)\psi_p(\Theta_2).$$

Finally, $\Theta_1 \circ \Theta_2 = \{p\colon (\Theta_1 \circ \Theta_2)|_p \mid p \in \mathcal{D}(X(n_1, \ldots, n_\gamma))\}$.

**Example 6.1.13** Let

$$\begin{aligned}
\Theta_1 &= \{X(\bar{0}, m) \leftarrow g(Y(n)), X(s(n), m) \leftarrow Y(n)\},\\
\Theta_2 &= \{Y(n) \leftarrow \hat{f}(Z(n), n), Y_1(\bar{0}, \bar{0}, k) \leftarrow a\},\\
\hat{f}(x, \bar{0}) &= x,\\
\hat{f}(x, s(n)) &= f(\hat{f}(x, n)).
\end{aligned}$$

As we have 3 parameters in $\Theta_1, \Theta_2$ we have 27 different states. We select 3 states $p_1, p_2, p_3$ of them and compute $(\Theta_1 \circ \Theta_2)_p$. Let

$$\begin{aligned}
&p_1 = n = \bar{0} \wedge m = \bar{0} \wedge k = \bar{0},\\
&p_2 = n \neq \bar{0} \wedge p(n) = \bar{0} \wedge m = \bar{0} \wedge k \neq \bar{0} \wedge p(k) \neq \bar{0},\\
&p_3 = n \neq \bar{0} \wedge p(n) \neq \bar{0} \wedge m \neq \bar{0} \wedge p(m) \neq \bar{0} \wedge k \neq \bar{0} \wedge p(k) \neq \bar{0}.
\end{aligned}$$

Then

$$\begin{aligned}
&(\Theta_1 \circ \Theta_2)|_{p_1} = \psi_{p_1}(\Theta_1)\psi_{p_1}(\Theta_2) =\\
&\{X(\bar{0}, \bar{0}) \leftarrow g(Y(\bar{0})), X(\bar{1}, \bar{0}) \leftarrow Y(\bar{0})\}\{Y(\bar{0}) \leftarrow \hat{f}(Z(\bar{0}), \bar{0}), Y_1(\bar{0}, \bar{0}, \bar{0}) \leftarrow a\} =\\
&\{X(\bar{0}, \bar{0}) \leftarrow g(\hat{f}(Z(\bar{0}), \bar{0})),\ X(\bar{1}, \bar{0}) \leftarrow \hat{f}(Z(\bar{0}), \bar{0}),\\
&Y(\bar{0}) \leftarrow \hat{f}(Z(\bar{0}), \bar{0}), Y_1(\bar{0}, \bar{0}, \bar{0}) \leftarrow a\} =\\
&\{X(\bar{0}, \bar{0}) \leftarrow g(Z(\bar{0}), \bar{0})),\ X(\bar{1}, \bar{0}) \leftarrow Z(\bar{0}), Y(\bar{0}) \leftarrow Z(\bar{0}),\ Y_1(\bar{0}, \bar{0}, \bar{0}) \leftarrow a\}.\\
&(\Theta_1 \circ \Theta_2)|_{p_2} = \psi_{p_2}(\Theta_1)\psi_{p_2}(\Theta_2) =\\
&\{X(\bar{0}, \bar{0}) \leftarrow g(Y(\bar{1})),\ X(\bar{2}, \bar{0}) \leftarrow Y(\bar{1})\}\{Y(\bar{1}) \leftarrow \hat{f}(Z(\bar{1}), \bar{1}), Y_1(\bar{0}, \bar{0}, k) \leftarrow a\} =\\
&\{X(\bar{0}, \bar{0}) \leftarrow g(\hat{f}(Z(\bar{1}), \bar{1})),\ X(\bar{2}, \bar{0}) \leftarrow \hat{f}(Z(\bar{1}), \bar{1}),\\
&Y(\bar{1}) \leftarrow \hat{f}(Z(\bar{1}), \bar{1}), Y_1(\bar{0}, \bar{0}, k) \leftarrow a\} =\\
&\{X(\bar{0}, \bar{0}) \leftarrow g(f(Z(\bar{1}))),\ X(\bar{2}, \bar{0}) \leftarrow f(Z(\bar{1})), Y(\bar{1}) \leftarrow f(Z(\bar{1})), Y_1(\bar{0}, \bar{0}, k) \leftarrow a\}.\\
&(\Theta_1 \circ \Theta_2)|_{p_3} = \psi_{p_3}(\Theta_1)\psi_{p_3}(\Theta_2) =\\
&\{X(\bar{0}, m) \leftarrow g(Y(n)), X(s(n), m) \leftarrow Y(n)\}\{Y(n) \leftarrow \hat{f}(Z(n), n), Y_1(\bar{0}, \bar{0}, k) \leftarrow a\}\\
&= \{X(\bar{0}, m) \leftarrow g(\hat{f}(Z(n), n)),\ X(s(n), m) \leftarrow \hat{f}(Z(n), n),\\
&Y(n) \leftarrow \hat{f}(Z(n), n),\ Y_1(\bar{0}, \bar{0}, k) \leftarrow a\}.
\end{aligned}$$

**Theorem 6.1.2** *The composition of standard s-substitutions is sound: let $\mathcal{D}$ be the set of parameters in $\Theta_1$ and $\Theta_2$. Then, for all $p \in \mathcal{D}$ and $\sigma \in \mathcal{S}(p)$ we have $\sigma((\Theta_1 \circ \Theta_2)|_p)\downarrow = \Theta_1[\sigma]\Theta_2[\sigma]$.*

***Proof*** Immediate by Definition 6.1.14, by Definition 6.1.11 and by Theorem 6.1.1.

It remains to extend substitution and unification to schematic formulas. We restrict our analysis to simple $\omega\iota o$-theories. Let $\Theta$ be a standard $s$-substitution and $\mathcal{D}$ a set of states containing all parameters of the formula. We first define the application of $\psi_p(\Theta)$ to $\mathbf{F}^o_b$, the basic formulas (see Definition 4.2.2).

**Definition 6.1.15** We assume that all formulas have parameters in $\mathcal{D}$ and $p \in \mathcal{D}$. Let $\Theta$ be a standard$s$-substitution and $T^*$ be an $\omega\iota$-theory. Then

- If $\xi \in \mathbf{V}^o$ then $Ap(\Theta, \xi)|_p = \xi\psi_p(\Theta) = \xi$.
- If $P \in \mathbf{P}^o_k$ and $t_1, \ldots, t_k \in \mathbf{F}^o_b(T^*)$ then $Ap(\Theta, P(t_1, \ldots, t_k))|_p = P(Ap(\Theta, t_1)|_p, \ldots, Ap(\Theta, t_k)|_p)$
- If $F \in \mathbf{F}^o_b(T^*)$ then $Ap(\Theta, \neg F)|_p = \neg Ap(\Theta, F)|_p$.
- If $F_1, F_2 \in \mathbf{F}^o_b(T^*)$ then $Ap(\Theta, F_1 \circ F_2)|_p = Ap(\Theta, F_1)|_p \circ Ap(\Theta, F_2)|_p$ for $\circ \in \{\wedge, \vee\}$.

The extension of an $s$-substitutions to schematic formulas is not completely trivial. It is not immediately clear how an $s$-substitution should be applied to an expression of the form $\hat{p}(X_1, \ldots, X_j, t_1, \ldots, t_k)$ where $\hat{p}$ is a schematic predicate symbol and $X_1, \ldots, X_j$ are variable classes. The solution consists in introducing a new symbol $\hat{p}(\Theta)$ which depends on $\hat{p}$ and $\Theta$.

**Definition 6.1.16** We assume that all formulas have parameters in $\mathcal{D}$ and $p \in \mathcal{D}$. Let $\Theta$ be a standard $s$-substitution. Then we define

- If $\xi \in \mathbf{V}^o$ then $Ap(\Theta, \xi)|_p = \xi\psi_p(\Theta) = \xi$.
- If $P \in \mathbf{P}^o_k$ and $t_1, \ldots, t_k \in \hat{\mathbf{T}}^\iota(T^*)$ then $Ap(\Theta, P(t_1, \ldots, t_k))|_p = P(Ap(\Theta, t_1)|_p, \ldots, Ap(\Theta, t_k)|_p)$.
- $Ap(\Theta, \neg F)|_p = \neg Ap(\Theta, F)|_p$.
- $Ap(\Theta, F_1 \circ F_2)|_p = Ap(\Theta, F_1)|_p \circ Ap(\Theta, F_2)|_p$ for $\circ \in \{\wedge, \vee\}$.
- if $\pi = (i, j, k)$ is a profile and $\hat{p} \in \hat{P}(\pi)$, $X_l \in \mathbf{V}^\iota_i$ for $l = 1, \ldots, j, t_1, \ldots, t_k \in \mathbf{T}^\omega$ then

$$Ap(\Theta, \hat{p}(X_1, \ldots, X_j, t_1, \ldots, t_k)|_p = \hat{p}(\Theta)(X_1, \ldots, X_j, t_1, \ldots, t_k)|_p.$$

where $\hat{p}(\Theta)$ is defined as follows: Let us assume that $D(\hat{p})$ is of the form

$$D(\hat{p}) = \{\hat{p}(X_1, \ldots X_j, n_1, \ldots, n_k) =_d \underline{\text{if}}\ n_k = \bar{0}\ \underline{\text{then}}\ \hat{q}_B \\ \underline{\text{else}}\ \hat{p}_S\{\xi \leftarrow \hat{p}(X_1, \ldots, X_j, n_1, \ldots, p(n_k))\}\}.$$

Then we define $D(\hat{p}(\Theta))|_p)$ as $D(\hat{p}(\Theta))_p =$

$$\{\hat{p}(\Theta)(X_1, \ldots, X_j, n_1, \ldots, n_k)|_p =_d \underline{\text{if}}\ n_k = \bar{0}\ \underline{\text{then}}\ Ap(\Theta, \hat{q}_B)|_p$$
$$\underline{\text{else}}\ Ap(\Theta, \hat{p}_S)|_p\{\xi \leftarrow \hat{p}(\Theta)(X_1, \ldots, X_j, n_1, \ldots, p(n_k))|_p\}\}.$$

**Example 6.1.14** Let $(T, T', \emptyset)$ the $\omega\iota o$-theory defined in Example 6.1.4. We had the following definitions of recursive predicate—and term symbols:

$$D(\hat{q}) = \{\hat{q}(X, n) =_d\ \underline{\text{if}}\ n = \bar{0}\ \underline{\text{then}}\ Q(X(\bar{0}))$$
$$\underline{\text{else}}\ \hat{q}(X, p(n)) \vee Q(\hat{t}(X(n), n))\}.$$
$$D(\hat{t}) = \{\hat{t}(x, n) =_d\ \underline{\text{if}}\ n = \bar{0}\ \underline{\text{then}}\ x\ \underline{\text{else}}\ h(\hat{t}(x, p(n)))\}.$$

and let

$$\Theta = \{X(n) \leftarrow \hat{t}(X(n), n)\},$$
$$p = n > \bar{0} \wedge p(n) > \bar{0},$$
$$F = \hat{q}(X, n).$$

Note that $\psi_p(\Theta) = \Theta$. Now we apply $\Theta$ to $F$:

$$Ap(\Theta, F)|_p = Ap(\Theta, \hat{q}(X, n))|p = \hat{q}(\Theta)(X, n)|_p,\ \text{where, as } n > \bar{0}$$
$$\hat{q}(\Theta)(X, n)|_p = Ap(\Theta, \hat{q}(X, p(n)) \vee Q(\hat{t}(X(n), n)))|p =$$
$$Ap(\Theta, \hat{q}(X, p(n)))|_p \vee Ap(\Theta, Q(\hat{t}(X(n), n)))|p =$$
$$Ap(\Theta, \hat{q}(X, p(n)))|_p \vee Q(\hat{t}(X(n), n))\psi_p(\Theta) =$$
$$\hat{q}(\Theta)(X, p(n))|_p \vee Q(\hat{t}(X(n), n))\Theta = \hat{q}(\Theta)(X, p(n))|p \vee Q(\hat{t}(\hat{t}(X(n), n), n)).$$

Now we also have $p(n) > \bar{0}$ and therefore

$$\hat{q}(\Theta)(X, p(n))|_p = Ap(\Theta, \hat{q}(X, p(p(n))) \vee Q(\hat{t}(X(p(n)), p(n))))|p =$$
$$Ap(\Theta, \hat{q}(X, p(p(n))))|_p \vee Q(\hat{t}(X(p(n)), p(n)))\psi_p(\Theta) =$$
$$\hat{q}(\Theta)(X, p(p(n))))|_p \vee Q(\hat{t}(X(p(n)), p(n))).$$

Note that $p$ does not give any further information to reduce $\hat{q}(\Theta)(X, p(p(n)))$; this expression cannot be evaluated unless we have a parameter assignment respecting $p$. Putting things together we obtain

$$Ap(\Theta, F)|_p = \hat{q}(\Theta)(X, p(p(n))))|_p \vee Q(\hat{t}(X(p(n)), p(n)))$$
$$\vee\ Q(\hat{t}(\hat{t}(X(n), n), n)).$$

**Theorem 6.1.3** *Schematic substitution is sound on schematic formulas: let $F$ be a formula in an $\omega\iota o$-theory, $\Theta$ be a standard s-substitution and $\sigma$ a parameter assigment. Then $\sigma(F){\downarrow}\ \Theta[\sigma] = \sigma(Ap(\Theta, F)|_p){\downarrow}$ for $p \in \mathcal{D}$ and $\sigma(p){\downarrow} = \top$.*

***Proof*** By Theorem 6.1.1 we know that, for schematic terms $t$ in $T^*$, $\sigma(t){\downarrow}\ \Theta[\sigma] = \sigma(Ap(\Theta, t)|_p){\downarrow}$ for $\sigma(p){\downarrow} = \top$. If $F$ is a formula in $T^*$ without schematic predicate symbols then, by Definition 6.1.16, $\sigma(F)\downarrow \Theta[\sigma] = \sigma(Ap(\Theta, F)|_p)\downarrow$ for $\sigma(p)\downarrow = \top$ (here we only have homorphic extensions of the application of $\Theta$). What remains to be shown is

$$\sigma(F)\downarrow \Theta[\sigma] = \sigma(Ap(\Theta(F))|p)\downarrow \text{ for } \sigma(p)\downarrow = \top$$

and for $F = \hat{p}(X_1, \dots, X_j, t_1, \dots, t_k)$ with the corresponding definitions

$$\{\hat{p}(X_1, \dots X_j, n_1, \dots, n_{k-1}, s(n_k)) = \hat{p}_S\{\xi \leftarrow \hat{p}(X_1, \dots, X_j, n_1, \dots, n_{k-1}, n_k)\},$$
$$\hat{p}(X_1, \dots X_j, n_1, \dots, n_{k-1}, \bar{0}) = \hat{q}_B\}.$$

Moreover, we assume $\sigma(t_i)\downarrow = \bar{\alpha}_i$ for $i = 1, \dots, k$.

Here we proceed by the depth of recursion. Let us assume that $\hat{p}$ is minimal and $\sigma(n_k) = \bar{0}$; then in the base case

$$\sigma(\hat{p}(X_1, \dots X_j, \bar{\alpha}_1, \dots, \bar{\alpha}_{k-1}, \bar{0})\downarrow = \hat{p}(X_1, \dots, X_j, \bar{\alpha}_1, \dots, \bar{\alpha}_{k-1}, \bar{0})\downarrow$$
$$= \sigma(\hat{p}_B)\downarrow.$$

$\hat{p}_B$ is a formula without schematic predicate symbols. In this case we have already shown that

$$\sigma(\hat{p}_B)\downarrow \Theta[\sigma] = \sigma(Ap(\Theta, \hat{p}_B)|_p)\downarrow \text{ for } \sigma(p)\downarrow = \top.$$

So we obtain

$$\sigma(\hat{p}(X_1, \dots, X_j, n_1, \dots, n_{k-1}, \bar{0}))\downarrow \Theta[\sigma] =$$
$$\sigma(Ap(\Theta, \hat{p}(X_1, \dots, X_j, n_1, \dots, \bar{\alpha}_{k-1}, \bar{0}))|p)\downarrow \text{ for } \sigma(p)\downarrow = \top.$$

Note that also $\hat{p}(\Theta)(X_1, \dots, X_j, n_1, \dots, \bar{0})|_p = Ap(\Theta, \hat{q}_B)|_p$.
For the step case of $\hat{p}$ ($\bar{\alpha}_k > \bar{0}$ we have that $\hat{p}_S$ does not contain schematic predicate symbols and we may assume that for $\sigma(n_k) = \bar{\alpha}_k$ we have

$$\sigma(\hat{p}(X_1, \dots, X_j, n_1, \dots, n_{k-1}, n_k))\downarrow \Theta[\sigma] =$$
$$\sigma(Ap(\Theta, \hat{p}(X_1, \dots, X_j, n_1, \dots, n_{k-1}, n_k)|_p)\downarrow$$

for $\sigma(p) = \top$. By

$$\hat{p}(X_1, \dots X_j, n_1, \dots, n_{k-1}, s(n_k)) = \hat{p}_S\{\xi \leftarrow \hat{p}(X_1, \dots, X_j, n_1, \dots, n_{k-1}, n_k)\}$$

we obtain for $\sigma(p) = \top$

$$\sigma(\hat{p}(X_1, \dots X_j, n_1, \dots, n_{k-1}, s(n_k)))\downarrow \Theta[\sigma] =$$
$$\sigma(\hat{p}_S)\downarrow \Theta[\sigma]\{\xi \leftarrow \sigma(\hat{p}(X_1, \dots, X_j, n_1, \dots, n_{k-1}, n_k))\downarrow \Theta[\sigma]\} =$$
$$\sigma(\hat{p}_S)\downarrow \Theta[\sigma]\{\xi \leftarrow \sigma(Ap(\Theta, \hat{p}(X_1, \dots, X_j, n_1, \dots, n_{k-1}, n_k))|_p)\}\downarrow =$$
$$\sigma(Ap(\Theta, \hat{p}_S)|p)\downarrow \{\xi \leftarrow \sigma(Ap(\Theta, \hat{p}(X_1, \dots, X_j, n_1, \dots, n_{k-1}, n_k))|_p)\downarrow\} =$$
$$\sigma(Ap(\Theta, \hat{p}(X_1, \dots X_j, n_1, \dots, n_{k-1}, s(n_k))|p))\downarrow.$$

The case for nonminimal $\hat{p}$ is analogous with the difference that the induction hypothesis holds for $\hat{p}_S$, $\hat{p}_b$ as they contain only schematic symbols smaller than $\hat{p}$.

## 6.2 Schematic Unification

Unification is crucial to any machine oriented logic, in particular to the resolution calculus [36]. In Sect. 7.2 we will present a refutational calculus for formula schemata, generalizing the first-order resolution principle to schemata. In particular we will concentrate on unification in $\omega\iota$-theories. Our first step consists in defining the concept of schematic unification.

**Definition 6.2.1** Let $T^*$ be a simple $\omega\iota$-theory $T^*$ and $t_1, \ldots, t_\alpha \in \hat{\mathbf{T}}^\iota_s(T^*)$. Then the set $\{t_1, \ldots, t_\alpha\}$ is ***unifiable*** if, for all $\sigma \in \mathcal{S}$, $\{\sigma(t_1)\downarrow_{T^*}, \ldots, \sigma(t_\alpha)\downarrow_{T^*}\}$ is first-order unifiable.

Unfortunately, unification in simple $\omega\iota$-theories is undecidable as we will show below. First we will show that all primitive recursive functions can be computed in such theories by translating the constructions in Theorem 4.1.1 to simple $\omega\iota$-theories. We define a constant *null* for zero (which is now an invidual constant, not a number constant) and a one-place function symbol *succ*. The successor function and the predecessor function are defined via the following schematic symbols $\hat{s}$, $\hat{t}$, $n\hat{u}m$ (all of them having the profile $(1, 1)$):

$$\begin{aligned} D(n\hat{u}m) &= \{n\hat{u}m(x,n) =_d \underline{\text{if}}\ n = \bar{0}\ \underline{\text{then}}\ null\ \underline{\text{else}}\ succ(n\hat{u}m(x, p(n)))\}, \\ D(\hat{s}) &= \{\hat{s}(x,n) = succ(n\hat{u}m(x,n))\}, \\ D(\hat{t}) &= \{\hat{t}(x,n) =_d \underline{\text{if}}\ n = \bar{0}\ \underline{\text{then}}\ null\ \underline{\text{else}}\ n\hat{u}m(x, p(n))\}. \end{aligned}$$

It is easy to see that for $\sigma(n) = \alpha\ \sigma(\hat{s}(x,n))\downarrow = succ^\alpha(null)$ which represents $s^\alpha(\bar{0})$ within the individual terms. In the same way $\hat{t}(x,n)$ represents the predecessor $p$:

$$\begin{aligned} &\hat{t}(x,\bar{0})\downarrow = null, \\ &\hat{t}(x,\bar{1})\downarrow = n\hat{u}m(x,\bar{0}) = null, \\ &\hat{t}(x,\bar{2})\downarrow = n\hat{u}m(x,\bar{1})\downarrow = succ(x,\bar{0})\downarrow = succ(null)\ \text{etc.} \end{aligned}$$

An easy induction argument yields that for all $\sigma(n) = \alpha \geq \bar{2}\ \sigma(\hat{t}(x,n))\downarrow = succ^{\alpha-1}(null)$ which represents $s^{\alpha-1}(\bar{0})$ in the individual terms. It is easy to see that the variable $x$ (which could be replaced by any first-order term) plays no role in the definition above. We illustrate the recursive definition of $+$ within the individual terms schemata and show how to restrict the individual terms such that the schema evaluates to the new "numerals". We start with

$$D(\oplus) = \{\oplus(x,n) =_d \underline{\text{if}}\ n = \bar{0}\ \underline{\text{then}}\ x\ \underline{\text{else}}\ succ(\oplus(x, p(n))\}.$$

This definition would be inappropriate as $x$, which is an individual variable, cannot be evaluated under a $\sigma \in \mathcal{S}$. Instead we have to evaluate $\oplus(\hat{s}(x, m), n)$ under $\sigma$ which gives the desired result. Using this trick we can simulate the computation of primitive recursive functions in simple $\omega\iota$-theories. It is easy to see that all operations needed in Theorem 4.1.1 can be carried out in this way; we don't give the rather tedious details here. We call the set of schemata $\omega\iota$ theories corresponding to the primitive recursive schemata in $T^{\omega}$ as $\mathrm{PRL}_{\iota}$. While, given a $\sigma \in \mathcal{S}$, every term in $T^{\omega}$ evaluates to a numeral, the terms in $\mathrm{PRL}_{\iota}$ evaluate to elements in $Num_{\iota}$ which is defined as

- $null \in Num_{\iota}$,
- if $t \in Num_{\iota}$ then $succ(t) \in Num_{\iota}$.

**Theorem 6.2.1** *The unification problem for schematic terms in* $\mathrm{PRL}_{\iota}$ *is undecidable.*

***Proof*** It is well known that the equivalence problem of primitive recursive programs is undecidable (see, e.g. [37]). Both the schematic terms defined in Theorem 4.1.1 and their modifications in indidual term schemata define a programming language for primitive recursive functions. In particular it is undecidable whether a primitive recursive program computes the function *zero* ($zero(n) = 0$ for all $n \in \mathbb{N}$. Therefore the equivalence problem $t = null$ where $t \in \mathrm{PRL}_{\iota}$ is undecidable; that means it is undecidable whether for all $\sigma \in \mathcal{S}$ $\sigma(t){\downarrow} = null$ (note that $\sigma(null){\downarrow} = null$ for all $\sigma \in \mathcal{S}$). We consider now the unfication problem $M\colon \{t, null\}$. According to Definition 6.2.1 $M$ is unifiable iff for all $\sigma \in \mathcal{S}$ the set $M(\sigma)\colon \{\sigma(t){\downarrow}, null{\downarrow}\} = \{\sigma(t){\downarrow}, null\}$ is first-order unifiable. But as $t \in \mathrm{PRL}_{\iota}$ we have $\sigma(t){\downarrow} \in Num_{\iota}$, i.e. $\sigma(t){\downarrow}$ is a normalized ground term. Therefore $M(\sigma)$ is only unifiable if $\sigma(t){\downarrow} = null$, and $M$ is unifiable iff for all $\sigma \in \mathcal{S}$ $\sigma(t){\downarrow} = null$ iff the equivalence problem $t = null$ holds. Hence the unification problem for $\mathrm{PRL}_{\iota}$ is undecidable.

As schematic unfication is not only undecidable but even $\Pi_1$-complete there exists no complete method for unfication, i.e. there are always unifiers which cannot be found algorithmic methods. However, we need a method for subclasses of the problem which does not only detect unifiability but also produces a syntactic representation of a unifier. Below we extend unification to standard terms.

**Definition 6.2.2** Let $\mathcal{T}\colon \{t_1, \ldots, t_\alpha\}$ be a set of standard terms over the parameters $\mathcal{N}_0\colon \{n_1, \ldots, n_\beta\}$ and $\mathcal{D}$ the set of states over $\mathcal{N}_0$. A standard $s$-substitution $\Theta$ is called a *standard unifier* of $\mathcal{T}$ if for all $p \in \mathcal{D}$

$$Ap(\Theta, t_1)|p = \cdots = Ap(\Theta, t_\alpha)|p.$$

$\mathcal{T}$ is called *uniformly standard unifiable* if there exists a standard unifier of $\mathcal{T}$.

**Example 6.2.1** Let $T = (T_1, \emptyset)$ be a $\omega\iota$-theory where $T_1 = (\hat{t}, \{\hat{t}, \hat{s}\}, D(\{\hat{t}, \hat{s}\}), \{x, y\}, <)$ and

- $\pi(\hat{t}) = (2, 2)$, $\pi(\hat{s}) = (1, 1)$,
- $x, y \in \mathbf{V}_0^{\iota}$,

- $\hat{s} < \hat{t}$,
- $g \in \mathbf{F}_2^\iota, h \in \mathbf{F}_1^\iota$.

Moreover,

$$D(\hat{t}) = \{\hat{t}(x, y, n_1, n_2) =_d \underline{\text{if}}\ n_2 = \bar{0}\ \underline{\text{then}}\ y$$
$$\underline{\text{else}}\ g(\hat{t}(x, y, n_1, p(n_2)), \hat{s}(x, n_1))\},$$
$$D(\hat{s}) = \{\hat{s}(x, n_1) =_d \underline{\text{if}}\ n_1 = \bar{0}\ \underline{\text{then}}\ x\ \underline{\text{else}}\ h(\hat{s}(x, p(n_1)))\}.$$

Here the set of parameters is $\{n_1, n_2\}$ and $\mathcal{D} = \{p_1, \ldots, p_9\}$ as in Example 6.1.11. Let $X, Y, Z \in \mathbf{V}_2^\iota$ and $\mathcal{T} =$

$$\{g(X(n_1, \bar{0}), \hat{t}(X(n_1, n_2), Y(n_1, n_2), n_1, n_2)), g(\hat{s}(X(\bar{0}, n_2), n_1), Z(n_1, n_2)))\}.$$

Then

$$\Theta = \{X(n_1, \bar{0}) \leftarrow \hat{s}(X(\bar{0}, n_2), n_1),\ Z(n_1, n_2) \leftarrow \hat{t}(X(n_1, n_2), Y(n_1, n_2), n_1, n_2))\}$$

is a standard unifier of $\mathcal{T}$ which is easily realizable and $\mathcal{T}$ is uniformly standard-unifiable. Note that $\mathcal{T}'$:

$$\{g(X(n_1, \bar{0}), \hat{t}(X(n_1, n_2), Y(n_1, n_2), n_1, n_2)), g(\hat{s}(X(\bar{0}, n_2), n_1), Y(\bar{0}, \bar{0})))\}$$

is not standard unifiable as for $\sigma(n_1) = \sigma(n_2) = \bar{0}$ we obtain

$$\sigma(\mathcal{T})\downarrow = \{g(X(\bar{0}, \bar{0}), \hat{t}(X(\bar{0}, \bar{0}), Y(\bar{0}, \bar{0}), \bar{0}, \bar{0})), g(\hat{s}(X(\bar{0}, \bar{0}), \bar{0}), Y(\bar{0}, \bar{0})))\}$$

which is not first-order unifiable due to occurs-check: $Y(\bar{0}, \bar{0})$ occurs in the term $\hat{t}(X(\bar{0}, \bar{0}), Y(\bar{0}, \bar{0}), \bar{0}, \bar{0}))$. Indeed, the standard substitution

$$\Theta' \colon \{X(n_1, \bar{0}) \leftarrow \hat{s}(X(\bar{0}, n_2), n_1),\ Y(\bar{0}, \bar{0}) \leftarrow \hat{t}(X(n_1, n_2), Y(n_1, n_2), n_1, n_2))\}$$

is not a standard unifier of $\mathcal{T}'$ as for $p_1 = n_1 = \bar{0} \wedge n_2 = \bar{0}$

$$Ap(\Theta', g(X(n_1, \bar{0}), \hat{t}(X(n_1, n_2), Y(n_1, n_2), n_1, n_2)))|_{p_1} \neq$$
$$Ap(\Theta', g(\hat{s}(X(\bar{0}, n_2), Y(\bar{0}, \bar{0})))|_{p_1}.$$

The unification problem $\mathcal{T}'$:

$$\{g(X(n_1, \bar{0}), \hat{t}(X(n_1, n_2), Y(n_1, n_2), n_1, n_2)), g(\hat{s}(X(\bar{0}, n_2), n_1), Y(\bar{0}, \bar{0})))\}$$

in Example 6.2.1 shows that we may not simply replace a variable expression by a terms without knowing that the variable expressions are not parameter-unifiable with variable expressions (in the example above $Y(n_1, n_2)$ is parameter-unifiable with $Y(\bar{0}, \bar{0})$ which generates an occurs-check). It also turns out that it is easier to define unifiers for all $p \in \mathcal{D}$ separately than starting with a substitution and testing whether it is actually a uniform standard-unifier. The example below shows that the construction of a standard unifier cannot be obtained by using the variable expressions of the terms to be unified.

**Example 6.2.2** Let $\mathcal{T} = \{g(X_1(n), X_1(\bar{0})),\ g(X_2(\bar{0}), X_2(p(n)))\}$. Neither

$$\Theta_1 = \{X_1(n) \leftarrow X_2(\bar{0}),\ X_1(\bar{0}) \leftarrow X_2(p(n))\} \text{ nor}$$
$$\Theta_2 = \{X_2(\bar{0}) \leftarrow X_1(n),\ X_2(p(n)) \leftarrow X_1(\bar{0})\}$$

are extended $s$-unifiers as the domain variables are parameter-unifiable. Is

$$\Theta_3 = \{X_1(n) \leftarrow X_2(\bar{0}),\ X_2(p(n)) \leftarrow X_1(\bar{0})\} \text{ an s-unifier?.}$$

No, it is not! consider $\Theta_3[\sigma]$ for $\sigma(n) = \bar{0}$. Then

$$\Theta_3[\sigma] = \{X_1(\bar{0}) \leftarrow X_2(\bar{0}), X_2(\bar{0}) \leftarrow X_1(\bar{0})\}$$

which is a standard $s$-substitution but no unfier of the set

$$\sigma(\mathcal{T})\downarrow = \{g(X_1(\bar{0}), X_1(\bar{0})),\ g(X_2(\bar{0}), X_2(\bar{0}))\}.$$

The remedy lies in the computation of unifiers for the sets

$$\mathcal{T}(p_i) = \psi_{p_i}(\mathcal{T}) = \{\psi_{p_i}(g(X_1(n), X_1(\bar{0}))),\ \psi_{p_i}(g(X_2(\bar{0}), X_2(p(n))))\}$$

for $i = 1, 2, 3$ and

$$p_1 = n = \bar{0},$$
$$p_2 = n \neq \bar{0} \wedge p(n) = \bar{0},$$
$$p_3 = n \neq \bar{0} \wedge p(n) \neq \bar{0}.$$

Here we get

$$\mathcal{T}(p_1) = \{g(X_1(\bar{0}), X_1(\bar{0})),\ g(X_2(\bar{0}), X_2(\bar{0}))\},$$
$$\mathcal{T}(p_2) = \{g(X_1(\bar{1}), X_1(\bar{0})),\ g(X_2(\bar{0}), X_2(\bar{0}))\},$$
$$\mathcal{T}(p_3) = \{g(X_1(n), X_1(\bar{0})),\ g(X_2(\bar{0}), X_2(p(n)))\}.$$

Let us define

$$\Theta_1 = \{X_1(\bar{0}) \leftarrow X_2(\bar{0})\} \text{ and}$$
$$\Theta_2 = \{X_1(\bar{1}) \leftarrow X_2(\bar{0}), X_1(\bar{0}) \leftarrow X_2(\bar{0})\}.$$

Obviously $\Theta_1$ is a unifier of $\mathcal{T}(p_1)$ and $\Theta_2$ of $\mathcal{T}(p_2)$. In case $\mathcal{T}(p_3)$ we seem to be lost as this set of terms is the original one, but we are not! The expression

$$\Theta_3\colon \{X_1(n) \leftarrow X_2(\bar{0}),\ X_1(\bar{0}) \leftarrow X_2(p(n))\}$$

is not an extended $s$-substitution, but for $\sigma \in \mathcal{S}(p_3)$ it is: indeed, for all $\sigma \in \mathcal{S}$ such that $\sigma(n) \neq \bar{0}$ and $\sigma(p(n))\downarrow \neq \bar{0}$ $\Theta_3[\sigma]$ is indeed a first-order substitution and $\Theta_3[\sigma]$ unifies $\sigma(\mathcal{T}(p_3))$.

The last observation above leads us to the following definition.

**Definition 6.2.3** Consider $\hat{\mathbf{T}}^{\iota}(T^*)$ for a theory $T^*$ over the parameter set $\mathcal{N}_\alpha$ for

- $\mathcal{N}_\alpha = \{n_1, \ldots, n_\alpha\}$

and $\mathcal{D}$ be the set of all states over $\mathcal{N}_\alpha$. An expression of the form

$$\Theta\colon \{V_1 \leftarrow t_1, \ldots, V_\beta \leftarrow t_\beta\}$$

for variable expressions $V_i$ over $\mathcal{N}_\alpha$ and $t_i \in \hat{\mathbf{T}}^{\iota}(T^*)$ is called a standard $s$-substitution over $p$ for $p \in \mathcal{D}$ if $\Theta[\sigma]$ is a first-order substitution for all $\sigma \in \mathcal{S}(p)$. $\Theta$ is called a standard $s$-unifier over $p$ of a set $\mathcal{T}$ of terms in $\hat{\mathbf{T}}^{\iota}(T^*)$ if $\Theta$ $s$-unifies $\mathcal{T}$, i.e. for all $\sigma \in \mathcal{S}(p)\colon |\sigma(\mathcal{T}\Theta))\downarrow| = 1$.

**Example 6.2.3** Let $\Theta_3 = \{X_1(n) \leftarrow X_2(\bar{0}),\ X_1(\bar{0}) \leftarrow X_2(p(n))\}$ and $p_1, p_2, p_3$ be as in Example 6.2.2. Then $\Theta_3$ is a standard $s$-unifier of $\mathcal{T}$ over $p_3$ for

$$\mathcal{T} = \{g(X_1(n), X_1(\bar{0})),\ g(X_2(\bar{0}), X_2(p(n)))\}.$$

Indeed, $\mathcal{T}\Theta_3 = \{g(X_2(\bar{0}), X_2(p(n))\}$ and $\Theta_3$ is a standard $s$-substitution over $p_3$. Note that $\Theta_3$ is *not* a standard $s$-substitution over $p_1$.

**Lemma 6.2.1** *Let $\mathcal{N}_\alpha$ and $\mathcal{D}$ as in Definition 6.2.3 and $p \in \mathcal{D}$. Let $\mathcal{T}$ be a set of standard terms such that $\psi_p(\mathcal{T}) = \mathcal{T}$ and $V_1, \ldots, V_\beta$ be arbitrary variable expressions in $\mathcal{T}$. Then for all standard terms $t_1, \ldots, t_\beta$ in $\hat{\mathbf{T}}^{\iota}(T^*)$ for a theory $T^*$ the expression*

$$\{V_1 \leftarrow t_1, \ldots, V_\beta \leftarrow t_\beta\}$$

*is a standard $s$-substitution if the $V_j$ are pairwise different.*

***Proof*** It is enough to show that for $i, j \in \{1, \ldots, \beta\}$ and $i \neq j$ $\{V_i, V_j\}$ is not parameter-unifiable. Let $X(r_1, \ldots, r_\gamma)$ and $X(s_1, \ldots, s_\gamma)$ be two different variable expressions occurring in $\mathcal{T}$ and $\psi_p(X(r_1, \ldots, r_\gamma)) = X(r_1, \ldots, r_\gamma)$, $\psi_p(X(s_1, \ldots, s_\gamma)) = X(s_1, \ldots, s_\gamma)$. In the proof of Lemma 6.1.2 we have shown that for the variable expressions above we have

- $X(r_1, \ldots, r_\gamma) = X(s_1, \ldots, s_\gamma)$ iff there exists a $\sigma \in \mathcal{S}(p)$ such that

$$\sigma(X(r_1, \ldots, r_\gamma))\downarrow = \sigma(X(s_1, \ldots, s_\gamma))\downarrow.$$

I.e., as $X(r_1, \ldots, r_\gamma)$ and $X(s_1, \ldots, s_\gamma)$ are different they are not parameter-unifiable.

Below we define a sound (though incomplete) unification algorithm for sets of terms in $\hat{\mathbf{T}}^\iota(T^*)$ for a theory $T^*$. Given a set of standard terms

$$\mathcal{T}\colon \{t_1, \ldots, t_\alpha\}$$

for $\alpha > 1$ (to be unified) we start with the following preprocessing:

- Check whether there are two different variable expressions which are parameter-unifiable.
- If there are none then define $\mathcal{T}^* = \{\mathcal{T}\}$.
- If there exist different parameter-unifiable variable expressions then select the variable expression with the highest arity, e.g. $X(s_1, \ldots, s_\beta)$ and compute the set of all states $p \in \mathcal{D}$ for $\mathcal{D}$ being the set of states corresponding to $X(n_1, \ldots, n_\alpha)$. Then define
$$\mathcal{T}^* = \{\mathcal{T}(p) \mid p \in \mathcal{D}\}.$$

For the set $\mathcal{T}$ (respectively $\mathcal{T}(p)$ for $p \in \mathcal{D}$) we define the set

$$\mathcal{U}(\mathcal{T}) = \{t_1 \simeq t_2, \ldots, t_1 \simeq t_\alpha\} \text{ respectively },$$
$$\mathcal{U}(\mathcal{T}(p)) = \{\psi_p(t_1) \simeq \psi_p(t_2), \ldots, \psi_p(t_1) \simeq \psi_p(t_\alpha)\}.$$

Now assume that $\mathcal{U} = \mathcal{U}(\mathcal{T})$ or $\mathcal{U} = \mathcal{U}(\mathcal{T}(p))$. Let *VEXP* be the set of all standard variable expressions. Note that we have have shown that the unification problem is undecidable and even that unifiability is not recursively enumerable. Therefore any unification algorithm for standard terms will be incomplete by the nature of the problem. When we arrive at an equation of the form

$$\hat{s}(t_1, \ldots, t_\beta, r_1, \ldots, r_\gamma) \simeq \hat{t}(t'_1, \ldots, t'_\delta, r'_1, \ldots, r'_\delta)$$

such that the terms are not syntactically equal then we stop the algorithm (to be defined) although the terms might be unifiable. The following example illustrates the problem.

**Example 6.2.4** Let us consider the set $\mathcal{U}$ consisting of the single equation

$$\{\hat{f}(x, y, m, n) \simeq \hat{f}(u, v, m, s(n))\}.$$

for

$$\begin{aligned} D(\hat{f}) = \{&\hat{f}(x, y, m, n) =_d \underline{\text{if}}\ n = \bar{0}\ \underline{\text{then}}\ \hat{g}(x, y, m) \\ &\underline{\text{else}}\ h(x, y, \hat{f}(x, y, m, p(n)))\}, \\ D(\hat{g}) = \{&\hat{g}(x, y, n) =_d \underline{\text{if}}\ n = \bar{0}\ \underline{\text{then}}\ h(x, y, c)\ \underline{\text{else}}\ h(\hat{g}(x, y, n), x, y)\}. \end{aligned}$$

The problem $\mathcal{U}$ is not solvable as $\hat{f}(x, y, m, \bar{0}) = \hat{g}(x, y, m)$, $\hat{f}(u, v, m, \bar{1}) = h(u, v, \hat{g}(u, v, m))$ and $\hat{g}(x, y, m)$ and $h(u, v, \hat{g}(u, v, m))$ are not unifiable.

For the following set of definitions $\mathcal{U}$ unsolvability does not appear so "soon":

$$\begin{aligned} D(\hat{f}) &= \{\hat{f}(x, y, m, n) =_d \underline{\text{if}}\ n = \bar{0}\ \underline{\text{then}}\ \hat{g}(x, y, m) \\ &\quad \underline{\text{else}}\ h(x, y, \hat{f}(x, y, m, p(n)))\}, \\ D(\hat{g}) &= \{\hat{g}(x, y, n) =_d x\}. \end{aligned}$$

Indeed,

$$\begin{aligned} &\hat{f}(x, y, m, \bar{0}) = x,\ \ \hat{f}(u, v, m, \bar{1}) = h(u, v, u),\ \hat{f}(x, y, m, \bar{1}) = h(x, y, x), \\ &\hat{f}(u, v, m, \bar{2}) = h(u, v, h(u, v, u)). \end{aligned}$$

While $\{\hat{f}(x, y, m, \bar{0}), \hat{f}(u, v, m, \bar{1})\}$ is unfiable $\{\hat{f}(x, y, m, \bar{1}), \hat{f}(u, v, m, \bar{2})\}$ is not (occurs check!)

On the other hand, for

$$\begin{aligned} D(\hat{f}) &= \{\hat{f}(x, y, m, n) =_d \underline{\text{if}}\ n = \bar{0}\ \underline{\text{then}}\ \hat{g}(x, y, m) \\ &\quad \underline{\text{else}}\ f(\hat{f}(x, y, m, p(n)))\}, \\ D(\hat{g}) &= \{\hat{g}(x, y, n) =_d x\}. \end{aligned}$$

we get (informally) $\hat{f}(x, y, m, k) = f^k(x)$, $\hat{f}(u, v, m, s(k)) = f^{k+1}(u)$ and $\mathcal{U}$ is unifiable by the simple unifier $\Theta$: $\{x \leftarrow f(u)\}$. But even for this simple case we would need a proof by induction that $\Theta$ is indeed a unifier.

By undecidability of the problem and by the fact that unifiability is a $\Pi_1$-complete property, the only property which we can obtain is soundness. We distinguish several different types of equations.

**Definition 6.2.4** An equation $s \simeq t$ is called

- a *clash-equation* if the head symbols of $s$ and $t$ are different function symbols in $\mathbf{F}^{\iota}$,
- a *occurs-equation* if one of $s$,$t$ is a variable expression properly occurring in the other term.
- a *complex equation* if $s \neq t$ and at least one of the head symbols is in $\hat{\mathbf{F}}^{\iota}$.
- an *admissible equation* if one of $s$, $t$ is a variable expression, $s \neq t$ and the variable expression does not occur in the term.

Below we define reduction relation $\rightarrow_{\mathcal{R}}$ inspired by the Martelli-Montanari unification algorithm in [33] working on finite sets of equations $\mathcal{U}$. Thereby we extend the syntax by the symbols $r$, $\perp$ and by the equation $r \simeq \perp$.

**Definition 6.2.5** $\mathcal{U}$ is irreducible if $r \simeq \perp \in \mathcal{U}$ or $\mathcal{U} = \emptyset$. Otherwise $\mathcal{U}$ is reducible under $\rightarrow_{\mathcal{R}}$. Now assume $\mathcal{U}$ is reducible, then

- (failure) If $s \simeq t \in \mathcal{U}$ and $s \simeq t$ is either an occurs-equation, a clash equation of a complex equation then $\mathcal{U} \rightarrow_{\mathcal{R}} \{r \simeq \bot\}$.
- (remove identity) If $s \simeq s \in \mathcal{U}$ then $\mathcal{U} \rightarrow_{\mathcal{R}} \mathcal{U} \backslash \{s \simeq s\}$.
- (decompose) If $f(s_1, \ldots, s_\alpha) \simeq f(r_1, \ldots, r_\alpha) \in \mathcal{U}$ for $f \in \mathbf{F}^\iota$ and $s_i, r_i \in \hat{\mathbf{T}}^\iota$ then

$$\mathcal{U} \rightarrow_{\mathcal{R}} \mathcal{U} \backslash \{f(s_1, \ldots, s_\alpha) \simeq f(r_1, \ldots, r_\alpha)\} \cup \{s_1 \simeq r_1, \ldots, s_\alpha \simeq r_\alpha\}.$$

If $\mathcal{U} \rightarrow^*_{\mathcal{R}} \mathcal{U}'$ and $\mathcal{U}'$ is irreducible (thus equal to $\emptyset$ or $\{r \simeq \bot\}$) then we write $\mathcal{U}' = \mathcal{U}{\downarrow_{\mathcal{R}}}$.

We will now define the unification algorithm UAL$(\mathcal{T}, \Theta)$ where $\mathcal{T}$ is a finite set of terms in $\hat{\mathbf{T}}^\iota$ and $\Theta$ is an $s$-substitution:

```
UAL(𝒯, Θ) =
begin
 if 𝒰(𝒯)↓ℛ = ∅ then return Θ
   else if 𝒰(𝒯)↓ℛ = {r ≃ ⊥} then return ⊥
    else begin
      select an admissible equation V ≃ t or t ≃ V in 𝒰;
       return UAL(𝒯Θ{V ← t}, Θ{V ← t});
      end;
end.
```

**Lemma 6.2.2** *Let $\mathcal{U}$ be a finite set of equations over $\hat{\mathbf{T}}^\iota$. Then*

*(1) $\rightarrow_{\mathcal{R}}$ is defined on $\mathcal{U}$ and $\mathcal{U}{\downarrow_{\mathcal{R}}}$ is finite.*
*(2) If $\mathcal{U}{\downarrow_{\mathcal{R}}} = \emptyset$ then all equations in $\mathcal{U}$ are of the form $s \simeq s$.*

***Proof*** (1) Assume that $\rightarrow_{\mathcal{R}}$ is nonterminating on $\mathcal{U}$. By Definition of $\rightarrow_{\mathcal{R}}$ the rules (decompose) and (remove identity) must be applied infinitely often. But this is impossible as the complexity decreases under application of $\rightarrow_{\mathcal{R}}$ via those rules. As complexity measure can choose e.g. the following one:

$$\begin{aligned} c(V) &= 0 \text{ for variable expressions V}, \\ c(f(t_1, \ldots, t_\alpha)) &= 2^{c(t_1)+\cdots+c(t_\alpha)} \text{ for } f \in \mathbf{F}^\iota, \\ c(\hat{f}(t_1, \ldots, t_\alpha, r_1, \ldots, r_\beta) &= 2^{c(t_1)+\cdot\cdot+c(t_\alpha)}, \\ c(s \simeq t) &= c(s) + c(t). \end{aligned}$$

(2) if there exists an equation $s \simeq t$ such that $s \neq t$ and $\rightarrow_{\mathcal{R}}$ does not terminate with $r \simeq \bot$ then there must exist an equation of the form $V \simeq r$ or $r \simeq V$ in $\mathcal{U}{\downarrow_{\mathcal{R}}}$ and $\mathcal{U}{\downarrow_{\mathcal{R}}}$ is not empty.

The unifier of a finite set of terms $\mathcal{T}$ in $\hat{\mathbf{T}}^\iota$ is then computed by calling UAL$(\mathcal{T}, \epsilon)$ where $\epsilon$ denotes the identical substitution $\emptyset$.

**Theorem 6.2.2** *(1) For all finite sets $\mathcal{T}$ of terms in $\hat{\mathbf{T}}^{\iota}$ such that no parameter-unifiable variable expressions exist in $\mathcal{T}$ we have*

- UAL$(\mathcal{T}, \epsilon)$ *terminates.*
- *If* UAL$(\mathcal{T}, \epsilon)$ *terminates with an s-substitution $\Theta$ then $\Theta$ is an s-unifier of $\mathcal{T}$.*

*(2) For all finite sets $\mathcal{T}$ of terms with parameter-unifiable variable expressions and appropriate sets of states $\mathcal{D}$ we get*

- UAL$(\mathcal{T}(p), \epsilon)$ *terminates for all $p \in \mathcal{D}$.*
- *If, for all $p \in \mathcal{D}$* UAL$(\mathcal{T}(p), \epsilon)$ *terminates with an s-substitution $\Theta(p)$ then $\mathcal{T}$ is unifiable and for $\sigma(p) = \top\ \Theta(p)$ is a unifier of $\mathcal{T}(p)$.*

***Proof*** In every call of UAL we first compute $\mathcal{U}(\mathcal{T})\downarrow_{\mathcal{R}}$, where the computation terminates and the result is finite. If $\mathcal{U}(\mathcal{T})\downarrow_{\mathcal{R}} = \emptyset$ the the algorithm returns a substitution $\Theta$. If $\mathcal{U}(\mathcal{T})\downarrow_{\mathcal{R}} = \{r \simeq \bot\}$ then UAL also terminates with $\bot$. If both cases do not apply then we call

$$\text{UAL}(\mathcal{T}\Theta\{V \leftarrow t\}, \Theta\{V \leftarrow t\}).$$

But, as $t$ does not contain $V$ the set of terms $\mathcal{T}\Theta\{V \leftarrow t\}$ contains one variable expression less. Therefore the number of calls is limited by the number of variable expressions in the input set $\mathcal{T}$. Thus UAL is always terminating. This holds also for case (2) where we only have several different problems of the same type.
It remains to show that, in case UAL returns a substitution $\Theta$, $\Theta$ is a unifier of $\mathcal{T}$. We proceed by induction on the number $\gamma$ of calls of UAL.

- $\gamma = 1$: Then $\mathcal{U}(\mathcal{T})\downarrow_{\mathcal{R}} = \emptyset$ and UAL terminates with $\Theta = \epsilon$. $\epsilon$ is indeed a unifier as by Lemma 6.2.2 all equations in $\mathcal{U}(\mathcal{T})$ are of type $s = s$ which implies that all terms in $\mathcal{T}$ are equal.
- (IH) Assume that, for all $\mathcal{T}$ where UAL$(\mathcal{T}, \epsilon)$ terminates after $\leq \gamma$ calls and yields a substitution $\Theta$, this $\Theta$ is a unifier of $\mathcal{T}$.
  Now Let $\mathcal{T}$ be a set of terms where UAL$(\mathcal{T}, \epsilon)$ terminates after $\gamma + 1$ steps and returns $\Theta$. Then by definition of UAL there exists a variable expression $V$ and and a term $t$ that
  $$\text{UAL}(\mathcal{T}, \epsilon) = \text{UAL}(\mathcal{T}\{V \leftarrow t\}, \{V \leftarrow t\}).$$
  Now UAL$(\mathcal{T}\{V \leftarrow t\}, \{V \leftarrow t\}$ must terminate after $\gamma$ steps and returns $\Theta$. By (IH) $\Theta$ is a unifier of $\mathcal{T}\{V \leftarrow t\}$ and so UAL returns $\Theta'$: $\{V \leftarrow t\}\Theta$ and, by $\mathcal{T}\Theta' = \mathcal{T}\{V \leftarrow t\}\Theta$, $\Theta'$ is a unifier of $\mathcal{T}$.

This concludes the induction proof.
In case (2) we must get a unifier *for all* $p \in \mathcal{D}$ as the $\mathcal{D}$ define a partition of the parameter space. For assume that for some $p \in \mathcal{D}$ UAL$(\mathcal{T}(p), \epsilon)$ terminates with $\bot$. As $\mathcal{S}'$: $\{\sigma \mid \sigma \in \mathcal{S}, \sigma(p)\downarrow = \top\} \neq \emptyset$ either $\mathcal{T}(p)\sigma$ for $\sigma \in \mathcal{S}'$ is not unifiable at all or no unifier can be found.

**Example 6.2.5** Let

$$\mathcal{T} = \{f(\hat{f}(X(\bar{0}, m), X(s(n), m), n, m), Y(\bar{0})), f(Z(\bar{0}), \hat{g}(Z(s(n)), n))\}$$

for $f \in \mathbf{F}^{\iota}$ and $\hat{f}, \hat{g} \in \hat{\mathbf{F}}^{\iota}$. No pair in the set of variable expressions

$$\{X(\bar{0}, m), X(s(n), m), Y(\bar{0}), Z(\bar{0}), Z(s(n))\}$$

is parameter-unifiable and so we can apply UAL to $\mathcal{T}$ only. For better legibility we use the variable names $x, y, z, u, v$ and write

$$\mathcal{T} = \{f(\hat{f}(x, y, n, m), z), f(u, \hat{g}(v, n))\}.$$

We now compute UAL$(\mathcal{T}, \epsilon)$:

- $\mathcal{U}(\mathcal{T}) = \{f(\hat{f}(x, y, n, m), z) \simeq f(u, \hat{g}(v, n))\}$.
- $\mathcal{U}(\mathcal{T}){\downarrow_{\mathcal{R}}} = \{\hat{f}(x, y, n, m) \simeq u, z \simeq \hat{g}(v, n)\}$.
- we choose the first equation, define $\Theta = \epsilon\{u \leftarrow \hat{f}(x, y, n, m)\} = \{u \leftarrow \hat{f}(x, y, n, m)\}$ and call
$$\text{UAL}(\mathcal{T}\Theta, \Theta).$$
- $\mathcal{U}(\mathcal{T}\Theta) = \{f(\hat{f}(x, y, n, m), z) \simeq f(\hat{f}(x, y, n, m), \hat{g}(v, n))\}$.
- $\mathcal{U}(\mathcal{T}\Theta){\downarrow_{\mathcal{R}}} = \{z \simeq \hat{g}(v, n)\}$.
- Now $\Theta \leftarrow \Theta\{z \leftarrow \hat{g}(v, n)\} = \{u \leftarrow \hat{f}(x, y, n, m), z \leftarrow \hat{g}(v, n)\}$ and we call again
$$\text{UAL}(\mathcal{T}\Theta, \Theta).$$

  Now $\mathcal{U}(\mathcal{T}\Theta) = \{f(\hat{f}(x, y, n, m), \hat{g}(v, n)) \simeq f(\hat{f}(x, y, n, m), \hat{g}(v, n))\}$ and
- $\mathcal{U}(\mathcal{T}\Theta){\downarrow_{\mathcal{R}}} = \emptyset$. So we return

$$\{u \leftarrow \hat{f}(x, y, n, m), z \leftarrow \hat{g}(v, n)\}$$

which is indeed an $s$-unifier of $\mathcal{T}$.

# Chapter 7
# Proof Schemata

So far we have seen that terms, formulas, and even substitutions can be defined schematically. This can also be done for proofs. In this chapter we will introduce proof schemata for **LK** derivations and resolution refutations.

## 7.1 LK-Proof Schemata

The general idea of a proof schema is to represent a proof by a finite description of an infinite sequence of proofs. Assume a proof $\varphi$ of the end-sequent $\vdash \forall x A(x)$ that uses an induction inference. Instead of considering the proof $\varphi$, we instead consider the infinite sequence of proofs $\varphi_0, \varphi_1, \varphi_2, \ldots$ of end-sequents $\vdash A(0), \vdash A(1), \vdash A(2)$ …. The task is to find a finite description of this infinite sequence of proofs, the proof schema. A proof schema always represents a parameterized sequence, and an evaluation under an assignment $\alpha$ for a parameter $n$ results in the proof $\varphi_\alpha$ of $\vdash A(\alpha)$. The underlying problem, that initially lead to the development of proof schemata, is to be able to analyze inductive proofs. Indeed, each of the proofs $\varphi_0, \varphi_1, \varphi_2, \ldots$ is a simple **LK**-proof without induction inferences, and thus enjoys cut-elimination resulting in an analytic proof. The development of a cut-elimination method for the proof schema representing the infinite sequence of proofs $\varphi_0, \varphi_1, \varphi_2, \ldots$ will be elaborated in Chap. 9. The notion of proof schema was introduced in [15, 31] for schemata handling one parameter only. The notion of proof schema in this book is capable of handling arbitrarily many parameters, as first introduced in [32].

Formally, proof schemata are constructed using proofs in an extension of **LK**. First, let us define the concept of schematic sequents.

**Definition 7.1.1** (*schematic sequent*) Let $T^+$ be a super simple $\omega\iota o$ - theory. A schematic sequent in $T^+$ is a sequent of the form $F_1, \ldots, F_\alpha \vdash G_1, \ldots, G_\beta$ where the $F_i$ and $G_j$ for $1 \leq i \leq \alpha$ and $1 \leq j \leq \beta$ are q-schemata. Let $S\colon F_1, \ldots, F_\alpha \vdash$

 
A. Leitsch et al., *First-Order Schemata and Inductive Proof Analysis*, Computer Science Foundations and Applied Logic, https://doi.org/10.1007/978-3-032-05741-9_7

$G_1, \ldots, G_\beta$ be a schematic sequent and $\sigma$ a parameter assignment. Then the evaluation of $S$ under $\sigma$ is $\sigma(S)\downarrow: \sigma(F_1)\downarrow, \ldots, \sigma(F_\alpha)\downarrow \vdash \sigma(G_1)\downarrow, \ldots, \sigma(G_\beta)\downarrow$.

Schematic sequents are defined over tuples of $\omega$-terms $\vec{r} = (r_1, \ldots, r_\alpha)$, where $r_i \in \{\bar{0}, n_i, p(n_i)\}$ for parameters $n_i$ and $i \in \{1, \ldots, \alpha\}$, which we will denote as $S(\vec{r})$. Hence, the schematic sequents $S(\bar{0})$ and $S(n)$ denote the same schematic sequent, where only the parameters are different.

In this chapter we restrict the end-sequent schemata to skolemized sequent schemata in prenex form.

**Definition 7.1.2** Let $T^+$ be a supersimple $\omega\iota o$-theory. Then the calculus $\mathbf{LK}_{T^+}$ is the calculus **LK** extended by the defining equations of $T^+$.

As in ordinary sequent calculus proofs, we allow non axiomatic schematic initial sequents. Therefore any set of schematic sequents that is closed under substitution can be considered as a schematic axiom set, the resulting proof schema is then a proof schema w.r.t. $\mathcal{A}$.

**Definition 7.1.3** (*schematic axiom set*) Let $T^+$ be a supersimple $\omega\iota o$-theory, and $\mathcal{A}$ a set of schematic sequents consisting of atomic $q$-schemata that is closed under substitution. $\mathcal{A}$ is called a schematic axiom set.

A special case of the above definition is the schematic standard axiom set $\mathcal{A}_s$.

**Definition 7.1.4** (*schematic standard axiom set*) Let $T^+$ be a supersimple $\omega\iota o$-theory, and $\mathcal{A}_s$ the smallest set of schematic sequents that is closed under substitution containing all schematic sequents of the form $A \vdash A$ for arbitrary atomic q-schemata $A$. Then $\mathcal{A}_s$ is called the schematic standard axiom set.

Schematic derivations can be understood as parameterized sequences of **LK**-derivations where new kinds of axioms and end-sequents (in the form of so-called link-variables) are included. Before formally introducing the concepts, let us have a look at an example.

**Example 7.1.1** In this example we will define a proof schema in $\mathbf{LK}_{T^+}$ of

$$S(n) = \forall x(P(x) \to P(f(x))) \vdash \forall x(P(x) \to P(\hat{f}(x, n))),$$

where $T^+ = (\emptyset, T^*, \emptyset)$ and

$$D(\hat{f}) = \{\hat{f}(x, n) =_d \underline{\text{if}}\ n = \bar{0}\ \underline{\text{then}}\ x\ \underline{\text{else}}\ f(\hat{f}(x, p(n)))\}.$$

To this aim, we extend the calculus $\mathbf{LK}_{T^+}$ by two rules for link-variables: we create link-variables $V_S(r)$ corresponding to the end-sequents $S(r)$ for $r \in \{0, n, p(n)\}$ and add rules for the elimination and the introduction of the variable $V_S(r)$:

$$\frac{S(r)}{V_S(r)}\ V_S I \quad \frac{V_S(r)}{S(r)}\ V_S E$$

The idea is to use link-variables to define recursive proofs. In our example, the proof of $S(\overline{0})$ will be used to prove $S(\overline{1})$. This proof will then be used to define the proof of $S(\overline{2})$ and so on. This recursive proof is defined for parameters $n > \overline{0}$ and denoted as $\rho(n, V_S(n)) =$

$$\dfrac{\dfrac{V_S(p(n))}{S(p(n))}\,V_S E \qquad \begin{array}{c}(\varphi(n))\\ \forall x(P(x) \to P(\hat{f}(a, p(n)))), A \vdash \forall x(P(x) \to P(\hat{f}(x, n)))\end{array}}{\dfrac{S(n)}{V_S(n)}\,V_S I}\,cut, c_l$$

where $A = \forall x(P(x) \to P(f(x)))$, and $\varphi(n) =$

$$\dfrac{\dfrac{\dfrac{P(a) \vdash P(a) \qquad \dfrac{\dfrac{P(\hat{f}(a, p(n))) \vdash P(\hat{f}(a, p(n))) \qquad \dfrac{P(\hat{f}(a, n)) \vdash P(\hat{f}(a, n))}{P(f(\hat{f}(a, p(n)))) \vdash P(\hat{f}(a, n))}\,T^+}{P(\hat{f}(a, p(n))), P(\hat{f}(a, p(n))) \to P(f(\hat{f}(a, p(n)))) \vdash P(\hat{f}(a, n))}\,\to_l}{P(\hat{f}(a, p(n))), \forall x(P(x) \to P(f(x))) \vdash P(\hat{f}(a, n))}\,\forall_l}{P(a), P(a) \to P(\hat{f}(a, p(n))), \forall x(P(x) \to P(f(x))) \vdash P(\hat{f}(a, n))}\,\to_l}{P(a) \to P(\hat{f}(a, p(n))), \forall x(P(x) \to P(f(x))) \vdash P(a) \to P(\hat{f}(a, n))}\,\to_r}{\dfrac{\forall x(P(x) \to P(\hat{f}(x, p(n)))), \forall x(P(x) \to P(f(x))) \vdash P(a) \to P(\hat{f}(a, n))}{\forall x(P(x) \to P(\hat{f}(a, p(n)))), \forall x(P(x) \to P(f(x))) \vdash \forall x(P(x) \to P(\hat{f}(x, n)))}\,\forall_r}\,\forall_l$$

Using the link-variables it is possible to append the proof above to itself until we arrive at $V_S(0)$. This is achieved by the following recursive definition, where $\hat{\rho}$ is a recursive proof symbol:

$$\begin{array}{l}\hat{\rho}(n, V_S(n)) =_d \underline{\text{if}}\ n = \bar{0}\ \underline{\text{then}}\ V_S(0)\\ \underline{\text{else}}\ \rho(n, V_S(n))\{V_S(p(n)) \leftarrow \hat{\rho}(p(n), V_S(p(n)))\}.\end{array}$$

In carrying out the substitution $\{V_S(p(n)) \leftarrow \hat{\rho}(p(n), V_S(p(n)))\}$ we identify the root of $\hat{\rho}(p(n), V_S(p(n)))$ with the uppermost leaf $V_S(p(n))$ in $\rho(n, V_S(n))$ provided $n > \overline{0}$, otherwise we end up with $V_S(0)$.

However, $V_S(0)$ is not an axiom, which means that the above proof is not an $\mathbf{LK}_{T^+}$-proof schema. Instead, $S(0)$ is provable from axioms by $\rho' =$

$$\dfrac{\dfrac{\dfrac{\dfrac{P(\hat{f}(a, 0)) \vdash P(\hat{f}(a, 0))}{P(a) \vdash P(\hat{f}(a, 0))}\,T^+}{\vdash P(a) \to P(\hat{f}(a, 0))}\,\to_r}{\vdash \forall x(P(x) \to P(\hat{f}(x, 0)))}\,\forall : r}{\dfrac{S(0)}{V_S(0)}\,V_S I}\,w_l$$

The proof $\hat{\rho}(n, V_S(n))\{V_S(0) \leftarrow \rho')\}$ is an $\mathbf{LK}_{T^+}$-proof schema of $S(n)$.

**Definition 7.1.5** A *parameter substitution* is an expression of the form $\{n_1 \leftarrow t_1, \ldots, n_\alpha \leftarrow t_\alpha\}$ where the $n_i$ are different parameters and the $t_i$ are terms in $T_0$.

The application of parameter substitutions to terms and formulas is straightforward:

**Definition 7.1.6** Let $\theta\colon \{n_1 \leftarrow t_1, \ldots, n_\alpha \leftarrow t_\alpha\}$ be a parameter substitution. We first define the application of $\theta$ to terms in $T_0$:
• $\mathbf{0}\theta = \mathbf{0}$,
• $n_i\theta = t_i$ for $i = 1, \ldots, \alpha$,
• $m\theta = m$ for $m \in \mathcal{N} \setminus \{n_1, \ldots, n_\alpha\}$,
• $s(t)\theta = s(t\theta)$,
• $p(t)\theta = p(t\theta) \in T_0$.

The extension to $\mathbf{T}^\omega$ is defined below:

- if $t \in \mathbf{T}^\omega$ then $s(t)\theta = s(t\theta)$ and $p(t)\theta = p(t\theta)$,
- if $t_1, \ldots, t_m \in \mathbf{T}^\omega$ and $f \in \mathbf{F}^\omega_m$ then $f(t_1, \ldots, t_m)\theta = f(t_1\theta, \ldots, t_m\theta)$.

We extend the application to $\mathbf{T}^\iota$:

- $c\theta = c$ for $c \in \mathbf{F}^\iota_0$.
- If $X(t_1, \ldots, t_m)$ is a variable expression then

$$X(t_1, \ldots, t_m)\theta = X(t_1\theta, \ldots, t_m\theta).$$

- If $t_1, \ldots, t_k \in \mathbf{T}^\iota$ and $f \in \mathbf{F}^\iota_k$ then $f(t_1, \ldots, t_k)\theta = f(t_1\theta, \ldots, t_k\theta)$.

The extension to simple schematic terms is defined below:

- If $t_1, \ldots, t_k \in \hat{\mathbf{T}}^\iota_s$ and $f \in \mathbf{F}^\iota_k$ then

$$f(t_1, \ldots, t_k)\theta = f(t_1\theta, \ldots, t_k\theta),$$

- if $\pi = (0, j, k)$ is a profile and $\hat{t} \in \hat{F}(\pi)$, $s_l \in \hat{\mathbf{T}}^\iota_s$ for $l = 1, \ldots, j, t_1, \ldots, t_k \in \mathbf{T}^\omega$ then
$$\hat{t}(s_1, \ldots, s_j, t_1, \ldots, t_k)\theta = \hat{t}(s_1, \ldots, s_j, t_1\theta, \ldots, t_k\theta).$$

It remains to extend the application of $\theta$ to schematic formulas:

- $\xi\theta = \xi$ for $\xi \in \mathbf{V}^o$.
- If $P \in \mathbf{P}^o_k$ and $t_1, \ldots, t_k \in \hat{\mathbf{T}}^\iota(T^*)$ then

$$P(t_1, \ldots, t_k)\theta = P(t_1\theta, \ldots, t_k\theta).$$

- if $\pi = (i, j, k)$ is a profile and $\hat{p} \in \hat{P}(\pi)$, $X_l \in \mathbf{V}^\iota_i$ for $l = 1, \ldots, j, t_1, \ldots, t_k \in \mathbf{T}^\omega$ then
$$\hat{p}(X_1, \ldots, X_j, t_1, \ldots, t_k)\theta = \hat{p}(X_1, \ldots, X_j, t_1\theta, \ldots, t_k\theta).$$

- $(\neg F)\theta = \neg F\theta$.
- If $F_1, F_2 \in \hat{F}$ then $(F_1 \circ F_2)\theta = F_1\theta \circ F_2\theta$ for $\circ \in \{\wedge, \vee\}$.

Finally, if $S$ is a schematic sequent of the form $A_1, \ldots, A_k \vdash B_1, \ldots, B_l$ then

$$S\theta = A_1\theta, \ldots, A_k\theta \vdash B_1\theta, \ldots, B_l\theta.$$

**Definition 7.1.7** Let $A$ be a schematic formula (let $S$ be a schematic sequent) then $A\theta$ ($S\theta$) is called a parameter instance of $A$ ($S$).

*Sequent link variables* are similar to the higher-order variables we used in the definition of term- and formula schemata.

**Definition 7.1.8** (*sequent-link-variables*) Let $S$ be a schematic sequent. To every schematic sequent $S$ and to all its parameter instances we assign an infinite set of *link-variables* $V(S) = \{V_i \mid i \in \mathbb{N}\}$. If $V \in V(S)$ we also say that $V$ is of type $S$. Let $V$ be in $V(S)$ and $n_1, \ldots, n_\alpha$ be the parameters occuring in $S$ (we denote $S$ also by $S(n_1, \ldots, n_\alpha)$) then $V(n_1, \ldots, n_\alpha)$ is called a *link expression representing* $S(n_1, \ldots, n_\alpha)$. Let $\theta$ be a parameter substitution such that $n_i\theta = r_i$ for $i = 1, \ldots, \alpha$; then we say that $V(r_1, \ldots, r_\alpha)$ is a *link expression representing* $S\theta$. We also write $V_S(r_1, \ldots, r_\alpha)$ for this link-expression to emphasize that $V$ is in $V(S)$. Two link-expressions $V_S(\vec{r})$ and $U_{S'}(\vec{s})$ are defined as equal if $S = S'$, $U = V$, and $\vec{r} = \vec{s}$.

Link-variables serve the purpose to define locations in a proof where they can be replaced by a proof. These locations can be either the leaves of a proof or the root. In constructing $\mathbf{LK}_{T^+}$-proof schemata, link-variables are defined for sequents, because the end-sequent or initial sequent of a proof is being replaced. However, it might also make sense to use link-variables as references to other schematic expressions than schematic sequents. Indeed, in Sect. 7.2 link-variables will be used for schematic sequents in defining refutation schemata.

In order to place into or to remove link-variables from proofs we extend $\mathbf{LK}_{T^+}$ by variable elimination rules and variable introduction rules.

**Definition 7.1.9** ($\mathbf{LK}V$) The calculus $\mathbf{LK}V$ contains the rules of $\mathbf{LK}_{T^+}$ with two additional rules. Let $V$ be a link-variable of type $S$ then we define the rules

$$\frac{S(\vec{r})}{V(\vec{r})}\ V_I \qquad \frac{V(\vec{r})}{S(\vec{r})}\ V_E$$

Elimination rules can only be applied to leaves in an $LKV$-derivation (if the leaf is a link-expression) and introduction rules to root nodes which are labeled by a sequent of the form $S(\vec{r})$.

**Example 7.1.2** The derivations $\rho(0, V_S(0))$ and $\rho(n, V_S(n))$ in Example 7.1.1 are $\mathbf{LK}V$-derivations.

The link-variables in an $\mathbf{LK}V$-derivation can be replaced by other $\mathbf{LK}V$-derivations:

**Definition 7.1.10** (*link substitution*) Let $\rho$ be an $\mathbf{LK}V$-derivation such that the link variable of the root differs from all link variables in the leaves (note that this can be always achieved). Let $\rho_i$ for $i = 1, \ldots, \alpha$ be $\mathbf{LK}V$-derivations with link-expressions $V_i(\vec{r}_i)$ occurring at their roots such that $V_i \neq V_j$ for $i \neq j$. Then

$$\rho\{V_i(\vec{r}_i) \leftarrow \rho_i \mid i \in \{1, \ldots, \alpha\}\}$$

is the $\mathbf{LK}V$-derivation obtained from $\rho$ by replacing the link variables $V_i(\vec{r}_i)$ occurring at the leaves of $\rho$ by $\rho_i$ without the last inference.

**Example 7.1.3** Let $\rho_1$ be

$$\dfrac{\dfrac{\dfrac{\dfrac{\dfrac{P(\hat{f}(a,0)) \vdash P(\hat{f}(a,0))}{P(a) \vdash P(\hat{f}(a,0))}\,T^+}{\vdash P(a) \to P(\hat{f}(a,0))}\,\to_r}{\vdash \forall x(P(x) \to P(\hat{f}(x,0)))}\,\forall : r}{\dfrac{S(0)}{V_S(0)}\,V_S I}\,w_l$$

and $\rho_2$ be

$$\dfrac{\dfrac{V_S(0)}{S(0)}\,V_S E \qquad \dfrac{\dfrac{\dfrac{\dfrac{P(a) \vdash P(a) \qquad \dfrac{\dfrac{P(\hat{f}(a,0)) \vdash P(\hat{f}(a,0)) \qquad \dfrac{B \vdash B}{P(f(\hat{f}(a,0))) \vdash B}\,T^+}{P(\hat{f}(a,0)), P(\hat{f}(a,0)) \to P(f(\hat{f}(a,0))) \vdash B}\,\to_l}{P(\hat{f}(a,0)), A \vdash B}\,\forall_l}{P(a), P(a) \to P(\hat{f}(a,0)), A \vdash B}\,\to_l}{P(a) \to P(\hat{f}(a,0)), A \vdash P(a) \to B}\,\to_r}{\forall x(P(x) \to P(\hat{f}(x,0))), A \vdash P(a) \to B}\,\forall_l}{\forall x(P(x) \to P(\hat{f}(x,0))), A \vdash \forall x(P(x) \to P(\hat{f}(x,s(0))))}\,\forall_r}{\dfrac{S(s(0))}{V_S(s(0))}\,V_S I}\,cut, c_l$$

for $A = \forall x(P(x) \to P(f(x)))$, and $B = P(\hat{f}(a, s(0)))$. Then $\rho_2\{V_S(0) \leftarrow \rho_1\}$ is

$$\dfrac{\dfrac{\dfrac{\dfrac{\dfrac{\dfrac{P(\hat{f}(a,0)) \vdash P(\hat{f}(a,0))}{P(a) \vdash P(\hat{f}(a,0))}\,T^+}{\vdash P(a) \to P(\hat{f}(a,0))}\,\to_r}{\vdash \forall x(P(x) \to P(\hat{f}(x,0)))}\,\forall : r}{S(0)}\,w_l \qquad \dfrac{(\varphi(0))}{\forall x(P(x) \to P(\hat{f}(x,0))), A \vdash \forall x(P(x) \to P(\hat{f}(x,s(0))))}}{\dfrac{S(s(0))}{V_S(s(0))}\,V_S I}\,*$$

where the inference $*$ is an application of $cut$, $c_l$, and $\varphi(0) =$

$$
\dfrac{\dfrac{\dfrac{\dfrac{P(a)\vdash P(a) \qquad \dfrac{\dfrac{P(\hat{f}(a,0))\vdash P(\hat{f}(a,0)) \qquad \dfrac{P(\hat{f}(a,s(0)))\vdash P(\hat{f}(a,s(0)))}{P(f(\hat{f}(a,0)))\vdash P(\hat{f}(a,s(0)))}\,T^{+}}{P(\hat{f}(a,0)), P(\hat{f}(a,0))\to P(f(\hat{f}(a,0)))\vdash P(\hat{f}(a,s(0)))}\,\to_l}{P(\hat{f}(a,0)), \forall x(P(x)\to P(f(x)))\vdash P(\hat{f}(a,s(0)))}\,\forall_l}{P(a), P(a)\to P(\hat{f}(a,0)), \forall x(P(x)\to P(f(x)))\vdash P(\hat{f}(a,s(0)))}\,\to_l}{P(a)\to P(\hat{f}(a,0)), \forall x(P(x)\to P(f(x)))\vdash P(a)\to P(\hat{f}(a,s(0)))}\,\to_r}{\forall x(P(x)\to P(\hat{f}(x,0))), \forall x(P(x)\to P(f(x)))\vdash P(a)\to P(\hat{f}(a,s(0)))}\,\forall_l}{\forall x(P(x)\to P(\hat{f}(x,0))), \forall x(P(x)\to P(f(x)))\vdash \forall x(P(x)\to P(\hat{f}(x,s(0))))}\,\forall_r
$$

In general, schematic derivations will be based on recursive definitions. We first define proofs admiting proof recursions, i.e. derivations which can be iterated recursively.

**Definition 7.1.11** (*admission of proof recursion*) Let $S(\vec{X},\vec{n},k)$, and $S^*(\vec{X},\vec{n},k)$ be schemata of sequents of the form

$$
\begin{aligned}
S(\vec{X},\vec{n},k) &= \ F(\vec{X},\vec{n},p(k)), \Gamma\vdash\Delta, F(\vec{X},\vec{n},k) \text{ for } k>\bar{0} \text{ and}\\
S^*(\vec{X},\vec{n},k) &= \ \Gamma\vdash\Delta, F(\vec{X},\vec{n},k) \text{ for all } k.
\end{aligned}
$$

where $F(\vec{X},\vec{n},k)$ is a schematic formula in $\vec{X},\vec{n},k$ and the (schematic) formulas in $\Gamma$, $\Delta$ do not contain the parameter $k$. $F(\vec{X},\vec{n},k)$ is called the recursion formula corresponding to $S(\vec{X},\vec{n},k)$. Sequents of this form are called *recursion sequents*. Then any $\mathbf{LK}_{T^+}$ derivation $\rho(\vec{X},\vec{n},k)$ for $k>0$ of the form

$$
\dfrac{\dfrac{\dfrac{V_{S^*}(\vec{n},p(k))}{S^*(\vec{X},\vec{n},p(k))} \qquad \dfrac{V_S E^{\rho'(\vec{X},\vec{n},k)}}{S(\vec{X},\vec{n},k)}}{\Gamma\vdash\Delta, F(\vec{X},\vec{n},k)}\,cut+c^*}{V_{S^*}(\vec{X},\vec{n},k)}\,V_S I
$$

and with the cut formula $F(\vec{X},\vec{n},p(k))$ (where $\rho'(\vec{X},\vec{n},k)$ is an $\mathbf{LK}_{T^+}$ derivation of $S(\vec{X},\vec{n},k)$) is said to *admit proof recursion based on* $S(\vec{X},\vec{n},k)$ and $S^*(\vec{X},\vec{n},k)$. The link expression $V_{S^*}(\vec{n},p(k))$ is called the *recursion link expression.*

Note that the form of the sequents $S$ and $S^*$ are closely related to the form of the induction rule as defined in [39]:

$$
\dfrac{\Gamma, A(k)\vdash\Delta, A(s(k))}{A(\bar{0}),\Gamma\vdash\Delta, A(t)}\,ind
$$

where $k$ may not occur free in $\Gamma$, $\Delta$. While $S$ corresponds to the premise of *ind*, $S^*$ corresponds to the conclusion–with the exception that the base case $A(0)$ is used separately as expressed in form of proof recursion below.

**Definition 7.1.12** (*proof recursion*) Consider a recursion sequent $S(\vec{X}, \vec{n}, k)$ and $\rho(\vec{X}, \vec{n}, k)$ for $k > 0$ be an $\mathbf{LK}_{T^+}$ derivation admitting proof recursion based on sequents $S(\vec{X}, \vec{n}, k)$ and $S^*(\vec{X}, \vec{n}, k)$ . We abbreviate $V_{S^*}(\vec{X}, \vec{n}, k)$ by $V(k)$ and define

$$D(\hat{\rho}) = \{\hat{\rho}(\vec{X}, \vec{n}, k) =_d \underline{\text{if}}\ k = \bar{0}\ \underline{\text{then}}\ \rho_0 \\ \underline{\text{else}}\ \rho(\vec{X}, \vec{n}, k)\{V(p(k)) \leftarrow \hat{\rho}(\vec{X}, \vec{n}, p(k))\}\}.$$

where $\rho_0$ is an $\mathbf{LK}_{T^+}$ derivation of $V_{S^*}(\vec{X}, \vec{n}, \bar{0})$. We say that $\hat{\rho}$ is the *inductive closure* of $\rho$, or $\hat{\rho}$ is *obtained via proof recursion* from $\rho$ based on $S(\vec{X}, \vec{n}, k)$ and $S^*(\vec{X}, \vec{n}, k)$.

We will see that in Chap. 8 proof recursion in schemata is closely related to the induction principle.

**Example 7.1.4** $\hat{\rho}$ is the inductive closure of $\rho$ for

$$\hat{\rho}(n, V_S(n)) \text{ and } \rho(n, V_S(n))$$

in Example 7.1.1.

Note that $\hat{\rho}$ is obtained from $\rho$ by a kind of primitive recursion on proofs.

**Definition 7.1.13** ($\mathbf{LK}_{T^+}$*-derivation schema*) We define $\mathbf{LK}_{T^+}$-derivation schemata inductively:

- Any $\mathbf{LK}V$-derivation of a schematic sequent $S$ (of a link variable $V_S$) is an $\mathbf{LK}_{T^+}$-derivation schema of $S$ (of $V_S$).
- Let $\rho$ be an $\mathbf{LK}V$-derivation of $S$ (of $V_S$) such that the link variable of the root differs from all link variables in the leaves, and $\rho_i$ for $i = 1, \ldots, \alpha$ $\mathbf{LK}V$-derivations with link-variables $V_i(\vec{r}_i)$ occurring at their roots such that $V_i \neq V_j$ for $i \neq j$. Then

$$\rho\{V_i(\vec{r}_i) \leftarrow \rho_i \mid i \in \{1, \ldots, \alpha\}\}$$

  is an $\mathbf{LK}_{T^+}$-derivation schema of $S$ (of $V_S$).
- If $\rho$ is an $\mathbf{LK}_{T^+}$-derivation schema of a sequent $S^*(\vec{X}, \vec{n}, k)$ admitting proof recursion based on $S(\vec{X}, \vec{n}, k)$ and $S^*(\vec{X}, \vec{n}, k)$, then the inductive closure of $\rho$ is an $\mathbf{LK}_{T^+}$-derivation schema of $V_{S^*}(\vec{X}, \vec{n}, k)$.

If $\rho$ is a derivation schema of a link variable $V_S(\vec{X}, \vec{r})$ we say that the *end-sequent* of $\rho$ is $S(\vec{X}, \vec{r})$.

**Example 7.1.5** The proofs $\rho$, $\hat{\rho}$, and $\hat{\rho}(n, V_S(n))\{V_S(0) \leftarrow \rho'\}$ in Example 7.1.1 are $\mathbf{LK}_{T^+}$-derivation schemata.

$\mathbf{LK}_{T^+}$-derivation schemata are schematic derivations, but not necessarily proofs as their initial sequents might not be axioms. Below we define $\mathbf{LK}_{T^+}$-proof schemata. Basically, $\mathbf{LK}_{T^+}$-proof schemata are just $\mathbf{LK}_{T^+}$-derivation schemata where the leaves are axiom schemata. In case of proof recursion we have no direct way to specify the axiom schemata of a derivation schema which motivates the definition below.

**Definition 7.1.14** Let $\rho(\vec{X}, \vec{n}, k)$ for $k > 0$ be a schematic derivation admitting proof recursion. $\rho(\vec{X}, \vec{n}, k)$ is called a *semi-proof schema* if all leaves of $\rho(\vec{X}, \vec{n}, k)$ except the recursion link expression are axiom schemata.

**Definition 7.1.15** (**LK**$_{T^+}$*-proof schema*) **LK**$_{T^+}$-proof schemata are defined inductively: let $A \in \mathcal{A}$ be a schematic sequent in a schematic axiom set $\mathcal{A}$. Then

$$A$$

is an **LK**$_{T^+}$-proof schema of $A$ w.r.t $\mathcal{A}$. If $\mathcal{A}$ is the schematic standard axiom set $\mathcal{A}_S$, the above proof schema is an **LK**$_{T^+}$-proof schema of $A$ w.r.t $\mathcal{A}_S$, or simply an **LK**$_{T^+}$-proof schema.[1]

- If $\rho$ is a proof schema of a schematic sequent $S(\vec{X}, \vec{n}, k)$ then

$$\frac{(\rho)}{V_S(\vec{X}, \vec{n}, k)}\ V_I$$

  is a proof schema of $V_S(\vec{X}, \vec{n}, k)$.
- If $\rho_1$ and $\rho_2$ are **LK**$_{T^+}$-proof schemata of $S_1$ and $S_2$ and $\rho =$

$$\frac{\begin{matrix}(\rho_1) & (\rho_2)\\ S_1 & S_2\end{matrix}}{S}\ \xi$$

  for a binary rule $\xi$ then $\rho$ is an **LK**$_{T^+}$-proof schema of $S$.
- If $\rho'$ is an **LK**$_{T^+}$-proof schema of $S'$ and $\rho =$

$$\frac{\begin{matrix}(\rho')\\ S'\end{matrix}}{S}\ \xi$$

  for a unary rule $\xi$ then $\rho$ is an **LK**$_{T^+}$-proof schema of $S$.
- Let $\rho$ be an **LK**$_{T^+}$-derivation schema of a schematic sequent $S$ (of a link expression $V_S(\vec{r})$) such that all leaves of $\rho$ which are not link expressions are axioms. Let $V_i(\vec{r_i})$ for $i = 1, \ldots, \alpha$ be the link expressions on the leaves of $\rho$ (we assume that all these link expressions are different from $V_S(\vec{r})$ and are different from each other) and $\rho_i$ for $i = 1, \ldots, \alpha$ be **LK**$_{T^+}$-proof-schemata of $V_i(\vec{r_i})$. Then $\rho\{V_i(\vec{r_i}) \leftarrow \rho_i \mid i \in \{1, \ldots, \alpha\}\}$ is an **LK**$_{T^+}$-proof schema of $S$ (of $V_S(\vec{r})$).
- Let $S(\vec{X}, \vec{n}, k)$ and $S^*(\vec{X}, \vec{n}, k)$ be a recursion sequents and

$$\rho(\vec{X}, \vec{n}, V_S(\vec{X}, \vec{n}, k))$$

[1] From now on we will refer to an **LK**$_{T^+}$-proof schema w.r.t $\mathcal{A}$ as an **LK**$_{T^+}$-proof schema and assume that the schematic axiom set is clear from the context.

for $k > 0$ be an $\mathbf{LK}_{T^+}$ derivation schema admitting proof recursion based on $S$ and $S^*$. Furthermore assume that $\rho(\vec{X}, \vec{n}, V_S(\vec{X}, \vec{n}, k))$ is a semi-proof schema and

$$D(\hat{\rho}) = \{\hat{\rho}(\vec{X}, \vec{n}, k) =_d \underline{\text{if}}\ k = \bar{0}\ \underline{\text{then}}\ \rho_0 \\ \underline{\text{else}}\ \rho^*(\vec{X}, \vec{n}, k)\{V(p(k)) \leftarrow \hat{\rho}(\vec{X}, \vec{n}, p(k))\}\}.$$

Assume that $\rho_0$ is an $\mathbf{LK}_{T^+}$-proof schema of $V(0)$ then $\hat{\rho}(\vec{n}, V(k))$ is an $\mathbf{LK}_{T^+}$-proof schema of $V(k)$.

Note that, as a consequence, $\hat{\rho}(\vec{n}, V(k))\{V(k) \leftarrow S^*(k)\}$ is a proof schema of $S^*(k)$.

**Example 7.1.6** $\hat{\rho}(n, V_S(n))\{V_S(0) \leftarrow \rho')\}$ in Example 7.1.1 is an $\mathbf{LK}_{T^+}$-proof schema of $S(n) = \forall x(P(x) \to P(f(x))) \vdash \forall x(P(x) \to P(\hat{f}(x,n)))$. We define $\rho_1 =$

$$\dfrac{\begin{matrix}(\hat{\rho}(n, V_S(n))) \\ \{V_S(0) \leftarrow \rho')\} \\ S(n)\end{matrix} \qquad \begin{matrix}(\varphi(n)) \\ A \vdash (P(\hat{f}(c,n)) \to P(g(c,n))) \to (P(c) \to P(g(c,n)))\end{matrix}}{\forall x(P(x) \to P(f(x))) \vdash (P(\hat{f}(c,n)) \to P(g(c,n))) \to (P(c) \to P(g(c,n)))}\ cut$$

for $A = \forall x(P(x) \to P(\hat{f}(x,n)))$, and $\varphi(n) =$

$$\dfrac{\dfrac{\dfrac{\dfrac{P(c) \to P(c) \qquad \dfrac{P(\hat{f}(c,n)) \vdash P(\hat{f}(c,n)) \qquad P(g(c,n)) \vdash P(g(c,n))}{P(\hat{f}(c,n)) \to P(g(c,n)), P(\hat{f}(c,n)) \vdash P(g(c,n))}\ \to_l}{P(c), P(\hat{f}(c,n)) \to P(g(c,n)), P(c) \to P(\hat{f}(c,n)) \vdash P(g(c,n))}\ \to_l}{P(\hat{f}(c,n)) \to P(g(c,n)), P(c) \to P(\hat{f}(c,n)) \vdash P(c) \to P(g(c,n))}\ \to_r}{\dfrac{P(c) \to P(\hat{f}(c,n)) \vdash (P(\hat{f}(c,n)) \to P(g(c,n))) \to (P(c) \to P(g(c,n)))}{\forall x(P(x) \to P(\hat{f}(x,n))) \vdash (P(\hat{f}(c,n)) \to P(g(c,n))) \to (P(c) \to P(g(c,n)))}\ \forall_l}\ \to_r$$

$\rho_1$ is an $\mathbf{LK}_{T^+}$-proof schema of

$$\forall x(P(x) \to P(f(x))) \vdash (P(\hat{f}(c,n)) \to P(g(c,n))) \to (P(c) \to P(g(c,n))).$$

When the parameters in an $\mathbf{LK}_{T^+}$-proof schema are instantiated with numerals, we obtain an $\mathbf{LK}$-derivation.

**Theorem 7.1.1** *Let $\rho$ be an $\mathbf{LK}_{T^+}$-proof schema of a schematic sequent that contains parameters $n_1, \ldots, n_\alpha$, denoted as $S(n_1, \ldots, n_\alpha)$. Then, for all numerals $\nu_1, \ldots, \nu_\alpha$, the evaluation of $\rho\{n_1 \leftarrow \nu_1, \ldots, n_\alpha \leftarrow \nu_\alpha\}$ is an $\mathbf{LK}$-derivation of $S(\nu_1, \ldots, \nu_\alpha)$.*

***Proof*** Let $\sigma(n_i) = \nu_i$ for $i = 1, \ldots, \alpha$ and $\rho' = \sigma(\rho)$. If $\rho$ is defined only via link-substitution then $\rho'$ is an $\mathbf{LK}V$-derivation of $S(\nu_1, \ldots, \nu_\alpha)$. Evaluating all schematic expressions like formula, predicate, and term schemata under $\sigma$, and eliminating link introduction and elimination rules will yield an $\mathbf{LK}$-derivation. If $\rho$ is defined via a condition $C$ then exactly one of $\rho_1$ or $\rho_2$ is chosen for the parameter assignment $\sigma$.

So $\rho'$ is either $\rho'_1$ or $\rho'_2$ and we may apply the induction hypothesis.
The only nontrivial case occurs when $\rho$ is defined via proof recursion. Here we must proceed by induction on the number of proof recursions $\delta(\rho)$ in $\rho$. If $\delta(\rho) = 0$ we have no inductive closures and we are done. Otherwise assume that $\hat{\psi}(\vec{m}, k)$ is a minimal inductive closure in $\rho$ via the active parameter $k$, i.e. $\psi$ itself does not contain proof recursions. Assume that $\hat{\psi}(\vec{m}, k)$ is an $\mathbf{LK}_{T^+}$-proof schema of a sequent $S$. We prove by induction that $\hat{\psi}(\vec{m}, k)$ evaluates to a derivation in **LK**. Under the parameter assignment $\sigma\colon \{n_1 \leftarrow \nu_1, \ldots, n_\alpha \leftarrow \nu_\alpha\}$ we obtain $\hat{\psi}(\vec{\mu}, \nu)$ for a vector of numerals $\vec{\mu}$ and a numeral $\nu$; $\hat{\psi}(\vec{\mu}, \nu)$ is a derivation of $\sigma(S)$. If $\nu = \bar{0}$ then by definition of inductive closure and by minimality of $\hat{\psi}$, $\hat{\psi}(\vec{\mu}, \bar{0})$ is a proof without proof recursion and we can argue as above. If $\nu > \bar{0}$ we obtain the derivation by link-substitution: the link $V(\vec{\mu}, p(\nu))$ occurring in the derivation $\psi(\vec{\mu}, \nu)$ is substituted

$$\hat{\psi}(\vec{\mu}, \nu) = \psi(\vec{\mu}, \nu)\{V(\vec{\mu}, p(\nu)) \leftarrow \hat{\psi}(\vec{\mu}, p(\nu))\}.$$

By induction hypothesis $\hat{\psi}(\vec{\mu}, p(\nu))$ evaluates to a derivation $\phi \in \mathbf{LK}$ and so $\hat{\psi}(\vec{\mu}, \nu)$ evaluates to

$$\psi' = \psi(\vec{\mu}, \nu)\{V(\vec{\mu}, p(\nu)) \leftarrow \phi\}.$$

$\psi'$ does not contain proof recursions and thus is a derivation in **LK**. When we replace $\hat{\psi}(\vec{\mu}, \nu)$ in $\sigma(\rho)$ by $\psi'$ we obtain a derivation $\rho'$ of the desired end-sequent.

In the example below we will illustrate how a particular proof schema can be evaluated for a given parameter assignment.

**Example 7.1.7** Consider the proof schema $\hat{\rho}(n, V_S(n))\{V_S(0) \leftarrow \rho')\}$ of $S(n)$ from Example 7.1.1, where the recursive proof symbol $\hat{\rho}$ is given by:

$$\begin{array}{l} D(\hat{\rho}) = \{\hat{\rho}(n, V_S(n)) =_d \underline{\text{if}}\ n = \bar{0}\ \underline{\text{then}}\ V_S(0) \\ \underline{\text{else}}\ \rho(n, V_S(n))\{V_S(p(n)) \leftarrow \hat{\rho}(p(n), V_S(p(n)))\}\end{array}$$

and $\rho(n, V_S(n)) =$

$$\dfrac{\dfrac{\dfrac{V_S(p(n))}{S(p(n))}\,V_SE \qquad \dfrac{\dfrac{\dfrac{\dfrac{P(a) \vdash P(a) \qquad \dfrac{\dfrac{C \vdash C \qquad \dfrac{B \vdash B}{P(f(\hat{f}(a, p(n)))) \vdash B}\,T^+}{C, C \to P(f(\hat{f}(a, p(n)))) \vdash B}\,\to_l}{C, A \vdash B}\,\forall_l}{P(a), P(a) \to C, A \vdash B}\,\to_l}{P(a) \to C, A \vdash P(a) \to B}\,\to_r}{\forall x(P(x) \to P(\hat{f}(x, p(n)))), A \vdash P(a) \to B}\,\forall_l}{\forall x(P(x) \to P(\hat{f}(x, p(n)))), A \vdash \forall x(P(x) \to P(\hat{f}(x, n)))}\,\forall_r}{S(n)}\,cut, c_l}{V_S(n)}\,V_SI$$

for $A = \forall x(P(x) \to P(f(x)))$, $B = P(\hat{f}(a, n))$, $C = P(\hat{f}(a, p(n)))$, and $\rho' =$

$$\dfrac{\dfrac{\dfrac{\dfrac{\dfrac{P(\hat{f}(a,0)) \vdash P(\hat{f}(a,0))}{P(a) \vdash P(\hat{f}(a,0))}\,T^{+}}{\vdash P(a) \to P(\hat{f}(a,0))}\,\to_r}{\vdash \forall x(P(x) \to P(\hat{f}(x,0)))}\,\forall : r}{S(0)}\,w_l$$

$$\dfrac{S(0)}{V_S(0)}\,V_S I$$

Now let the parameter assignment be $\bar{0}$ for $n$. Then, $\hat{\rho}(0, V_S(0))\{V_S(0) \leftarrow \rho'\} = \rho'$.

For the parameter assignment $\bar{1}$ for $n$ we obtain $\hat{\rho}(1, V_S(1))\{V_S(0) \leftarrow \rho'\}$, where

$$\begin{aligned} \hat{\rho}(1, V_S(1)) &= \rho(1, V_S(1))\{V_S(0) \leftarrow \hat{\rho}(0, V_S(0))\} \\ &= \rho(1, V_S(1))\{V_S(0) \leftarrow V_S(0)\} \\ &= \rho(1, V_S(1)). \end{aligned}$$

Thus, $\hat{\rho}(1, V_S(1))\{V_S(0) \leftarrow \rho'\} = \rho(1, V_S(1))\{V_S(0) \leftarrow \rho'\} =$

$$\dfrac{\dfrac{\dfrac{\dfrac{\dfrac{P(\hat{f}(a,0)) \vdash P(\hat{f}(a,0))}{P(a) \vdash P(\hat{f}(a,0))}\,T^{+}}{\vdash P(a) \to P(\hat{f}(a,0))}\,\to_r}{\vdash \forall x(P(x) \to P(\hat{f}(x,0)))}\,\forall : r}{S(0)}\,w_l \qquad \begin{array}{c}(\varphi(0)) \\ \forall x(P(x) \to P(\hat{f}(x,0))), A \vdash \forall x(P(x) \to P(\hat{f}(x,1)))\end{array}}{\dfrac{S(1)}{V_S(1)}\,V_S I}\,cut, c_l$$

where $A = \forall x(P(x) \to P(f(x)))$, and $\varphi(0) =$

$$\dfrac{\dfrac{\dfrac{P(a) \vdash P(a) \qquad \dfrac{\dfrac{P(\hat{f}(a,0)) \vdash P(\hat{f}(a,0)) \qquad \dfrac{P(\hat{f}(a,1)) \vdash P(\hat{f}(a,1))}{P(f(\hat{f}(a,0))) \vdash P(\hat{f}(a,1))}\,T^{+}}{P(\hat{f}(a,0)), P(\hat{f}(a,0)) \to P(f(\hat{f}(a,0))) \vdash P(\hat{f}(a,1))}\,\to_l}{P(\hat{f}(a,0)), \forall x(P(x) \to P(f(x))) \vdash P(\hat{f}(a,1))}\,\forall_l}{\dfrac{P(a), P(a) \to P(\hat{f}(a,0)), \forall x(P(x) \to P(f(x))) \vdash P(\hat{f}(a,1))}{P(a) \to P(\hat{f}(a,0)), \forall x(P(x) \to P(f(x))) \vdash P(a) \to P(\hat{f}(a,1))}\,\to_r}\,\to_l}{\dfrac{\forall x(P(x) \to P(\hat{f}(x,0))), \forall x(P(x) \to P(f(x))) \vdash P(a) \to P(\hat{f}(a,1))}{\forall x(P(x) \to P(\hat{f}(x,0))), \forall x(P(x) \to P(f(x))) \vdash \forall x(P(x) \to P(\hat{f}(x,1)))}\,\forall_r}\,\forall_l$$

The final evaluation of the above proof to a real **LK**-proof without schematic symbols is

$$\dfrac{\dfrac{\dfrac{\dfrac{P(a) \vdash P(a)}{\vdash P(a) \to P(a)}\,\to_r}{\vdash \forall x(P(x) \to P(x))}\,\forall : r}{\forall x(P(x) \to P(f(x))) \vdash \forall x(P(x) \to P(x))}\,w_l \qquad \begin{array}{c}(\varphi) \\ S_\varphi\end{array}}{\forall x(P(x) \to P(f(x))) \vdash \forall x(P(x) \to P(f(x)))}\,cut, c_l$$

where

$$S_\varphi\colon \forall x(P(x) \to P(x)), \forall x(P(x) \to P(f(x))) \vdash \forall x(P(x) \to P(f(x)))$$

and $\varphi =$

$$\dfrac{\dfrac{\dfrac{\dfrac{P(a) \vdash P(a) \qquad \dfrac{\dfrac{P(a) \vdash P(a) \qquad P(f(a)) \vdash P(f(a))}{P(a), P(a) \to P(f(a)) \vdash P(f(a))}\ \to_l}{P(a), \forall x(P(x) \to P(f(x))) \vdash P(f(a))}\ \forall_l}{P(a), P(a) \to P(a), \forall x(P(x) \to P(f(x))) \vdash P(f(a))}\ \to_l}{P(a) \to P(a), \forall x(P(x) \to P(f(x))) \vdash P(a) \to P(f(a))}\ \to_r}{\dfrac{\forall x(P(x) \to P(x)), \forall x(P(x) \to P(f(x))) \vdash P(a) \to P(f(a))}{\forall x(P(x) \to P(x)), \forall x(P(x) \to P(f(x))) \vdash \forall x(P(x) \to P(f(x)))}\ \forall_r}\ \forall_l$$

For (schematic) cut-elimination we need *regular proofs*. To ensure that the schemata will evaluate to regular proofs we have to regularize the schemata themselves. As we are describing infinite sequences of proofs (instead of single ones) we will use variables in $V_k^\iota$ for $k > 0$.

**Definition 7.1.16** (*regular proof schema*) Consider an $\mathbf{LK}_{T^+}$-proof schema $\rho$ of a schematic sequent $S(\vec{n})$, where $\vec{n}$ contains all parameters occurring in $\rho$. We will inductively regularize $\rho$ by replacing eigenvariables by new variable expressions.

- If $\rho$ is not defined via link substitution, inductive closure, or a condition, then $\rho$ is simply an $\mathbf{LK}V$-proof. Assume $x_1, \ldots, x_\alpha$ are all the eigenvariables occurring in $\rho$, and replace them by new variable classes $X_1^\rho(\vec{n}), \ldots, X_\alpha^\rho(\vec{n})$, thus obtaining the regular proof schema $\rho^R$.
- If $\rho = \rho'\{V_i(\vec{r}_i) \leftarrow \rho_i \mid i \in \{1, \ldots, \alpha\}$, then we first regularize $\rho'$ and the $\rho_i$ for $i \in \{1, \ldots, \alpha\}$, thus obtaining $\rho'^R$ containing variable expressions $X_j^{\rho'}$ for the $j$ eigenvariables in $\rho'$, and $X_{k_i}^{\rho_i}$ for the $k_i$ eigenvariables in $\rho_i$. The regular proof schema $\rho^R$ is then $\rho'^R\{V_i(\vec{r}_i) \leftarrow \rho_i^R \mid i \in \{1, \ldots, \alpha\}$.
- If $\rho$ is defined via proof recursion, i.e. $\rho$ is some $\hat{\rho}$, where

$$D(\hat{\rho}) = \{\hat{\rho}(\vec{n}, V(k)) =_d \underline{\text{if}}\ k = \bar{0}\ \underline{\text{then}}\ V(0)\{V(0) \leftarrow \rho_0\}$$
$$\underline{\text{else}}\ \rho'(\vec{n}, V(k))\{V(p(k)) \leftarrow \hat{\rho}((\vec{n}\{k \leftarrow p(k)\}, V(p(k))))\}$$

  then we proceed inductively and define the regularization $\hat{\rho}^R$ via

$$D(\hat{\rho}^R) = \{\hat{\rho}^R(\vec{n}, V(k)) =_d \underline{\text{if}}\ k = \bar{0}\ \underline{\text{then}}\ \rho_0^R$$
$$\underline{\text{else}}\ \rho'^R(\vec{n}, V(k))\{V(p(k)) \leftarrow \hat{\rho}^R((\vec{n}\{k \leftarrow p(k)\}, V(p(k))))\}.$$

- If $\rho$ is defined via a condition $C$

$$\underline{\text{if}}\ C\ \underline{\text{then}}\ \rho_1\ \underline{\text{else}}\ \rho_2$$

  then the regular proof schema $\rho^R$ is given by

$$\underline{\text{if}}\ C\ \underline{\text{then}}\ \rho_1^R\ \underline{\text{else}}\ \rho_2^R$$

for the regularization $\rho_{1,2}^R$ of $\rho_{1,2}$ containing the respective variable expressions instead of eigenvariables.

**Example 7.1.8** Consider the proof schema $\rho_1$ from Example 7.1.6. The regularization of $\rho_1$ is $\rho_1^R =$

$$\dfrac{\begin{array}{c}(\hat{\rho}^R(n, V_S(n))\\ \{V_S(0) \leftarrow \rho'^R)\})\\ S(n)\end{array} \qquad \dfrac{\dfrac{\dfrac{\dfrac{P(c) \to P(c) \qquad \dfrac{P(\hat{f}(c,n)) \vdash P(\hat{f}(c,n)) \qquad P(g(c,n)) \vdash P(g(c,n))}{P(\hat{f}(c,n)) \to P(g(c,n)), P(\hat{f}(c,n)) \vdash P(g(c,n))}\,\to_l}{P(c), P(\hat{f}(c,n)) \to P(g(c,n)), P(c) \to P(\hat{f}(c,n)) \vdash P(g(c,n))}\,\to_l}{P(\hat{f}(c,n)) \to P(g(c,n)), P(c) \to P(\hat{f}(c,n)) \vdash P(c) \to P(g(c,n))}\,\to_r}{P(c) \to P(\hat{f}(c,n)) \vdash (P(\hat{f}(c,n)) \to P(g(c,n))) \to (P(c) \to P(g(c,n)))}\,\to_r}{\forall x(P(x) \to P(\hat{f}(x,n))) \vdash (P(\hat{f}(c,n)) \to P(g(c,n))) \to (P(c) \to P(g(c,n)))}\,\forall_l}{\forall x(P(x) \to P(f(x))) \vdash (P(\hat{f}(c,n)) \to P(g(c,n))) \to (P(c) \to P(g(c,n)))}\,cut$$

where

$$S(n) = \forall x(P(x) \to P(f(x))) \vdash \forall x(P(x) \to P(\hat{f}(x,n)))$$

and $\hat{\rho}^R(n, V_S(n))\{V_S(0) \leftarrow \rho'^R)\}$ is given by

$$\begin{array}{l}D(\hat{\rho}^R) = \{\hat{\rho}^R(n, V_S(n)) =_d \underline{\text{if}}\ n = \bar{0}\ \underline{\text{then}}\ \rho'^R\\ \underline{\text{else}}\ \rho^R(n, V_S(n))\{V_S(p(n)) \leftarrow \hat{\rho}^R(p(n), V_S(p(n)))\}.\end{array}$$

$\rho^R(n, V_S(n))$ (the regularization of $\rho(n, V_S(n))$) is

$$\dfrac{\dfrac{\dfrac{V_S(p(n))}{S(p(n))}\,V_S E \qquad \begin{array}{c}(\varphi_1(n))\\ \forall x(P(x) \to P(\hat{f}(x, p(n)))), \forall x(P(x) \to P(f(x))) \vdash \forall x(P(x) \to P(\hat{f}(x,n)))\end{array}}{S(n)}\,cut, c_l}{V_S(n)}\,V_S I$$

for $\varphi_1(n) =$

$$\dfrac{\dfrac{\dfrac{\dfrac{P(X^\rho(n)) \vdash P(X^\rho(n)) \qquad \begin{array}{c}(\varphi_2(n))\\ P(\hat{f}(X^\rho(n), p(n))), \forall x(P(x) \to P(f(x))) \vdash P(\hat{f}(X^\rho(n), n))\end{array}}{P(X^\rho(n)), P(X^\rho(n)) \to P(\hat{f}(X^\rho(n), p(n))), \forall x(P(x) \to P(f(x))) \vdash P(\hat{f}(X^\rho(n), n))}\,\to_l}{P(X^\rho(n)) \to P(\hat{f}(X^\rho(n), p(n))), \forall x(P(x) \to P(f(x))) \vdash P(X^\rho(n)) \to P(\hat{f}(X^\rho(n), n))}\,\to_r}{\forall x(P(x) \to P(\hat{f}(x, p(n)))), \forall x(P(x) \to P(f(x))) \vdash P(X^\rho(n)) \to P(\hat{f}(X^\rho(n), n))}\,\forall_l}{\forall x(P(x) \to P(\hat{f}(x, p(n)))), \forall x(P(x) \to P(f(x))) \vdash \forall x(P(x) \to P(\hat{f}(x,n)))}\,\forall_r$$

and $\varphi_2(n) =$

$$\dfrac{\dfrac{P(\hat{f}(X^\rho(n), p(n))) \vdash P(\hat{f}(X^\rho(n), p(n))) \qquad \dfrac{P(\hat{f}(X^\rho(n), n)) \vdash P(\hat{f}(X^\rho(n), n))}{P(f(\hat{f}(X^\rho(n), p(n)))) \vdash P(\hat{f}(X^\rho(n), n))}\,T^+}{P(\hat{f}(X^\rho(n), p(n))), P(\hat{f}(X^\rho(n), p(n))) \to P(f(\hat{f}(X^\rho(n), p(n)))) \vdash P(\hat{f}(X^\rho(n), n))}\,\to_l}{P(\hat{f}(X^\rho(n), p(n))), \forall x(P(x) \to P(f(x))) \vdash P(\hat{f}(X^\rho(n), n))}\,\forall_l$$

and $\rho'^R$ (the regularization of $\rho'$) is

$$\cfrac{\cfrac{\cfrac{\cfrac{P(\hat{f}(X^{\rho'}(0),0)) \vdash P(\hat{f}(X^{\rho'}(0),0))}{P(X^{\rho'}(0)) \vdash P(\hat{f}(X^{\rho'}(0),0))}\ T^{+}}{\vdash P(X^{\rho'}(0)) \to P(\hat{f}(X^{\rho'}(0),0))}\ \to_r}{\vdash \forall x(P(x) \to P(\hat{f}(x,0)))}\ \forall_r}{\cfrac{S(0)}{V_S(0)}\ V_S I}\ w_l$$

## 7.2 Refutation Schemata

In the previous section we defined **LK** derivations schematically. This can also be done with $\mathrm{RPL}_0$ derivations (see Definition 2.2.1). Many of the concepts in this section are similar to the ones in the previous section, however, they are a little more involved given the nature of schematic refutations.

First, we will extend $\mathrm{RPL}_0$ by rules handling schematic formula definitions. But we have to consider another aspect as well: in inductive proofs the use of lemmas is vital, i.e. an ordinary refutational calculus, which has just a weak capacity of lemma generation, may fail to derive the desired invariant. To this aim we will add introduction rules for the connectives, giving us the potential to derive more complex formulas. Note that our aim is to use the calculi in an interactive way and not fully automatic, which justifies this process of "anti-refinement".

Another aspect is that we have to ensure that the formulas, on which the resolution rule is applied, have pairwise disjoint variables. We need a corresponding concept of disjointness for the schematic case.

**Definition 7.2.1** Let $A, B$ be finite sets of schematic variables. $A$ and $B$ are called essentially disjoint if for all $\sigma \in \mathcal{S}$ $A[\sigma] \cap B[\sigma] = \emptyset$.

**Definition 7.2.2** ($\mathrm{RPL}_0{}^{\Psi}$) Let $\Psi\colon ((\hat{q}, \mathcal{P}, D(\mathcal{P}), \vec{Y}, <, T', T'')$ be a theory, then for all schematic predicate symbols $\hat{p} \in \mathcal{P}$ for

$$D(\hat{p}) = \{\hat{p}(\vec{Y}, \vec{n}, m) =_d \ \underline{\text{if}}\ m = \bar{0}\ \underline{\text{then}}\ \hat{p}_B \\ \underline{\text{else}}\ \hat{p}_S\{\xi \leftarrow \hat{p}(\vec{Y}, \vec{n}, p(m))\}\}\},$$

we define the elimination of defined symbols

$$\frac{\Gamma \vdash \Delta, \hat{p}(\vec{Y}, \vec{n}, 0)}{\Gamma \vdash \Delta, \hat{p}_B}\ B\hat{p}r \qquad \frac{\Gamma \vdash \Delta, \hat{p}(\vec{Y}, \vec{n}, s(m))}{\Gamma \vdash \Delta, \hat{p}_S\{\xi \leftarrow \hat{p}(\vec{Y}, \vec{n}, m)\}}\ S\hat{p}r$$

$$\frac{\hat{p}(\vec{Y}, \vec{n}, 0), \Gamma \vdash \Delta}{\hat{p}_B, \Gamma \vdash \Delta}\ B\hat{p}l \qquad \frac{\hat{p}(\vec{Y}, \vec{n}, s(m)), \Gamma \vdash \Delta}{\hat{p}_S\{\xi \leftarrow \hat{p}(\vec{Y}, \vec{n}, m)\}, \Gamma \vdash \Delta}\ S\hat{p}l$$

and the introduction of defined symbols

$$\frac{\Gamma \vdash \Delta, \hat{p}_B}{\Gamma \vdash \Delta, \hat{p}(\vec{Y}, \vec{n}, 0)}\ B\hat{p}r^+ \quad \frac{\Gamma \vdash \Delta, \hat{p}_S\{\xi \leftarrow \hat{p}(\vec{Y}, \vec{n}, m)\}}{\Gamma \vdash \Delta, \hat{p}(\vec{Y}, \vec{n}, s(m))}\ S\hat{p}r^+$$

$$\frac{\hat{p}_B, \Gamma \vdash \Delta}{\hat{p}(\vec{Y}, \vec{n}, 0), \Gamma \vdash \Delta}\ B\hat{p}l^+ \quad \frac{\hat{p}_S\{\xi \leftarrow \hat{p}(\vec{Y}, \vec{n}, m)\}, \Gamma \vdash \Delta}{\hat{p}(\vec{Y}, \vec{n}, s(m)), \Gamma \vdash \Delta}\ S\hat{p}l^+$$

As the theory $\Psi$ contains the $\omega\iota$-theory $T'$ and, due to the defining equations for schematic term symbols, $T'$ is an equational theory we also need inference rules *within* schematic formulas. Let $t$ be a term in $T'$ of the form

$$(1)\colon \hat{s}(s_1, \ldots, s_i, t_1, \ldots, t_{j-1}, s(w))$$

or of the form

$$(2)\colon \hat{s}(s_1, \ldots, s_i, t_1, \ldots, t_{j-1}, \bar{0})$$

where

$$D(\hat{s}) = \{\hat{s}(x_1, \ldots, x_i, n_1, \ldots, n_{j-1}, n_j) =_d \underline{\text{if}}\ n_j = \bar{0}\ \underline{\text{then}}\ \hat{s}_B$$
$$\underline{\text{else}}\ \hat{s}_S\{z \leftarrow \hat{s}(x_1, \ldots, x_i, n_1, \ldots, n_{j-1}, p(n_j))\}\}\},$$

Then, in case (1) we define $t \sim_{T'} t'$ for

$$t' = \hat{s}_S\theta\{z \leftarrow \hat{s}(s_1, \ldots, s_i, t_1, \ldots, t_{j-1}, w)\},$$

where $\theta = \{x_1 \leftarrow s_1, \ldots, x_i \leftarrow s_i\}\{n_1 \leftarrow t_1, \ldots, n_{j-1} \leftarrow t_{j-1}, n_j \leftarrow w\}$.
In case (2) we define $t \sim_{T'} t'$ for

$$t' = \hat{s}_B\{x_1 \leftarrow s_1, \ldots, x_i \leftarrow s_i\}\{n_1 \leftarrow t_1, \ldots, n_{j-1} \leftarrow t_{j-1}\}.$$

We also define $\sim_{T'}$ as the reflexive and symmetric closure of the relation defined above. Now let $F$ be a schematic formula occurring in $\Psi$ such that $t$ occurs at position $\lambda$ in $F$ and $t \sim_{T'} t'$. Then $F[t']_\lambda$ is called a *$T'$-variant of $F$*. Now let $S\colon F_1, \ldots, F_i \vdash G_1, \ldots, G_j$ be a sequent, the $F'_\alpha$ be $T'$-variants of $F_\alpha$ and the $G'_\beta$ be $T'$-variants of $G_\beta$. Then we define the inference

$$\frac{F_1, \ldots, F_i \vdash G_1, \ldots, G_j}{F'_1, \ldots, F'_i \vdash G'_1, \ldots, G'_j}\ T'$$

We also adapt the resolution rule to the schematic case:
Let

$$T^\iota_V(\{A_1, \ldots, A_\alpha\}),\ T^\iota_V(\{B_1, \ldots, B_\beta\})$$

be essentially disjoint sets of schematic variables and $\Theta$ be an s-unifier of $\{A_1, \ldots, A_\alpha, B_1, \ldots, B_\beta\}$. Then the resolution rule is defined as

$$\frac{\Gamma \vdash \Delta, A_1, \ldots, A_\alpha \quad B_1, \ldots, B_\beta, \Pi \vdash \Lambda}{\Gamma\Theta, \Pi\Theta \vdash \Delta\Theta, \Lambda\Theta}\ res\{\Theta\}$$

The refutational completeness of $\mathrm{RPL}_0^\Psi$ is not an issue as already $\mathrm{RPL}_0$ is refutationally complete for $\mathrm{PL}_0$ formulas [3, 31]. Note that this is not the case anymore if parameters occur in formulas. Indeed, due to the usual theoretical limitations, the logic is not semi-decidable for schematic formulas [2]. $\mathrm{RPL}_0^\Psi$ is sound if the defining equations are considered.

**Proposition 7.2.1** *Let $S$ be a sequent which is derivable in $\mathrm{RPL}_0^\Psi$ for $\Psi\colon (\hat{q}, \mathcal{P}, D(\mathcal{P}), \vec{Y}, <), T', T'')$. Then $D(\mathcal{P}) \models S$.*

***Proof*** The introduction and elimination rules for defined predicate symbols are sound (we have to consider the equations $D(\mathcal{P})$), as are the $T'$ rules; also the resolution rule (involving s-unification ) is sound.

Before giving a formal definition of refutation schemata let us have a look at the following example.

**Example 7.2.1** We define a refutation schema for the $\omega\iota o$-theory $T^+\colon (T, T', T'')$ where

$$T = (\hat{p}, \{\hat{p}, \hat{q}\}, D(\{\hat{p}, \hat{q}\}), V, <),$$

$$T' = (\hat{s}, \{\hat{s}\}, D(\{\hat{s}\}), \emptyset),$$

$X\colon \omega \to \iota$, $Y, Z\colon \iota$ and

$$\begin{aligned}
D(\hat{p}) = {} & \{\hat{p}(X, n) =_d \underline{\text{if}}\ n = \bar{0}\ \underline{\text{then}}\ \neg P(X(\bar{0}), \hat{s}(a, \bar{0})) \\
& \underline{\text{else}}\ \hat{p}(X, p(n)) \vee \neg P(X(n), \hat{s}(a, n))\}, \\
D(\hat{q}) = {} & \{\hat{q}(X, Y, Z, n, m) =_d P(\hat{s}(Y, m), Z) \wedge \hat{p}(X, n)\}, \\
D(\hat{s}) = {} & \{\hat{s}(x, n) =_d \underline{\text{if}}\ n = \bar{0}\ \underline{\text{then}}\ x\ \underline{\text{else}}\ h(\hat{s}(x, p(n)))\}.
\end{aligned}$$

To this aim, we will extend the calculus $\mathrm{RPL}_0^\Psi$ by two rules involving link-variables. We create link-expressions $V_S(X, r)$ corresponding to $S\colon \vdash \hat{p}(X, r)$ for $r \in \{\bar{0}, n, p(n)\}$ and add rules for the elimination and introduction of $V_S(X, r)$:

$$\frac{V_S(X, r)}{\vdash \hat{p}(X, r)}\ V_S E \qquad \frac{\vdash \hat{p}(X, r)}{V_S(X, r)}\ V_S I$$

Analogously to **LK**-proof schemata, the idea is to use the sequent link-variables to define recursive refutations. We start with a recursive proof which is defined for $n > \bar{0}$ and denoted as $\rho(X, Y, Z, m, p(n), V_S(X, p(n))) =$

$$
\cfrac{\cfrac{\cfrac{\cfrac{\cfrac{\cfrac{V_S(X,n)}{\vdash \hat{p}(X,n)}\,V_S E}{\vdash \hat{p}(X,p(n)) \vee \neg P(X(n),\hat{s}(a,n))}}{\vdash \hat{p}(X,p(n)), \neg P(X(n),\hat{s}(a,n))}}{P(X(n),\hat{s}(a,n)) \vdash \hat{p}(X,p(n))} \quad \cfrac{V_{S'}(Y,Z,m)}{\vdash P(\hat{s}(Y,m),Z)}\,V_{S'}E}{\vdash \hat{p}(X,p(n))}\,R(\Theta(n,m))}{V_S(X,p(n))}\,V_S I
$$

where $S'(Y,Z,m) = \vdash P(\hat{s}(Y,m),Z)$ and $\Theta(n,m) = \{X(n) \leftarrow \hat{s}(Y,m), Z \leftarrow \hat{s}(a,n)\}$ is an s-substitution. Note that this substitution can be computed by UAL.

The idea is now to append $\rho(X,Y,Z,m,p(n),V_S(X,p(n)))$ to itself recursively and to use the resulting iteration to derive the sequent $V_S(X,\bar{0})$. We first define the recursive iteration of $\rho(X,Y,Z,m,n,V_S(X,n))$:

$$
\begin{aligned}
D(\hat{\rho}) = \; & \{\hat{\rho}(X,Y,Z,m,n) =_d \underline{\text{if}}\; n = \bar{0}\; \underline{\text{then}}\; \rho(X,Y,Z,m,\bar{0},V_S(X,\bar{0})) \\
& \underline{\text{else}}\; \hat{\rho}(X,Y,Z,m,p(n))\{V_S(X,n) \leftarrow \rho(X,Y,Z,m,n,V_S(X,n))\}\}.
\end{aligned}
$$

In carrying out the substitution

$$
\hat{\rho}(X,Y,Z,m,p(n))\{V_S(X,n) \leftarrow \rho(X,Y,Z,m,n,V_S(X,n))\}
$$

we identify the root of $\rho(X,Y,Z,m,n,V_S(X,n))$ with the uppermost leaf $\vdash V_S(X,n)$ in $\hat{\rho}(X,Y,Z,m,p(n))$ provided $p(n) > \bar{0}$.
In the next step we use a case-distinction such that for $n = \bar{0}$ we obtain $V_S(X,\bar{0})$ and $\hat{\rho}(X,Y,Z,m,p(n))$ otherwise. We achieve this by defining $\rho'$ as

$$
\begin{aligned}
D(\rho') = \; & \{\rho'(X,Y,Z,m,n) =_d \underline{\text{if}}\; n = \bar{0}\; \underline{\text{then}}\; V_S(X,\bar{0}) \\
& \underline{\text{else}}\; \hat{\rho}(X,Y,Z,m,p(n))\}.
\end{aligned}
$$

Note that, for any values of $n$ and $m$, $\rho'(X,Y,Z,m,n)$ evaluates to a proof of $V_{\hat{p}}(X,\bar{0})$. But still we do not have a proof with the right axioms, as one axiom in $\rho'(X,Y,Z,m,n)$ is $V_S(X,n)$.

To achieve a proof we have to apply a link substitution

$$
\theta(X,m,n)\colon \{V_S(X,n) \leftarrow \psi(X,m,n)\}
$$

and obtain

$$
\rho'(X,Y,Z,m,n)\{V_S(X,n) \leftarrow \psi(X,n)\}
$$

where $\psi(X,m,n)$ is a proof of $\vdash V_S(X,n)$ from $\hat{q}(X,Y,Z,n,m)$.

But still we are not done: in the proof schema

$$
\rho(X,Y,Z,m,p(k),V_S(X,p(k)))
$$

we have the link expressions $V_{S'}(Y, Z, m)$ for $\vdash P(\hat{s}(Y, m), Z)$. Therefore we still have no derivation from $\hat{q}(X, Y, Z, n, m)$. To avoid applying several link substitutions as a postprocessing we first "complete" the proof

$$\rho(X, Y, Z, m, p(k), V_S(X, p(k)))$$

to a proof

$$\rho^*(X, Y, Z, m, p(k), V_S(X, p(k)))$$

where $\rho^*(X, Y, Z, m, p(k), V_S(X, p(k))) =$

$$\dfrac{\dfrac{\dfrac{\dfrac{\dfrac{\dfrac{V_S(X,k)}{\vdash \hat{p}(X,k)}\,V_S E}{\vdash \hat{p}(X,p(k)) \lor \neg P(X(k), \hat{s}(a,k))}}{\vdash \hat{p}(X,p(k)), \neg P(X(k), \hat{s}(a,k))}}{P(X(k), \hat{s}(a,k)) \vdash \hat{p}(X,p(k))} \quad \dfrac{\{V_{S'}(Y,Z,m) \leftarrow \tau(Y,Z,m)\}}{\vdash P(\hat{s}(Y,m), Z)}\,V_{S'}E}{\vdash \hat{p}(X,p(k))}\,R(\Theta(k,m))}{V_S(X,p(k))}\,V_S I$$

where

$$\{V_{S'}(Y, Z, m) \leftarrow \tau(Y, Z, m, n)\}$$

is a link substitution replacing $V_{S'}(Y, Z, m)$ by a derivation $\tau(Y, Z, m)$ of $V_{S'}(Y, Z, m)$ from $\hat{q}(X, Y, Z, n, m)$. To this aim we have to replace $\rho$ by $\rho^*$ *before* defining $\hat{\rho}$ and $\rho'$. Then the following derivation below is a refutation of $\vdash \hat{q}(X, Y, Z, n, m)$: $\rho_0(X, Y, Z, n, m) =$

$$\dfrac{\dfrac{\dfrac{\dfrac{\dfrac{\rho'(X,Y,Z,m,n)\{V_S(X,n) \leftarrow \psi(X,m,n)\}}{V_S(X,\bar{0})}}{\hat{p}(X,\bar{0})}}{\vdash \neg P(X(\bar{0}), \hat{s}(a,\bar{0}))}}{P(X(\bar{0}), f(a,\bar{0})) \vdash} \quad \dfrac{\{V_{S'}(Y,Z,m) \leftarrow \tau(Y,Z,m,n)\}}{\vdash P(\hat{s}(Y,m), Z)}\,V_{S'}E}{\vdash}\,R(\Theta(\bar{0},m))$$

where $\Theta(\bar{0}, m) = \{X(\bar{0}) \leftarrow \hat{s}(Y, m), Z \leftarrow \hat{s}(a, \bar{0})\}$ is an s-substitution. It remains to define the schematic derivations $\psi(X, n)$ and $\tau(Y, Z, m, k)$:

$\psi(X, m, n) =$

$$\dfrac{\dfrac{\dfrac{\vdash \hat{q}(X, Y, Z, n, m)}{\vdash P(\hat{s}(Y, m), Z) \land \hat{p}(X, n)}}{\vdash \hat{p}(X, n)}}{V_S(X, n)}$$

and $\tau(Y, Z, m, k) =$

$$\dfrac{\dfrac{\vdash \hat{q}(X, Y, Z, n, m)}{\vdash P(\hat{s}(Y, m), Z) \land \hat{p}(X, n)}}{\vdash P(\hat{s}(Y, m), Z)}$$

Sequent link variables for the refutational calculus are defined like in Definition 7.1.8; the only extension consists in potentially adding parameters which are not contained in the corresponding sequent. To keep the formalism simple we write $S(\vec{X},\vec{r})$ for a sequent $S$ even if $\vec{r}$ contains strictly more parameters than $S$ itself. These additional parameters may enable more complex recursions in defining schematic derivations. Analogously to link-variables in **LK**-proof schemata, link-variables $V$ in refutation schemata serve the purpose to define locations in a proof where $V$ can be replaced by a proof; these locations can be either the leaves of a proof or the root. In order to place into or to remove variables from proofs we extend the $\mathrm{RPL}_0^\Psi$-calculus by variable elimination rules and variable introduction rules.

**Definition 7.2.3** ($\mathrm{RPL_0}^\Psi V$) The calculus $\mathrm{RPL}_0^\Psi V$ contains the rules of the calculus $\mathrm{RPL}_0^\Psi$ with two additional rules. Let $V$ be a variable of type $S$, we define the rules

$$\frac{S(\vec{X},\vec{r})}{V_S(\vec{X},\vec{r})}\ V_I \qquad \frac{V_S(\vec{X},\vec{r})}{S(\vec{X},\vec{r})}\ V_E$$

Elimination rules can only be applied to leaves in an $\mathrm{RPL}_0^\Psi V$-derivation if the leaf is a link-expression, and introduction rules to root nodes which are labeled by a sequent $S(\vec{X},\vec{r})$. Or expressed in another way: any $\mathrm{RPL}_0^\Psi V$-derivation can be obtained from an $\mathrm{RPL}_0^\Psi$-derivation $\rho$ by appending variable elimination rules on some leaves of $\rho$ and (possibly) a variable introduction rule on the root (provided the sequents on the nodes are of an appropriate form).

**Example 7.2.2** The derivation $\rho(X, Y, Z, m, p(n), V_{\hat{p}}(X, p(n)))$ in Example 7.2.1 is a $\mathrm{RPL}_0^\Psi V$-derivation.

The link-variable-expressions in an $\mathrm{RPL}_0^\Psi V$-derivation can be replaced by other $\mathrm{RPL}_0^\Psi V$-derivations:

**Definition 7.2.4** ($\mathrm{RPL_0}^\Psi V$*-link substitutions*) Let $\rho$ be a $\mathrm{RPL}_0^\Psi V$-derivation such that the link variable of the root differs from all link variables in the leaves. Let $\rho_i$ for $i = 1,\ldots,\alpha$ be $\mathrm{RPL}_0^\Psi V$-derivations with root-link-expressions $V_i(\vec{X},\vec{r}_i)$ such that $V_i \neq V_j$ for $i \neq j$. Then

$$\rho\{V_i(\vec{X},\vec{r}_i) \leftarrow \rho_i \mid i \in \{1,\ldots,\alpha\}\}$$

is the $\mathrm{RPL}_0^\Psi V$-derivation obtained from $\rho$ by replacing the link-expressions $V_i(\vec{X},\vec{r}_i)$ occurring at the leaves of $\rho$ by $\rho_i$ without the last inference.

**Definition 7.2.5** ($\mathrm{RPL_0}^\Psi V$*-recursion*) Let $\vec{n} = (n_1,\ldots,n_\alpha)$ and $\rho(\bar{X}, (\vec{n}, p(k)), V_S(\vec{X}, (\vec{n}, p(k)))$ for $k > 0$ be a proof with root $V_S(\vec{X}, (\vec{n}, p(k))$ and $m$ leaves

$$U_S^i(\vec{X}, (\psi_1^i(n_1),\ldots,\psi_\alpha^i(n_\alpha), k)$$

where for all $i, j$ $\psi_j^i(n_j) \in \{p(n_j), n_j, s(n_j), \bar{0}\}$. If there are leaves of this form we say that $\rho$ admits proof recursion and the $U_S^i(\vec{X}, (\psi_1^i(n_1),\ldots,\psi_\alpha^i(n_\alpha), k)$ are called the *recursion link expressions* of $\rho$. We define for $i = 1,\ldots,m$ and $k > 0$

$$\eta_i((n_1, \ldots, n_\alpha, p(k)) =$$
$$(\psi_1^i(n_1), \ldots, \psi_\alpha^i(n_\alpha), k).$$

Let $V_i(k)$ be an abbreviation for $U_S^i(\vec{X}, \eta_i((n_1, \ldots, n_\alpha, p(k))))$ and $\rho_i(k)$ for

$$\rho(\bar{X}, \eta_i(\vec{n}, p(k)), V_S(\vec{X}, \eta_i(\vec{n}, p(k))).$$

Then we define

$$\hat{\rho}(\bar{X}, \vec{n}, k)) =_d \underline{\text{if}}\ k = \bar{0}\ \underline{\text{then}}\ \rho(\bar{X}, (\vec{n}, \bar{0}), V_S(\vec{X}, (\vec{n}, \bar{0}))$$
$$\underline{\text{else}}\ \hat{\rho}((\bar{X}, \vec{n}, p(k))(\{V_i(k) \leftarrow \rho_i(k) \mid i \in \{1, \ldots, m\}\}.$$

$\hat{\rho}$ is called an *inductive closure* of $\rho$. If there exists only one recursive link expression $U(\vec{X}, \vec{r})$ in $\rho$ (which may occur several times as leaves) then we speak about a standard proof recursion and $\hat{\rho}$ is called a *standard inductive closure*.

**Example 7.2.3** Take $\rho(X, Y, Z, m, p(n), V_S(X, p(n)))$ and $\hat{\rho}(X, Y, Z, m, n)$ from Example 7.2.1. Then $\hat{\rho}$ is a standard inductive closure of $\rho$. Note that $\rho(X, Y, Z, m, p(n), V_S(X, p(n)))$ in Example 7.2.1 has the root $V_S(X, p(n))$ and only the link variable $V_S(X, n)$ in the leaves.

$\hat{\rho}$ is obtained from $\rho$ by a kind of primitive recursion on proofs. While in primitive recursion only one parameter is changed and the others are invariant, several parameters may be changed in our definition of inductive closure. An example for a change of two parameters can be found in the inductive closure $\hat{\rho}_1(k, l)$ in Example 9.2.6.

**Definition 7.2.6** ($\text{RPL}_0{}^\Psi V$-*schema*) $\text{RPL}_0{}^\Psi V$-derivation schemata are defined inductively:

- Any $\text{RPL}_0{}^\Psi V$-derivation of a schematic sequent $S$ is a $\text{RPL}_0{}^\Psi V$-derivation schema of $S$.
- Let $\rho$ be a $\text{RPL}_0{}^\Psi V$-derivation schema of $S$ (or of a link expression $V_S$) such that the link variable of the root differs from all link variables in the leaves, $\rho_i$ for $i = 1, \ldots, \alpha$ be $\text{RPL}_0{}^\Psi V$-derivations with root-link-variables $V_i(\vec{X}, \vec{r_i})$ such that $V_i \neq V_j$ for $i \neq j$. Then

$$\rho\{V_i(\vec{X}, \vec{r_i}) \leftarrow \rho_i \mid i \in \{1, \ldots, \alpha\}\}$$

  is a $\text{RPL}_0{}^\Psi V$-derivation schema of $S$ ($V_S$).
- If $\rho(\vec{X}, n, k)$ is an $\text{RPL}_0{}^\Psi V$-derivation schema which admits proof recursion and $\rho(\vec{X}, n, \bar{0})$ is an $\text{RPL}_0{}^\Psi V$-derivation schema of $S$ then the inductive closure of $\rho$ is an $\text{RPL}_0{}^\Psi V$-derivation schema of $S$.
- Let $\rho_1(\vec{X_1}, \vec{n})$, $\rho_2(\vec{X_2}, \vec{n})$ be $\text{RPL}_0{}^\Psi V$-derivation schemata over the parameter tuple $\vec{n}$ of a schematic sequent $S$ and let $C$ be a condition on the parameters in $\vec{n}$ then

$$\underline{\text{if}}\ C\ \underline{\text{then}}\ \rho_1(\vec{X_1}, \vec{n})\ \underline{\text{else}}\ \rho_2(\vec{X_2}, \vec{n})$$

  is a $\text{RPL}_0{}^\Psi V$-derivation schema of $S$.

**Example 7.2.4** Consider the proofs $\rho'$ and $\rho_0$ in Example 7.2.1. Both are $\mathrm{RPL_0}^\Psi V$-schemata; the $\mathrm{RPL_0}^\Psi V$-schema $\rho_0$ is also a *refutation schema* of $\hat{q}(X, Y, Z, n, m)$, a concept which will be formally defined below.

We call $\mathrm{RPL_0}^\Psi V$-schemata which are schematic derivations from sequents of the form $\vdash F$ *schematic* $F$*-proofs*. A schematic $F$-proof of $\vdash$ is called a *refutation schema* of $F$. In defining schematic $F$-proofs in terms of schematic $F$-refutations we face the same problem as in the case of input schemata. Like in Definition 7.1.14 we define here

**Definition 7.2.7** Let Let $F\colon \hat{q}(\vec{X}, \vec{n}, m)$ be the main schematic atom in a schematic definition $\Psi$ and $\rho(\vec{X}, \vec{n}, p(k))$ for $k > 0$ be a $\mathrm{RPL_0}^\Psi V$-derivation schema of $S(\vec{X}, (\vec{n}, p(k)))$ admitting proof recursion. $\rho(\vec{X}, \vec{n}, p(k))$ is called a *F-semi-proof schema* of $S(\vec{X}, (\vec{n}, p(k)))$ if all leaves of $\rho(\vec{X}, \vec{n}, p(k))$ except the recursion link expressions are of the form $\vdash \hat{q}(\vec{X}, \vec{n}, m)$.

In order to guarantee that an inductive closure $\hat{\rho}(\vec{X}, \vec{n}, k)$ of a semi-proof schema $\rho(\vec{X}, \vec{n}, k)$ is actually a schematic proof we have to replace all recursion link expressions $V_S(\vec{X}, \vec{r})$ on the leaves of $\hat{\rho}(\vec{X}, \vec{n}, k)$ by schematic proofs of $V_S(\vec{X}, \vec{r})$. To achive this aim we have to define the set of all recursion link expressions occurring at the leaves of $\hat{\rho}(\vec{X}, \vec{n}, k)$.

**Definition 7.2.8** Let $\vec{n} = (n_1, \dots, n_\alpha)$ and $\rho(\bar{X}, (\vec{n}, p(k)), V_S(\vec{X}, (\vec{n}, p(k)))$ for $k > 0$ be a proof with root $V_S(\vec{X}, (\vec{n}, p(k))$ and $m$ leaves

$$U^i_S(\vec{X}, (\psi^i_1(n_1), \dots, \psi^i_\alpha(n_\alpha), k)$$

where for all $i, j$ $\psi^i_j(n_j) \in \{p(n_j), n_j, s(n_j), \bar{0}\}$ and $\rho$ admits proof recursion such that the $U^i_S(\vec{X}, (\psi^i_1(n_1), \dots, \psi^i_\alpha(n_\alpha), k)$ are the recursion link expressions of $\rho$. We define for $i = 1, \dots, m$ and $k > 0$

$$\eta_i((n_1, \dots, n_\alpha, p(k)) = (\psi^i_1(n_1), \dots, \psi^i_\alpha(n_\alpha), k).$$

Let $V_i(k)$ be an abbreviation for $U^i_S(\vec{X}, \eta_i((n_1, \dots, n_\alpha, p(k))))$ and $\rho_i(k)$ for

$$\rho(\bar{X}, \eta_i(\vec{n}, p(k)), V_S(\vec{X}, \eta_i(\vec{n}, p(k))).$$

Then let

$$\begin{aligned}\hat{\rho}(\bar{X}, \vec{n}, k)) =_d\ &\underline{\text{if}}\ k = \bar{0}\ \underline{\text{then}}\ \rho(\bar{X}, (\vec{n}, \bar{0}), V_S(\vec{X}, (\vec{n}, \bar{0}))\\ &\underline{\text{else}}\ \hat{\rho}((\vec{X}, \vec{n}, p(k))\{V_i(k) \leftarrow \rho_i(k) \mid i \in \{1, \dots, m\}\}.\end{aligned}$$

be the inductive closure of $\rho$. We define the set of recursion link expressions $\mathcal{L}(\hat{\rho}, k)$ and the set of *link tuples* $\mathcal{T}(\hat{\rho}, k)$ in $\hat{\rho}(\vec{X}, \vec{n}, k)$ inductively:

$$\begin{aligned}\mathcal{L}(\hat{\rho}, \bar{0}) &= \{U^i_S(\vec{X}, \eta_i((n_1, \dots, n_\alpha, \bar{0}))) \mid i = 1, \dots, m\},\\ \mathcal{T}(\hat{\rho}, \bar{0}) &= \{\eta_i((n_1, \dots, n_\alpha, \bar{0})) \mid i = 1, \dots, m\}\end{aligned}$$

is the set of recursion link expressions in $\hat{\rho}(\vec{X}, \vec{n}, \bar{0})$. Let $\mathcal{L}(\hat{\rho}, k)$ be the set of recursion link expressions in $\hat{\rho}(\vec{X}, \vec{n}, k)$ and $\mathcal{T}(\hat{\rho}, k)$ be the set of link tuples in $\hat{\rho}(\vec{X}, \vec{n}, k)$. Then

$$\mathcal{L}(\hat{\rho}, s(k)) = \bigcup_{i=1}^{m} \bigcup_{\vec{r} \in \mathcal{T}(\hat{\rho},k)} \{U_S^i(\vec{X}, \eta_i \vec{r})\},$$

$$\mathcal{T}(\hat{\rho}, s(k)) = \bigcup_{i=1}^{m} \bigcup_{\vec{r} \in \mathcal{T}(\hat{\rho},k)} \{\eta_i \vec{r}\}.$$

It is provable by a straightforward induction that $\mathcal{L}(\hat{\rho}, k)$ is indeed the set of all recursion link expression occurring on the leaves of $\hat{\rho}(\vec{X}, \vec{n}, k)$.

**Example 7.2.5** Consider the derivation schemata

$$\rho(X, Y, Z, m, p(k), V_S(X, p(k))) \quad \text{and} \quad \rho^*(X, Y, Z, m, p(k), V_S(X, p(k)))$$

of $(\vec{X}, (\vec{n}, p(k)))$ for $k > 0$ in Example 7.2.1. $\rho(X, Y, Z, m, p(k), V_S(X, p(k)))$ is a $\mathrm{RPL_0}^\Psi V$-derivation schema of $S(\vec{X}, (\vec{n}, p(k)))$ but not a F-semi-proof schema of $S(\vec{X}, (\vec{n}, p(k)))$, but $\rho^*(X, Y, Z, m, p(k), V_S(X, p(k)))$ is a F-semi-proof schema of $S(\vec{X}, (\vec{n}, p(k)))$ for $F = \hat{q}(X, Y, Z, n, m)$.

Now consider

$$\begin{aligned} D(\hat{\rho}) = \ & \{\hat{\rho}(X, Y, Z, m, n) =_d \underline{\text{if}}\ n = \bar{0}\ \underline{\text{then}}\ \rho^*(X, Y, Z, m, \bar{0}, V_S(X, \bar{0})) \\ & \underline{\text{else}}\ \hat{\rho}(X, Y, Z, m, p(n))\{V_S(X, n) \leftarrow \rho^*(X, Y, Z, m, n, V_S(X, n))\}. \end{aligned}$$

in Example 7.2.1. We compute $\mathcal{L}(\hat{\rho}, k)$ and $\mathcal{T}(\hat{\rho}, k)$ for arbitrary $k$. Here we have $\eta(p(k)) = k$ for $k > 0$. For $k = \bar{0}$ we get

$$\begin{aligned} \mathcal{L}(\hat{\rho}, \bar{0}) &= \{V_S(X, \bar{1})\}, \\ \mathcal{T}(\hat{\rho}, \bar{0}) &= \{\bar{1}\} \end{aligned}$$

Assume that we already defined $\mathcal{L}(\hat{\rho}, k)$, $\mathcal{T}(\hat{\rho}, k)$; then

$$\mathcal{L}(\hat{\rho}, s(k)) = \bigcup_{r \in \mathcal{T}(\hat{\rho},k)} \{V_S(\vec{X}, \eta(r))\},$$

$$\mathcal{T}(\hat{\rho}, s(k)) = \bigcup_{r \in \mathcal{T}(\hat{\rho},k)} \{\eta(r)\}.$$

An easy induction argument shows that

$$\mathcal{L}(\hat{\rho}, k) = \{V_S(X, s(k))\}$$

and

$$\mathcal{T}(\hat{\rho}, k) = \{s(k)\}$$

for all $k$.

**Definition 7.2.9** (*schematic $F$-proofs*) Let $F : \hat{q}(\vec{X}, \vec{n})$ be the main schematic atom in a schematic definition $\Psi$. We define schematic $F$-proofs below. The $\mathrm{RPL_0}^\Psi V$-proof

$$\vdash \hat{q}(\vec{Y}, \vec{n})$$

for a variable tuple $\vec{Y}$ of the same length as $\vec{X}$ is a schematic $F$-proof.

- If $\rho_1$ and $\rho_2$ are schematic $F$-proofs of $S_1$ and $S_2$ from $\vdash \hat{q}(\vec{X}, \vec{n})$ and $\rho =$

$$\cfrac{\begin{matrix}(\rho_1) & (\rho_2)\\ S_1 & S_2\end{matrix}}{S}\ \xi$$

  for a binary rule $\xi$ then $\rho$ is a schematic $F$-proof of $S$.
- If $\rho'$ is a schematic $F$-proof of $S'$ and $\rho =$

$$\cfrac{\begin{matrix}(\rho')\\ S'\end{matrix}}{S}\ \xi$$

  for a unary rule $\xi$ then $\rho$ is a schematic $F$-proof of $S$.
- Let $\rho$ be a $\mathrm{RPL_0}^\Psi V$-derivation schema of a schematic sequent $S$ (or of a link expression $V_S$) such that the link expression of the root differs from all link expressions in the leaves. Let us assume that $V_{S_i}(\vec{X}_i, \vec{r}_i)$ for $i = 1, \ldots, \alpha$ are all the the link expressions occurring at the leaves of $\rho$ and every leaf which is not a link expression is of the form $\hat{q}(\vec{Y}, \vec{n})$. If there are schematic $F$-proofs $\rho_i$ of $V_{S_i}(\vec{X}_i, \vec{r}_i)$ then

$$\rho\{V_{S_i}(\vec{X}, \vec{r}_i) \leftarrow \rho_i \mid i \in \{1, \ldots, \alpha\}\}$$

  is a schematic $F$-proof.
- Let $\vec{n} = (n_1, \ldots, n_\alpha)$ and $\rho(\bar{X}, (\vec{n}, p(k)), V_S(\vec{X}, (\vec{n}, p(k)))$ for $k > 0$ be a proof with root $V_S(\vec{X}, (\vec{n}, p(k))$ and $m$ leaves

$$U_S^i(\vec{X}, (\psi_1^i(n_1), \ldots, \psi_\alpha^i(n_\alpha), k)$$

  where for all $i, j$ $\psi_j^i(n_j) \in \{p(n_j), n_j, s(n_j), \bar{0}\}$ and $\rho$ admits proof recursion such that the $U_S^i(\vec{X}, (\psi_1^i(n_1), \ldots, \psi_\alpha^i(n_\alpha), k)$ are the recursion link expressions of $\rho$. Moreover we assume that $\rho$ is a semi-proof schema. We define for $i = 1, \ldots, m$ and $k > 0$

$$\eta_i((n_1, \ldots, n_\alpha, p(k)) = (\psi_1^i(n_1), \ldots, \psi_\alpha^i(n_\alpha), k).$$

Let $V_i(k)$ be an abbreviation for $U_S^i(\vec{X}, \eta_i((n_1, \ldots, n_\alpha, p(k))))$ and $\rho_i(k)$ for

$$\rho(\bar{X}, \eta_i(\vec{n}, p(k)), V_S(\vec{X}, \eta_i(\vec{n}, p(k))).$$

Then let

$$\hat{\rho}(\bar{X}, \vec{n}, k)) =_d \underline{\text{if}}\ k = \bar{0}\ \underline{\text{then}}\ \rho(\bar{X}, (\vec{n}, \bar{0}), V_S(\vec{X}, (\vec{n}, \bar{0}))$$
$$\underline{\text{else}}\ \hat{\rho}((\bar{X}, \vec{n}, p(k))\{V_i(k) \leftarrow \rho_i(k) \mid i \in \{1, \ldots, m\}\}.$$

be the inductive closure of $\rho$. $\hat{\rho}$ is not a schematic $F$-proof of

$$V_S(\vec{X}, \vec{n}, \bar{0})$$

as the recursion link expressions in $\mathcal{L}(\hat{\rho}, k)$ (see Definition 7.2.8) at the leaves are not of the form $\vdash F$. But assume that for all $V \in \mathcal{L}(\hat{\rho}, k)$ there exists a schematic $F$-proof $\tau(V)$ of $V$, then

$$\hat{\rho}(\bar{X}, \vec{n}, k))\{V \leftarrow \tau(V) \mid V \in \mathcal{L}(\hat{\rho}, k)\}$$

is an $F$-proof of $V_S(\vec{X}, \vec{n}, \bar{0})$.

- Let $\hat{\rho}$ be given by the (standard) $\text{RPL}_0{}^\Psi V$-recursion

$$\hat{\rho}(\bar{X}, \vec{n}, k) =_d \underline{\text{if}}\ k = \bar{0}\ \underline{\text{then}}\ \rho(\bar{0})$$
$$\underline{\text{else}}\ \hat{\rho}(\bar{X}, \vec{n}, p(k))\{V(k) \leftarrow \rho(k)\}.$$

with the abbreviations $V(k)$ for $V_S(\vec{X}, (\vec{n}, k))$, $S(k)$ for $S(\vec{X}, (\vec{n}, k))$ and $\rho(k)$ for $\rho(\bar{X}, (\vec{n}, k), V_S(\vec{X}, (\vec{n}, k)))$. Let us assume that $\rho(p(k))$ admits proof recursion for $k > 0$ and $\rho(p(k))$ is a $F$-semi-proof schema. Let $\psi(k)$ be a schematic $F$-proof of $V(s(k))$. Then

$$\hat{\rho}(\bar{X}, \vec{n}, k)\{V(s(k)) \leftarrow \psi(k)\}$$

is a schematic $F$-proof of $V_S(\vec{X}, \vec{n}, \bar{0})$. Here, $\mathcal{L}(\hat{\rho}, k) = \{V(s(k))\}$ and there is just one single recursion link expression in $\hat{\rho}(\bar{X}, \vec{n}, k)$, namely $V(s(k))$.

- Let $\rho_1(\vec{X}_1, \vec{n})$, $\rho_2(\vec{X}_2, \vec{n})$ be schematic $F$-proofs of $S$ over the parameter tuple $\vec{n}$ and let $C$ be a condition on the parameters in $\vec{n}$; let $\rho(\vec{X}_1, \vec{X}_2, \vec{n}) =$

$$\underline{\text{if}}\ C\ \underline{\text{then}}\ \rho_1(\vec{X}_1, \vec{n})\ \underline{\text{else}}\ \rho_2(\vec{X}_2, \vec{n}).$$

Then $\rho(\vec{X}_1, \vec{X}_2, \vec{n})$ is a schematic $F$-proof of $S$.

A schematic $F$-proof of the empty sequent is called a *schematic $F$-refutation.*

When the parameters in a refutation schema are instantiated with numerals, we obtain a $\text{RPL}_0$ refutation. Below, we will use two evaluation methods:$\downarrow_1$ (to be defined in Lemma 7.2.1), which is a proof evaluation mapping schemata into $\mathrm{RPL}_0{}^{\Psi}V$ refutations, and $\downarrow_2$ which transforms $\mathrm{RPL}_0{}^{\Psi}V$ refutations into $\mathrm{RPL}_0$-refutations.

**Lemma 7.2.1** *Let $\rho$ be a refutation schema of $\hat{q}(\vec{X}, n_1, \ldots, n_\alpha)$. Then, for all numerals $\nu_1, \ldots, \nu_\alpha$, the evaluation $\rho\{n_1 \leftarrow \nu_1, \ldots, n_\alpha \leftarrow \nu_\alpha\}\downarrow_1$ is a $\mathrm{RPL}_0{}^{\Psi}V$ refutation of $\hat{q}(\vec{X}, \nu_1, \ldots, \nu_\alpha)$.*

***Proof*** Let $\sigma(n_i) = \nu_i$ for $i = 1, \ldots, \alpha$ and $\rho' = \sigma(\rho)$. If $\rho$ is defined only via applications of link substitututions then $\rho'$ is an $\mathrm{RPL}_0{}^{\Psi}V$ derivation of $\vdash$ from axioms $\vdash \hat{q}(\vec{X}, \nu_1, \ldots, \nu_\alpha)$ and thus an $\mathrm{RPL}_0{}^{\Psi}V$ refutation of $\hat{q}(\vec{X}, \nu_1, \ldots, \nu_\alpha)$. If $\rho$ is defined via a condition $C$ then exactly one of $\rho_1$ or $\rho_2$ is chosen for the parameter assignment $\sigma$. So $\rho'$ is either $\rho_1'$ or $\rho_2'$ and we may apply the induction hypothesis.

The only nontrivial case occurs when $\rho$ is defined via proof recursion. Here we must proceed by induction on the number of proof recursions $\delta(\rho)$ in $\rho$. If $\delta(\rho) = 0$ we have no inductive closures and we are done. Otherwise assume that $\hat{\psi}(\vec{Y}, \vec{m}, k)$ is a minimal inductive closure in $\rho$ via the active parameter $k$, i.e. $\psi$ itself does not contain proof recursions. Assume that $\hat{\psi}(\vec{Y}, \vec{m}, k)$ is a schematic $F$-proof of a sequent $S$. We prove by induction that $\hat{\psi}(\vec{Y}, \vec{m}, k)$ evaluates to a derivation in $\mathrm{RPL}_0{}^{\Psi}V$. Under the parameter assignment $\sigma$: $\{n_1 \leftarrow \nu_1, \ldots, n_\alpha \leftarrow \nu_\alpha\}$ we obtain $\hat{\psi}(\vec{Y}, \vec{\mu}, \nu)$ for a vector of numerals $\vec{\mu}$ and a numeral $\nu$; $\hat{\psi}(\vec{Y}, \vec{\mu}, \nu)$ is a derivation of $\sigma(S)$. If $\nu = \bar{0}$ then by definition of inductive closure and by minimality of $\hat{\psi}$ $\hat{\psi}(\vec{Y}, \vec{\mu}, \bar{0})$ is a proof without proof recursion and we can argue as above. If $\nu > \bar{0}$ we have

$$\hat{\psi}(\vec{Y}, \vec{\mu}, \nu) = \hat{\psi}(\vec{Y}, \vec{\mu}, p(\nu))\{V(p(\nu)) \leftarrow \psi(\vec{Y}, \vec{\mu}, \nu)\}.$$

By induction hypothesis $\hat{\psi}(\vec{Y}, \vec{\mu}, p(\nu))$ evaluates to a derivation $\phi \in \mathrm{RPL}_0{}^{\Psi}V$ and so $\hat{\psi}(\vec{Y}, \vec{\mu}, \nu)$ evaluates to

$$\psi' = \phi\{V(p(\nu)) \leftarrow \psi(\vec{Y}, \vec{\mu}, \nu)\}.$$

As $\phi$ and $\psi(\vec{Y}, \vec{\mu}, \nu)$ derivations without proof recursions and thus derivations in $\mathrm{RPL}_0{}^{\Psi}V$, also $\psi'$ is a derivation in $\mathrm{RPL}_0{}^{\Psi}V$. Replacing $\hat{\psi}(\vec{Y}, \vec{\mu}, \nu)$ in $\sigma(\rho)$ by $\psi'$ we obtain a derivation $\rho'$ of $\vdash$ from $\hat{q}(\vec{X}, \nu_1, \ldots, \nu_\alpha)$ with $\delta(\rho') < \delta(\rho)$.

In the theorem below, $\downarrow_2$ is actually not much more than evaluations of sequents directly based on the normalization of sequents.

**Theorem 7.2.1** *Let $\rho$ be a refutation schema of $\hat{q}(\vec{X}, n_1, \ldots, n_\alpha)$. Then, for all numerals $\nu_1, \ldots, \nu_\alpha$, the evaluation $\rho\{n_1 \leftarrow \nu_1, \ldots, n_\alpha \leftarrow \nu_\alpha\}\downarrow_2$ is a $\mathrm{RPL}_0$-refutation of the evaluation of $\hat{q}(\vec{X}, \nu_1, \ldots, \nu_\alpha)$.*

***Proof*** From $\rho$ we first obtain an $\mathrm{RPL}_0{}^{\Psi}V$ refutation of $\hat{q}(\vec{X}, \nu_1, \ldots, \nu_\alpha)$ by Lemma 7.2.1. To obtain an $\mathrm{RPL}_0$ refutation from an $\mathrm{RPL}_0{}^{\Psi}V$ refutation, we have to evaluate all schematic expressions like formula, predicate and term schemata (thus, $\hat{q}(\vec{X}, \nu_1, \ldots, \nu_\alpha)$) under $\sigma$. Link introduction and elimination rules can be omitted.

Our aim is to extract a global schematic substitution from $F$-refutations. Therefore parts of the proof containing no resolution rules do not contribute to the computation of such substitutions. It means that for the computation of substitutions parts of the proof can be cut off, saving the expense of regularizing parts of the proof which do not contribute to compute substitutions.

**Example 7.2.6** Let us consider Example 7.2.8 and $\tau(Y, Z, m, k) =$

$$\cfrac{\cfrac{\vdash \hat{q}(X, Y, Z, n, m)}{\vdash P(\hat{s}(Y, m), Z) \wedge \hat{p}(X, n)}}{\vdash P(\hat{s}(Y, m), Z)}$$

The only relevant part of $\tau(Y, Z, m, k)$ for a computation of a substitution is $\vdash P(\hat{s}(Y, m), Z)$ which is a premise of a resolution rule. We can therefore eliminate $\tau(Y, Z, m, k)$ from the proof ans replace it by $\vdash P(\hat{s}(Y, m), Z)$. We can prune the proof $\rho(X, Y, Z, m, p(k), V_S(X, p(k)))$ to the proof $\rho_r(X, Y, Z, m, p(k), V_S(X, p(k))) =$

$$\cfrac{\cfrac{\cfrac{\cfrac{\cfrac{V_S(X,k)}{\vdash \hat{p}(X,k)}\,V_S E}{\vdash \hat{p}(X, p(k)) \vee \neg P(X(k), \hat{s}(a, k))}}{\vdash \hat{p}(X, p(k)), \neg P(X(k), \hat{s}(a, k))}}{P(X(k), \hat{s}(a, k)) \vdash \hat{p}(X, p(k))} \quad \vdash P(\hat{s}(Y, m), Z)}{\cfrac{\vdash \hat{p}(X, p(k))}{V_S(X, p(k))}\,V_S I}\,R(\Theta(k, m))$$

We cannot prune the derivation $\rho_r(X, Y, Z, m, p(k), V_S(X, p(k)))$ futher because it admits proof recursion and the link variable $V_S(X, k)$ is vital to the recursion (and of course, in computing the recursion, there are resolution rules above $V_S(X, k)$). Also the derivation $\psi(X, m, n)$:

$$\cfrac{\cfrac{\cfrac{\vdash \hat{q}(X, Y, Z, n, m)}{\vdash P(\hat{s}(Y, m), Z) \wedge \hat{p}(X, n)}}{\vdash \hat{p}(X, n)}}{V_S(X, n)}$$

can be reduced to $V_S(X, n)$ because no substitutions take place in $\psi(X, n, m)$. as this is a derivation admitting proof recursion and contains a resolution rule with nonempty substitution. Any derivation which starts in a sequent $\vdash \hat{q}(X, Y, Z, n, m)$ which consists only of definition rules and propositional rules can be ignored to the point where the derived sequent becomes an ancestor sequent $S$ of a resolution rule. The derivation is only necessary to ensure that $S$ is indeed provable from the main sequent $\vdash \hat{q}(X, Y, Z, n, m)$. We are left with the following reduced procedures:

$\rho_r(X, Y, Z, m, p(k), V_S(X, p(k)))$ as defined above, $D(\hat{\rho}_r) =$

$$\{\hat{\rho}_r(X, Y, Z, m, n) =_d \underline{\text{if}}\ n = \bar{0}\ \underline{\text{then}}\ \rho_r(X, Y, Z, m, \bar{0}, V_S(X, \bar{0}))$$
$$\underline{\text{else}}\ \hat{\rho}_r(X, Y, Z, m, p(n))\{V_S(X, n) \leftarrow \rho_r(X, Y, Z, m, n, V_S(X, n))\}.$$

$$D(\rho'_r) = \{\rho'(X, Y, Z, m, n) =_d \underline{\text{if}}\ n = \bar{0}\ \underline{\text{then}}\ V_S(X, \bar{0})$$
$$\underline{\text{else}}\ \hat{\rho}_r(X, Y, Z, m, p(n))\}.$$

$\rho_{0r}(X, Y, Z, n, m) =$

$$\dfrac{\dfrac{\dfrac{\dfrac{\dfrac{\rho'_r(X, Y, Z, m, n)}{V_S(X, \bar{0})}}{\hat{p}(X, \bar{0})}}{\vdash \neg P(X(\bar{0}), \hat{s}(a, \bar{0}))}}{P(X(\bar{0}), f(a, \bar{0})) \vdash} \quad \vdash P(\hat{s}(Y, m), Z)}{\vdash}\ R(\Theta(\bar{0}, m))$$

Note that our pruning of substitution-free parts of the proof is closely related to the midsequent theorem of Gentzen [20]; there a proof of a prenex end-sequent can be divided in two parts, the propositional above and the quantificational below; these two parts are separated by the so-called *midsequent*. This sequent is propositionally valid and contains the instantiations of all quantifiers, the *Herbrand instances*. We call such a sequent *Herbrand sequent*. For a formal definition see Definitions 3.1.7 and 3.1.9.

**Definition 7.2.10** (*substitution-free $F$-derivation*) A $\mathrm{RPL_0}^{\Psi}V$–derivation schema $\varphi$ is called *substitution-free* if no nonground resolution rule occurs in $\varphi$. In particular a derivation schema is substitution-free if it contains no resolutions at all.

**Example 7.2.7** $\tau(Y, Z, m, k)$ in Example 7.2.6 is substitution-free.

**Definition 7.2.11** (*schematic reduced $F$-proof*) Roughly speaking, a *reduced $F$-proof* is a proof with initial sequents $S$ which are derivable from $\vdash F$ without substitutions and are premises of a nonground resolution rule. Below we give a formal definition of a reduction method for $F$-proofs.

- If $\varphi$ is a substitution-free $\mathrm{RPL_0}^{\Psi}V$-proof of $S$ ($V_S$) then $\varphi_r = S$ ($\varphi_r = V_S$), i.e. the proof is replaced by its end-node. If $\varphi$ contains nonground resolution rules then $\varphi_r$ is constructed in a way that all nodes above the premises of resolution rules are eliminated.
- Let $\rho$ be a schematic $F$-proof of the form

$$\dfrac{\begin{matrix}(\rho') & (\rho'') \\ S_1 & S_2\end{matrix}}{S}\ \xi$$

such that $\xi$ is not a nonground resolution rule. We distinguish two cases:

1. $\rho'_r = S_1$ and $\rho''_r = S_2$; then $\rho_r = S$.
2. otherwise $\rho_r =$

$$\frac{\begin{array}{cc}(\rho'_r) & (\rho''_r)\\ S_1 & S_2\end{array}}{S}\ \xi$$

- Let $\rho$ be a schematic $F$-proof of the form

$$\frac{\begin{array}{cc}(\rho') & (\rho'')\\ S_1 & S_2\end{array}}{S}\ R$$

such that $R$ is a nonground resolution rule. Then $\rho_r =$

$$\frac{\begin{array}{cc}(\rho'_r) & (\rho''_r)\\ S_1 & S_2\end{array}}{S}\ R$$

- $\rho =$

$$\frac{\begin{array}{c}(\rho')\\ S'\end{array}}{S}\ \xi$$

for a unary rule $\xi$. We distinguish two cases:

1. $\rho'_r = S'$; then $\rho_r = S$.
2. $\rho'_r \neq S'$; then $\rho_r =$

$$\frac{\begin{array}{c}(\rho'_r)\\ S'\end{array}}{S}\ \xi$$

- Let $\rho$ be of the form

$$\rho'\{V_{S_i}(\vec{X}, \vec{r}_i) \leftarrow \rho^i \mid i \in \{1, \ldots, \alpha\}\}$$

We distinguish two cases

1. $\rho^i_r = V_{S_i}(\vec{X}, \vec{r}_i)$ for all $i \in \{1, \ldots, \alpha\}$. Then $\rho_r = \rho'_r$.
2. otherwise

$$\rho_r = \rho'\{V_{S_i}(\vec{X}, \vec{r}_i) \leftarrow \rho^i_r \mid i \in \{1, \ldots, \alpha\}\}.$$

- Let $V_i(k)$ be an abbreviation for $U_S^i(\vec{X}, \eta_i((n_1, \ldots, n_\alpha, p(k))))$ and $\rho^i(k)$ for

$$\rho(\bar{X}, \eta_i(\vec{n}, p(k)), V_S(\vec{X}, \eta_i(\vec{n}, p(k))).$$

Then let

$$\begin{aligned}\hat{\rho}(\bar{X}, \vec{n}, k)) =_d\ & \underline{\text{if}}\ k = \bar{0}\ \underline{\text{then}}\ \rho(\bar{X}, (\vec{n}, \bar{0}), V_S(\vec{X}, (\vec{n}, \bar{0}))\\ & \underline{\text{else}}\ \hat{\rho}((\bar{X}, \vec{n}, p(k))\{V_i(k) \leftarrow \rho_i(k) \mid i \in \{1, \ldots, m\}\}.\end{aligned}$$

If $\rho(\bar{X}, (\vec{n}, k), V_S(\vec{X}, (\vec{n}, k))$ is substitution-free then

$$\hat{\rho}_r(\bar{X}, \vec{n}, k)) = V_S(\vec{X}, \vec{n}, \bar{0})$$

(the proof collapses). Otherwise

$$\begin{aligned}\hat{\rho}_r(\bar{X}, \vec{n}, k)) =_d\ & \underline{\text{if}}\ k = \bar{0}\ \underline{\text{then}}\ \rho_r(\bar{X}, (\vec{n}, \bar{0}), V_S(\vec{X}, (\vec{n}, \bar{0}))\\ & \underline{\text{else}}\ \hat{\rho}_r((\bar{X}, \vec{n}, p(k))\{V_i(k) \leftarrow \rho_r^i(k) \mid i \in \{1, \ldots, m\}\}.\end{aligned}$$

- Let $\rho(\vec{X}_1, \vec{X}_2, \vec{n}) =$

$$\underline{\text{if}}\ C\ \underline{\text{then}}\ \rho'(\vec{X}_1, \vec{n})\ \underline{\text{else}}\ \rho''(\vec{X}_2, \vec{n}).$$

Then $\rho_r(\vec{X}_1, \vec{X}_2, \vec{n}) =$

$$\underline{\text{if}}\ C\ \underline{\text{then}}\ \rho_r'(\vec{X}_1, \vec{n})\ \underline{\text{else}}\ \rho_r''(\vec{X}_2, \vec{n}).$$

We have seen that by regularization of an $\mathrm{RPL}_0$ refutation we can obtain a so-called *global substitution* of the derivation which turns a resolution refutation into a refutation using a cut rule and collecting all instances of the formula $F$ to be refuted we obtain an unsatisfiable set, the set of *Herbrand instances*. The example below shows that regularization in presence of inductive closures is a more delicate matter.

**Example 7.2.8** Let us consider the reduced schematic refutation described in Example 7.2.6. Consider $\rho_r(X, Y, Z, m, p(k), V_S(X, p(k)))$:

$$\dfrac{\dfrac{\dfrac{\dfrac{\dfrac{\dfrac{V_S(X, k)}{\vdash \hat{p}(X, k)}\ V_S E}{\vdash \hat{p}(X, p(k)) \vee \neg P(X(k), \hat{s}(a, k))}}{\vdash \hat{p}(X, p(k)), \neg P(X(k), \hat{s}(a, k))}}{P(X(k), \hat{s}(a, k)) \vdash \hat{p}(X, p(k))} \qquad \vdash P(\hat{s}(Y, m), Z)}{\vdash \hat{p}(X, p(k))}\ R(\Theta(k, m))}{V_S(X, p(k))}\ V_S I$$

This schematic proof contains one resolution with the $s$-substitution

$$\Theta(k,m)\colon \{X(k) \leftarrow \hat{s}(Y,m), Z \leftarrow \hat{s}(a,k)\}.$$

In the substitution $\Theta(k,m)\colon \{X(k) \leftarrow \hat{s}(Y,m), Z \leftarrow \hat{s}(a,k)\}$ $X$ depends on $k$, while $Y$ and $Z$ do not. This creates a nonregularity in the proof $\hat{\rho}_r$: $D(\hat{\rho}_r) =$

$$\{\hat{\rho}_r(X,Y,Z,m,k) =_d \underline{\text{if}}\ k=\bar{0}\ \underline{\text{then}}\ \rho_r(X,Y,Z,m,\bar{0},V_S(X,\bar{0}))$$
$$\underline{\text{else}}\ \hat{\rho}_r(X,Y,Z,m,p(k))\{V_S(X,k) \leftarrow \rho_r(X,Y,Z,m,k,V_S(X,k))\}.$$

where the schematic proof $\rho_r$ is appended recursively to itself for different values of $k$, so $Z$ would be replaced by various $\hat{s}(a,k)$ for different $k$ which prevents, given a parameter assignment to $m$ and $k$, the construction of a regular $RPL_0$ deduction corresponding to $\hat{\rho}$. As already shown for first-order proofs, a global substitution (and thus a Herbrand substitution) can only be constructed from a regular $RPL_0$-proof (see Definitions 2.2.3 and 3.2.3). Therefore we have to rewrite $\rho_r(X,Y,Z,m,p(k),V_S(X,p(k)))$ to the proof $\rho'_r(X,Y,Z,m,p(k),V_S(X,p(k)))$:

$$\cfrac{\cfrac{\cfrac{\cfrac{\cfrac{V_S(X,k)}{\vdash \hat{p}(X,k)}\,V_SE}{\vdash \hat{p}(X,p(k)) \vee \neg P(X(k),\hat{s}(a,k))}}{\vdash \hat{p}(X,p(k)), \neg P(X(k),\hat{s}(a,k))}}{P(X(k),\hat{s}(a,k)) \vdash \hat{p}(X,p(k))} \quad \vdash P(\hat{s}(Y(k),m),Z(k))}{\cfrac{\vdash \hat{p}(X,p(k))}{V_S(X,p(k))}\,V_SI}\,R(\Theta(k,m))$$

This gives us a new $\Theta(k,m)$ of the form

$$\Theta(k,m)\colon \{X(k) \leftarrow \hat{s}(Y(k),m), Z(k) \leftarrow \hat{s}(a,k)\}$$

By this variable renaming the schematic proof schema $\hat{\rho}_r$ becomes regular in the sense that for any parameter assignment $\sigma$ the proof $\sigma(\hat{\rho}(X,Y,Z,m,k))\downarrow$ is regular in the sense of Definition 2.2.2. Still we have to take account of the substitutions above $\rho(X,Y,Z,m,k,V_S(X,k))$ in $\hat{\rho}$. Therefore we have to start at the top and obtain

$$\Phi(X,Y,Z,k,m) = \{X(k) \leftarrow \hat{s}(Y(k),m),\ Z(k) \leftarrow \hat{s}(a,k)\}.$$

The recursive global substitution for $\hat{\rho}_r(X,Y,Z,m,p(k))$ then is given by $\Phi^*(X,Y,Z,p(k),m)$ where

$$\Phi^*(X,Y,Z,k,m) = \underline{\text{if}}\ k=\bar{0}\ \underline{\text{then}}\ \Phi(X,Y,Z,s(k),m)\ \underline{\text{else}}$$
$$\Phi^*(X,Y,Z,p(k),m) \circ \Phi(X,Y,Z,s(k),m).$$

For the final refutation $\rho_{0r}(X, Y, Z, n, m)$ we have to perform a final resolution which yields the additional substitution

$$\Psi(X, Y, Z, m) = \{X(0) \leftarrow \hat{s}(Y', m),\ Z' \leftarrow \hat{s}(a, 0)\}.$$

Note that we changed $Y, Z$ in $\rho_{0r}$ to $Y', Z'$ as $Y, Z$ got one place variables in the recursion loop and we have to change the name of these variables (we could use $Y(0)$ for $Y'$ and $Z(0)$ for $Z'$ but there is non necessity to do so). Therefore the global substitution schema for $\rho_{0r}$ is

$$\begin{aligned}\Theta^*(X, Y, Z, n, m) =_d\ & \underline{\text{if}}\ n = \bar{0}\ \underline{\text{then}}\ \Psi(X, Y, Z, m) \\ & \underline{\text{else}}\ \Psi(X, Y, Z, m) \cup \Phi^*(X, Y, Z, p(n), m).\end{aligned}$$

Given a parameter assignment $\sigma$ ($\sigma(n) = \bar{\alpha}$ and $\sigma(m) = \bar{\beta}$), $\sigma(\Theta^*(X, Y, Z, n, m))$ applied to $\sigma(\rho_{0r}(X, Y, Z, n, m))\!\downarrow$ results in a refutation of

$$\sigma(\hat{q}(X, Y, Z, n, m)\!\downarrow$$

in which the resolution rules of $\sigma(\rho_{0r}(X, Y, Z, n, m))\!\downarrow$ are replaced by cut rules. The leaves of this refutation are of the form

$$\begin{aligned}&\vdash \sigma(P(\hat{s}(Y(k), m), Z(k))\Theta^*(X, Y, Z, n, m))\!\downarrow\ \text{ for } \sigma(k) \in \{\bar{0}, \ldots, \bar{\alpha}\} \text{ and} \\ &\vdash \sigma(\hat{p}(X, n)\Theta^*(X, Y, Z, n, m))\!\downarrow .\end{aligned}$$

and thus the set of these leaves is propositionally unsatisfiable as they are refuted by propositional rules only; so we obtain a Herbrand set for any parameter assignment $\sigma$.

We don't need to insert $\hat{q}(X, Y, Z, n, m)$ instead of $P(\hat{s}(Y(k), m), Z(k))$ as $P(\hat{s}(Y, m), Z)$ follows from $\hat{q}(X, Y, Z, n, m)$ via a substitution-free derivation and $P(\hat{s}(Y(k), m), Z(k))$ is logically equivalent to $P(\hat{s}(Y, m), Z)$ - when they are considered as universal formulas. Therefore also

$$P(\hat{s}(Y(k), m), Z(k))\Theta^*(n, m)$$

logically follows from $\hat{q}(X, Y, Z, n, m)$. Therefore proof reduction gives us a mean to compactify information in the spirit of the midsequent theorem of Gentzen. For illustration we compute $\sigma(\Theta^*(X, Y, Z, n, m)\!\downarrow$ and $\sigma(\rho_{0r}(X, Y, Z, n, m))\!\downarrow$ for $\sigma(n) = \sigma(m) = \bar{2}$.

We first compute $\sigma(\Theta^*(X, Y, Z, n, m)\!\downarrow$. By

$$\begin{aligned}\Theta^*(X, Y, Z, n, m) =_d\ & \underline{\text{if}}\ n = \bar{0}\ \underline{\text{then}}\ \Psi(X, Y, Z, m) \\ & \underline{\text{else}}\ \Psi(X, Y, Z, m) \cup \Phi^*(X, Y, Z, p(n), m).\end{aligned}$$

We get

$$\begin{aligned}
&\Theta^*(X, Y, Z, \bar{2}, \bar{2})\downarrow = \Psi(X, Y, Z, \bar{2}) \cup \Phi^*(X, Y, Z, \bar{1}, \bar{2}) = \\
&\Psi(X, Y, Z, \bar{2}) \cup (\Phi^*(X, Y, Z, \bar{0}, \bar{2}) \circ \Phi(X, Y, Z, \bar{2}, \bar{2}) = \\
&\Psi(X, Y, Z, \bar{2}) \cup (\Phi(X, Y, Z, \bar{1}, \bar{2}) \circ \Phi(X, Y, Z, \bar{2}, \bar{2})) = \\
&\{X(0) \leftarrow \hat{s}(Y', \bar{2}),\ Z' \leftarrow \hat{s}(a, 0)\} \cup \\
&(\{X(\bar{1}) \leftarrow \hat{s}(Y(\bar{1}), \bar{2}),\ Z(\bar{1}) \leftarrow \hat{s}(a, \bar{1})\} \circ \\
&\{X(\bar{2}) \leftarrow \hat{s}(Y(\bar{2}), \bar{2}),\ Z(\bar{2}) \leftarrow \hat{s}(a, \bar{2})\}) = \\
&\{X(0) \leftarrow \hat{s}(Y', \bar{2}),\ Z' \leftarrow \hat{s}(a, 0), X(\bar{1}) \leftarrow \hat{s}(Y(\bar{1}), \bar{2}),\ Z(\bar{1}) \leftarrow \hat{s}(a, \bar{1}), \\
&X(\bar{2}) \leftarrow \hat{s}(Y(\bar{2}), \bar{2}),\ Z(\bar{2}) \leftarrow \hat{s}(a, \bar{2})\}.
\end{aligned}$$

By the definition

$$D(\hat{s}) = \{\hat{s}(x, n) =_d \underline{\text{if}}\ n = \bar{0}\ \underline{\text{then}}\ x\ \underline{\text{else}}\ h(\hat{s}(x, s(n)))\}$$

we can rewrite $\Theta^*(X, Y, Z, \bar{2}, \bar{2})\downarrow$ to the ordinary first-order substitution

$$\begin{aligned}
&\{X(0) \leftarrow h(h(Y')),\ Z' \leftarrow a, X(\bar{1}) \leftarrow h(h((Y(\bar{1})))),\ Z(\bar{1}) \leftarrow h(a), \\
&X(\bar{2}) \leftarrow h(h((Y(\bar{2})))),\ Z(\bar{2}) \leftarrow h(h(a))\}.
\end{aligned}$$

We now compute $\sigma(\rho_{0r}(X, Y, Z, n, m))\downarrow$. Remember that $D(\hat{\rho}_r) =$

$$\begin{aligned}
&\{\hat{\rho}_r(X, Y, Z, m, n) =_d \underline{\text{if}}\ n = \bar{0}\ \underline{\text{then}}\ \rho_r(X, Y, Z, m, \bar{0}, V_S(X, \bar{0})) \\
&\underline{\text{else}}\ \hat{\rho}_r(X, Y, Z, m, p(n))\{V_S(X, n) \leftarrow \rho_r(X, Y, Z, m, n, V_S(X, n))\}.
\end{aligned}$$

$$\begin{aligned}
D(\rho'_r) = \ &\{\rho'_r(X, Y, Z, m, n) =_d \underline{\text{if}}\ n = \bar{0}\ \underline{\text{then}}\ V_S(X, \bar{0}) \\
&\underline{\text{else}}\ \hat{\rho}(X, Y, Z, m, p(n))\}.
\end{aligned}$$

$\rho_{0r}(X, Y, Z, n, m) =$

$$\cfrac{\cfrac{\cfrac{\cfrac{\begin{array}{c}\rho'_r(X, Y, Z, m, n)\\ V_S(X, \bar{0})\end{array}}{\hat{p}(X, \bar{0})}}{\vdash \neg P(X(\bar{0}), \hat{s}(a, \bar{0}))}}{P(X(\bar{0}), f(a, \bar{0})) \vdash} \quad \vdash P(\hat{s}(Y', m), Z')}{\vdash}\ R(\Theta(\bar{0}, m))$$

We first compute $\sigma(\rho'_r(X, Y, Z, m, n))\downarrow = \rho'_r(X, Y, Z, \bar{2}, \bar{2})\downarrow$. By definition of $\rho'_r$ we obtain

$$\rho'_r(X, Y, Z, \bar{2}, \bar{2})\downarrow = \hat{\rho}(X, Y, Z, \bar{2}, \bar{1})\downarrow .$$

By definition of $\hat{\rho}_r$ we obtain

$$\hat{\rho}(X, Y, Z, \bar{2}, \bar{1})\downarrow = \hat{\rho}_r(X, Y, Z, \bar{2}, \bar{0})\{V_S(X, \bar{1}) \leftarrow \rho_r(X, Y, Z, \bar{2}, \bar{1}, V_S(X, \bar{1}))\} =$$
$$\rho_r(X, Y, Z, \bar{2}, \bar{0}, V_S(X, \bar{0}))\{V_S(X, \bar{1}) \leftarrow \rho_r(X, Y, Z, \bar{2}, \bar{1}, V_S(X, \bar{1}))\}.$$

Therefore $\hat{\rho}_r(X, Y, Z, \bar{2}, \bar{1})$ evaluates to

$$\dfrac{\dfrac{\dfrac{\dfrac{\dfrac{\begin{array}{c}\varphi\\ \vdash \hat{p}(X, \bar{1})\end{array}}{\vdash \hat{p}(X, \bar{0}) \vee \neg P(X(\bar{1}), \hat{s}(a, \bar{1}))}}{\vdash \hat{p}(X, \bar{0}), \neg P(X(\bar{1}), \hat{s}(a, \bar{1}))}}{P(X(\bar{1}), \hat{s}(a, \bar{1})) \vdash \hat{p}(X, \bar{0})} \quad \vdash P(\hat{s}(Y(\bar{1}), \bar{2}), Z(\bar{1}))}{\vdash \hat{p}(X, \bar{0})}\, R(\Theta(\bar{1}, \bar{2}))}{V_S(X, \bar{0})}\, V_S I$$

For $\varphi =$

$$\dfrac{\dfrac{\dfrac{\dfrac{\vdash \hat{p}(X, \bar{2})}{\vdash \hat{p}(X, \bar{1}) \vee \neg P(X(\bar{2}), \hat{s}(a, \bar{2}))}}{\vdash \hat{p}(X, \bar{1}), \neg P(X(\bar{2}), \hat{s}(a, \bar{2}))}}{P(X(\bar{2}), \hat{s}(a, \bar{2})) \vdash \hat{p}(X, \bar{1})} \quad \vdash P(\hat{s}(Y(2), \bar{2}), Z(\bar{2}))}{\vdash \hat{p}(X, \bar{1})}\, R(\Theta(\bar{2}, \bar{2}))$$

Note that we eliminated the uppermost link variable $V_S(X, \bar{2})$ to obtain a derivation from $\hat{p}(X, \bar{2})$ directly. Finally $\rho_{0r}(X, Y, Z, \bar{2}, \bar{2}) =$

$$\dfrac{\dfrac{\dfrac{\dfrac{\begin{array}{c}\hat{\rho}_r(X, Y, Z, \bar{2}, \bar{1})\\ V_S(X, \bar{0})\end{array}}{\hat{p}(X, \bar{0})}}{\vdash \neg P(X(\bar{0}), \hat{s}(a, \bar{0}))}}{P(X(\bar{0}), f(a, \bar{0})) \vdash} \quad \vdash P(\hat{s}(Y', \bar{2}), Z')}{\vdash}\, R(\Theta(\bar{0}, \bar{2}))$$

We see that in all resolution rules the premises are completely variable-disjoint (they are the only binary rules in this proof), thus $\rho_{0r}(X, Y, Z, \bar{2}, \bar{2})$ is regular. It is easy to verify that the for the set of leaves $S$ and the for instance

$$S\Theta^*(X, Y, Z, \bar{2}, \bar{2})\downarrow = \{\vdash P(h(h(Y')), a), \vdash P(h(h((Y(\bar{1})))), h(a)),$$
$$\vdash P(h(h((Y(\bar{2})))), h(h(a))), \vdash \hat{p}(X, \bar{2})\Theta^*(X, Y, Z, \bar{2}, \bar{2})\downarrow\}$$

where

$$\vdash \hat{p}(X, \bar{2})\Theta^*(X, Y, Z, \bar{2}, \bar{2})\downarrow =$$
$$\vdash \neg P(h(h((Y(\bar{2})))), h(h(a))) \vee \neg P(h(h((Y(\bar{1})))), h(a)) \vee \neg P(h(h(Y')), a)$$

the set $(S\Theta^*(X, Y, Z, \bar{2}, \bar{2}))\downarrow$ is unsatisfiable and

$$(\rho_{0r}(X, Y, Z, \bar{2}, \bar{2})\Theta^*(X, Y, Z, \bar{2}, \bar{2}))\downarrow$$

is a cut-refutation of $(S\Theta^*(X, Y, Z, \bar{2}, \bar{2}))\downarrow$ in the sense of Definition 3.2.3.

Below, we will give a formal definition of a global substitution schema of a schematic $F$-refutation. The first step consists in transforming a schematic $F$-refutation into a regular one. To this aim we need a slight extension of the concept of schematic $F$-proofs which serves only the purpose to obtain regularization and–finally–the global substitution schema. We need some technical notions to define the extension of $F$-proofs.

**Definition 7.2.12** ($V$*-renamings*) A schematic substitution is a *simple* $V$*-renaming* if it is of the form

$$\{U_i(\vec{t_i}) \leftarrow U_i'(\vec{t_i}) \mid i = 1, \ldots, \alpha\}$$

where $U_i$ and $U_i'$ are of appropriate types, for $i \neq j$ $U_i \neq U_j$, and

$$\{U_1, \ldots, U_\alpha\} \cap \{U_1', \ldots, U_\alpha'\} = \emptyset.$$

Now let the $U_i$ for $i = 1, \ldots, \alpha$ be of type $\omega^\alpha \to \iota$ (and for $i \neq j$ $U_i \neq U_j$) and the $U_i'$ for $i = 1, \ldots, \alpha$ be of type $\omega^{\alpha+1} \to \iota$ (and for $i \neq j$ $U_i' \neq U_j'$) . Let $k$ be a parameter not occurring in the set of term tuples $\{\vec{t_i} \mid i = 1, \ldots, \alpha\}$. Then the schematic substitution

$$\{U_i(\vec{t_i}) \leftarrow U_i'(\vec{t_i}, k) \mid i = 1, \ldots, \alpha\}$$

is called an *extended* $V$*-renaming*.

**Definition 7.2.13** We consider reduced proof schemata and extend them by unary rules which perform simple $V$-renaming. The rule we add to schematic $\mathrm{RPL_0}^\Psi V$-derivations is of the form

$$\frac{S}{S\Theta}\ s$$

where $\Theta$ is a simple $V$-renaming. Proof schemata containing this substitution rule are called *extended proof schemata*.

Note that the addition of the substitution rule is just a technical tool for computing the global substitution of a refutation. This rule is not part of the specification of a derivation.

**Definition 7.2.14** A reduced extended $\text{RPL}_0{}^{\Psi}V$-derivation schema is called *regular* if

- $\rho$ is of the form $S$ for a schematic sequent $S$.
- $\rho$ is of the form

$$\frac{\begin{array}{c}(\rho')\\ S'\end{array}}{S}\ \xi$$

  and $\rho'$ is regular. If $\rho'$ is also an extended schematic $F$-proof then $\rho$ is a regular extended $F$-proof .
- $\rho$ is of the form

$$\frac{\begin{array}{cc}(\rho_1) & (\rho_2)\\ \Gamma_1 \vdash \Delta_1 & \Gamma_2 \vdash \Delta_2\end{array}}{\Gamma_1', \Gamma_2' \vdash \Delta_1', \Delta_2'}\ \xi$$

  such that $\rho_1$ and $\rho_2$ are regular and $\rho_1, \rho_2$ are variable-disjoint.
- Let $\rho'$ be an extended $\text{RPL}_0{}^{\Psi}V$-derivation schema of a schematic sequent $S$ (or of a link expression $V_S$) such that the link expression of the root differs from all link expressions in the leaves. Let us assume that $V_{S_i}(\vec{X}_i, \vec{r}_i)$ for $i = 1, \ldots, \alpha$ are all the the link expressions occurring at the leaves of $\rho'$ and every leaf which is not a link expression is of the form $\hat{q}(\vec{Y}, \vec{n})$. Let us assume that there are extended $\text{RPL}_0{}^{\Psi}V$-derivation schema $\rho_i$ of $V_{S_i}(\vec{X}_i, \vec{r}_i)$ such that

$$\rho = \rho'\{V_{S_i}(\vec{X}, \vec{r}_i) \leftarrow \rho_i \mid i \in \{1, \ldots, \alpha\}\}$$

  Then $\rho$ is regular if $\rho'$ is regular, all the $\rho_i$ are regular and pairwise variable disjoint.
- $\rho$ is of the form

$$\begin{array}{l}\rho(\bar{X}, \vec{n}, k)) =_d \underline{\text{if}}\ k = \bar{0}\ \underline{\text{then}}\ \rho'(\bar{X}, (\vec{n}, \bar{0}), V_S(\vec{X}, (\vec{n}, \bar{0}))\\ \underline{\text{else}}\ \rho((\bar{X}, \vec{n}, p(k))(\{V_i(k) \leftarrow \rho_i'(k) \mid i \in \{1, \ldots, m\}\}.\end{array}$$

  as in Definition 7.2.5 and $\rho'$ is regular, the $\rho_i'(k)$ are regular, pairwise variable disjoint and all the variable expressions occurring in $\rho_i'(k)$ contain the parameter $k$.

**Example 7.2.9** The modified proof schema (after extending the arity of variables) in Example 7.2.8 is regular.

**Proposition 7.2.2** *Let $\rho$ be a schematic extended reduced $F$-proof of a schematic sequent $S$. Then there exists a schematic extended reduced $F$-proof $\rho^R$ of $S$ which is regular.*

***Proof*** We proceed by induction on the defining proof operations.

- Let $\rho = \vdash F$ then $\rho^R = \rho$.
- Let $\rho =$

$$\cfrac{\begin{array}{c}(\rho_1)\\ S'\end{array}}{S}\ \xi$$

where $\xi$ is a unary inference and $\rho_1$ is an extended reduced $F$-proof then $\rho^R =$

$$\cfrac{\begin{array}{c}(\rho_1^R)\\ S'\end{array}}{S}\ \xi$$

Note that $\xi$ can be also a $V$-renaming.
- Let $\rho =$

$$\cfrac{\begin{array}{cc}(\rho_1) & (\rho_2)\\ S_1 & S_2\end{array}}{S}\ \xi$$

where $\xi$ is a binary rule different from resolution. Assume that $\rho_1^R, \rho_2^R$ are the regularized versions of $\rho_1, \rho_2$. If $\rho_1^R$ and $\rho_2^R$ are variable-disjoint then $\rho^R =$

$$\cfrac{\begin{array}{cc}(\rho_1^R) & (\rho_2^R)\\ S_1 & S_2\end{array}}{S}\ \xi$$

Otherwise, let $\{V_1, \ldots, V_m\}$ be the set of variables which occur in both $\rho_1^R, \rho_2^R$. Let

$$U_1^1, \ldots, U_m^1,\ U_1^2, \ldots, U_m^2$$

be different variables which neither occur in $\rho_1^R$ nor in $\rho_2^R$ such that the types of $U_i^1, U_i^2$ coincide with the type of $V_i$. Let

$$\begin{aligned}\Theta_1 &= \{V_1 \leftarrow U_1^1, \ldots, V_m \leftarrow U_m^1\},\\ \Theta_2 &= \{V_1 \leftarrow U_1^2, \ldots, V_m \leftarrow U_m^2\},\\ \Phi &= \bigcup_{i=1}^{m}\{U_i^1 \leftarrow V_i, U_i^2 \leftarrow V_i\}.\end{aligned}$$

Then $\rho^R =$

$$\cfrac{\cfrac{\begin{array}{cc}(\rho_1^R\Theta_1) & (\rho_2^R\Theta_2)\\ S_1\Theta_1 & S_2\Theta_2\end{array}}{S'}\ \xi}{S}\ \Phi$$

- $\rho =$

$$\dfrac{\begin{array}{cc}(\rho_1) & (\rho_2)\\ \Gamma \vdash \Delta, A_1, \ldots, A_\alpha & B_1, \ldots, B_\beta, \Pi \vdash \Lambda\end{array}}{\Gamma\Theta, \Pi\Theta \vdash \Delta\Theta, \Lambda\Theta}\ res\{\Theta\}$$

As the rule is a resolution rule the sets of variables $\mathcal{V}_1$ in $\{A_1, \ldots, A_\alpha\}$ and $\mathcal{V}_2$ in $\{B_1, \ldots B_\beta\}$ are disjoint. However, $\Gamma$, $\Delta$ and $\Pi$, $\Lambda$ may have common variables not occurring in $\mathcal{V}_1 \cup \mathcal{V}_2$. Let $\{V_1, \ldots, V_m\}$ be the set of these variables. The substitutions $\Theta_1, \Theta_2$ and $\Phi$ are defined exactly as above. But here we have to emphasize that the set of variables in the domain and range of $\Theta$ are disjoint to the set $\{V_1, \ldots, V_m\}$ when we compute $\Theta$ via the algorithm UAL in Sect. 6.2. So $\rho^R =$

$$\dfrac{\dfrac{\begin{array}{cc}(\rho_1^R\Theta_1) & (\rho_2^R\Theta_2)\\ \Gamma\Theta_1 \vdash \Delta\Theta_1, A_1, \ldots, A_\alpha & B_1, \ldots, B_\beta, \Pi\Theta_2 \vdash \Lambda\Theta_2\end{array}}{\Gamma\Theta_1 \circ \Theta, \Pi\Theta_2 \circ \Theta \vdash \Delta\Theta_1 \circ \Theta, \Lambda\Theta_1 \circ \Theta}\ res\{\Theta\}}{\Gamma\Theta, \Pi\Theta \vdash \Delta\Theta, \Lambda\Theta}\ \Phi$$

- Let

$$\rho = \rho_0\{V_{S_i}(\vec{X}, \vec{r}_i) \leftarrow \rho_i \mid i \in \{1, \ldots, \alpha\}\}$$

We first regularize $\rho_0$ to $\rho_0^R$ such that the leaves $V_{S_i}(\vec{X}, \vec{r}_i)$ are renamed to $V_{S_i}(\vec{X}_j, \vec{r}_i\{\vec{X} \to \vec{X}_j)$ for variable tuples $\vec{X}_i$ for $i = 1, \ldots, \alpha$ of the same type as $\vec{X}$; moreover the variables in $\vec{X}_i$ and $\vec{X}_j$ are disjoint for $i \neq j$. Let

$$\Theta_j = \{\vec{X} \leftarrow \vec{X}_j\}.$$

Then we first obtain

$$\rho' = \rho_0^R\{V_{S_i}(\vec{X}, \vec{r}_i)\Theta_i \leftarrow \rho_i\Theta_i \mid i \in \{1, \ldots, \alpha\}\}.$$

The end-sequent of $\rho'$ can still be different from that of $\rho$, but $\rho^R$ can be obtained by appropriate renamings of variables in $\rho'$.

- Let $V_i(k)$ be an abbreviation for

$$U_S^i(\vec{X}, \eta_i((n_1, \ldots, n_\alpha, p(k))))$$

and $\rho_i(k)$ for

$$\rho(\bar{X}, \eta_i(\vec{n}, p(k)), V_S(\vec{X}, \eta_i(\vec{n}, p(k))).$$

Then let

$$\begin{array}{l}\hat{\rho}(\bar{X}, \vec{n}, k)) =_d \underline{\text{if}}\ k = \bar{0}\ \underline{\text{then}}\ \rho(\bar{X}, (\vec{n}, \bar{0}), V_S(\vec{X}, (\vec{n}, \bar{0}))\\ \underline{\text{else}}\ \hat{\rho}((\bar{X}, \vec{n}, p(k))\{V_i(k) \leftarrow \rho_i(k) \mid i \in \{1, \ldots, m\}\}.\end{array}$$

We first regularize the proof $\rho(\bar{X}, (\vec{n}, k), V_S(\vec{X}, (\vec{n}, k))$ by the regularization steps defined above (note that $\rho$ may also contain proof substitutions) and obtain $\rho^R(\bar{X}, (\vec{n}, k), V_S(\vec{X}, (\vec{n}, k)))$. Then we apply an extended $V$-renaming by replacing every variable expression $U(t_1, \ldots, t_m)$ in $\rho^R$ by an expression $U'(t_1, \ldots, t_m, k)$. We denote the resulting proof schema by $\rho^{RR}(\bar{X}, (\vec{n}, k), V_S(\vec{X}, (\vec{n}, k)))$. We apply the same procedure to the proofs $\rho_i(k)$ and obtain proof schemata $\rho_i^{RR}(k)$ (here we take care that the schemata $\rho_i^{RR}(k)$ are pairwise variable disjoint). Then we define

$$\begin{aligned}\hat{\rho}^R(\bar{X}, \vec{n}, k)) =_d\ & \underline{\text{if}}\ k = \bar{0}\ \underline{\text{then}}\ \rho^{RR}(\bar{X}, (\vec{n}, \bar{0}), V_S(\vec{X}, (\vec{n}, \bar{0}))\\ & \underline{\text{else}}\ \hat{\rho}^r((\bar{X}, \vec{n}, p(k))\{V_i(k) \leftarrow \rho_i^{RR}(k) \mid i \in \{1, \ldots, m\}\}.\end{aligned}$$

**Definition 7.2.15** (*global substitution schema*) Let $\rho$ be an extended regular reduced $F$-proof of a sequent $S$. We define the *global substitution schema* for $\rho$.

- Let $S$ be a leaf sequent in $\rho$ be the extended $F$-proof then $\Theta(S) = \emptyset$ (which is the identical substitution).
- $\rho =$

$$\cfrac{\begin{array}{c}\rho'\\ S'\end{array}}{S}\ \xi$$

  where $\xi$ is a unary rule different from a $V$-renaming rule. Then $\Theta(\rho) = \Theta(\rho')$.
- $\rho =$

$$\cfrac{\begin{array}{c}\rho'\\ S'\end{array}}{S}\ \Theta_0$$

  where $\Theta_0$ is a $V$-renaming. Then $\Theta(\rho) = \Theta(\rho') \circ \Theta_0$
- Let $\rho$ be an extended regular $F$-proof of the form

$$\cfrac{\begin{array}{cc}(\rho_1) & (\rho_2)\\ S_1 & S_2\end{array}}{S}\ \xi$$

  for a binary rule $\xi$ which is not the resolution rule. Then

$$\Theta(\rho) = \Theta(\rho_1) \cup \Theta(\rho_2).$$

- Let $\rho$ be an extended regular $F$-proof of the form

$$\cfrac{\begin{array}{cc}(\rho_1) & (\rho_2)\\ \Gamma \vdash \Delta, A_1, \ldots, A_\alpha & B_1, \ldots, B_\beta, \Pi \vdash \Lambda\end{array}}{\Gamma\Theta', \Pi\Theta' \vdash \Delta\Theta', \Lambda\Theta'}\ res\{\Theta'\}$$

  Then

$$\Theta(\rho) = (\Theta(\rho_1) \cup \Theta(\rho_2)) \circ \Theta'.$$

- Let $\rho'$ be a regular extended $\mathrm{RPL_0}^\Psi V$-derivation schema of a schematic sequent $S$ (or of a link expression $V_S$) such that the link expression of the root differs from all link expressions in the leaves. Let us assume that $V_{S_i}(\vec{X}_i, \vec{r}_i)$ for $i = 1, \ldots, \alpha$ are all the the link expressions occurring at the leaves of $\rho'$ and every leaf which is not a link expression is of the form $\vdash F$. Let $\rho_i$ be regular extended $F$-proofs $\rho_i$ of $V_{S_i}(\vec{X}_i, \vec{r}_i)$ such that

$$\rho\colon \rho'\{V_{S_i}(\vec{X}, \vec{r}_i) \leftarrow \rho_i \mid i \in \{1, \ldots, \alpha\}\}$$

is a regular extended $F$-proof. Then

$$\Theta(\rho) = \Theta(\rho)(\Theta(\rho_1) \cup \cdots \cup \Theta(\rho_\alpha)).$$

- Let $\rho$ be a regular extended $\mathrm{RPL_0}^\Psi V$-derivation schema (an inductive closure) of the form

$$\begin{aligned} \rho(\bar{X}, \vec{n}, k)) =_d\ & \underline{\text{if}}\ k = \bar{0}\ \underline{\text{then}}\ \rho'(\bar{X}, (\vec{n}, \bar{0}), V_S(\vec{X}, (\vec{n}, \bar{0})) \\ & \underline{\text{else}}\ \rho((\bar{X}, \vec{n}, p(k))\{V_i(k) \leftarrow \rho'_i(k) \mid i \in \{1, \ldots, \alpha\}\}. \end{aligned}$$

Then $\Theta(\rho(\bar{X}, \vec{n}, k)) =_d$

$$\begin{aligned} & \underline{\text{if}}\ k = \bar{0}\ \underline{\text{then}}\ \Theta(\rho'(\bar{X}, (\vec{n}, \bar{0}), V_S(\vec{X}, (\vec{n}, \bar{0})))) \\ & \underline{\text{else}}\ \Theta(\rho((\bar{X}, \vec{n}, p(k))) \circ (\Theta(\rho'_1(k)) \cup \cdots \cup \Theta(\rho'_\alpha(k))). \end{aligned}$$

- Let $\rho$ be an extended regular $F$-proof of the form

$$\underline{\text{if}}\ C\ \underline{\text{then}}\ \rho_1(\vec{X}_1, \vec{n})\ \underline{\text{else}}\ \rho_2(\vec{X}_2, \vec{n}).$$

Then

$$\Theta(\rho) =_d\ \underline{\text{if}}\ C\ \underline{\text{then}}\ \Theta(\rho_1(\vec{X}_1, \vec{n}))\ \underline{\text{else}}\ \Theta(\rho_2(\vec{X}_2, \vec{n})).$$

If $\rho$ is schematic regular, extended and reduced $F$-refutation then $\Theta(\rho)$ is called the *Herbrand system* of $\rho$.

**Example 7.2.10** Let $\Theta^*(n, m)$ be the substitution schema defined in Example 7.2.8. Then $\Theta^*(n, m)$ is a Herbrand system for the reduced refutation $\rho_{0r}$ of $\hat{q}(X, Y, X, n, m)$ defined in Example 7.2.6.

## 7.3 Variants and Extensions of Proof Schemata

In this section we compare our definitions of proof schemata with that from earlier publications (e.g. [12, 31, 32]) and define some new variants which will turn out useful in the comparison of schemata and proofs in theories of arithmetic. We fist address input schemata.

### 7.3.1 *Variants of* LK*-Schemata*

We will first-define the variant of *condition-free schemata* (or cf-schemata) which do not make use of if-then-else and have no explicit case distinctions (a property shared by schemata as defined in [12, 31, 32] ) but otherwise use the same formalism of link variables introduced in Definition 7.1.8. In the second step we will simulate the input schemata in [12, 31, 32] by our schemata as defined in Definition 7.1.15.

Recall the definition of proof recursion in **LK**-proofs (Definitions 7.1.11 and 7.1.12), first this of recursion sequents:

Let $S(\vec{X}, \vec{n}, k)$, $S^*(\vec{X}, \vec{n}, k)$ be schemata of sequents of the form

$$\begin{aligned} S(\vec{X}, \vec{n}, k) &= F(\vec{X}, \vec{n}, p(k)), \Gamma \vdash \Delta, F(\vec{X}, \vec{n}, k) \text{ for } k > \bar{0} \text{ and} \\ S^*(\vec{X}, \vec{n}, k) &= \Gamma \vdash \Delta, F(\vec{X}, \vec{n}, k) \text{ for all } k. \end{aligned}$$

where $F(\vec{X}, \vec{n}, k)$ is a schematic formula in $\vec{X}, \vec{n}, k$ and the (schematic) formulas in $\Gamma$, $\Delta$ do not contain the parameter $k$. Sequents of this form are called *recursion sequents*. Then any $\mathbf{LK}_{T^+}$ derivation $\rho(\vec{X}, \vec{n}, k)$ for $k > 0$ of the form

$$\dfrac{\dfrac{\dfrac{V_{S^*}(\vec{n}, p(k))}{S^*(\vec{X}, \vec{n}, p(k))}\, V_S E \qquad \dfrac{\rho'(\vec{X}, \vec{n}, k)}{S(\vec{X}, \vec{n}, k)}}{\Gamma \vdash \Delta, F(\vec{X}, \vec{n}, k)}\, cut + c^*}{V_{S^*}(\vec{X}, \vec{n}, k)}\, V_S I$$

and with the cut formula $F(\vec{X}, \vec{n}, p(k))$ (where $\rho'(\vec{X}, \vec{n}, k)$ is an $\mathbf{LK}_{T^+}$ derivation of $S(\vec{X}, \vec{n}, k)$) is said to *admit proof recursion based on* $S(\vec{X}, \vec{n}, k)$ *and* $S^*(\vec{X}, \vec{n}, k)$. The link expression $V_{S^*}(\vec{n}, p(k))$ is called the *recursion link expression.*

Second, proof recursion itself:
Let $S(\vec{X}, \vec{n}, k)$ be a recursion sequent and $\rho(\vec{X}, \vec{n}, k)$ for $k > 0$ be an $\mathbf{LK}_{T^+}$ derivation admitting proof recursion based on $S(\vec{X}, \vec{n}, k)$ and $S^*(\vec{X}, \vec{n}, k)$. We abbreviate $V_{S^*}(\vec{X}, \vec{n}, k)$ by $V(k)$ and define

$$\begin{aligned} D(\hat{\rho}) = \{\hat{\rho}(\vec{X}, \vec{n}, k) =_d\ & \underline{\text{if}}\ k = \bar{0}\ \underline{\text{then}}\ \rho_0 \\ & \underline{\text{else}}\ \rho(\vec{X}, \vec{n}, k)\{V(p(k)) \leftarrow \hat{\rho}(\vec{X}, \vec{n}, p(k))\}\}. \end{aligned}$$

where $\rho_0$ is an $\mathbf{LK}_{T^+}$ derivation of $V_{S^*}(\vec{X}, \vec{n}, \bar{0})$. We say that $\hat{\rho}$ is the *inductive closure* of $\rho$, or $\hat{\rho}$ is *obtained via proof recursion* from $\rho$ based on $S(\vec{X}, \vec{n}, k)$ and $S^*(\vec{X}, \vec{n}, k)$.

We will now modify the definitions in replacing the condition

$$\underline{\text{if}}\ k = \bar{0}\ \underline{\text{then}}\ \ldots\ \underline{\text{else}}$$

by an implicit case distinction between the terms **0** and $k + 1$ for a parameter $k$. To this aim we have to replace $p(k)$ by $k$ in the definitions below. We first give a new definition of recursion sequents:

**Definition 7.3.1** (*admission of cf-proof recursion*) Let $S(\vec{X}, \vec{n}, k)$ and $S^*(\vec{X}, \vec{n}, k)$ be schemata of sequents of the form

$$\begin{aligned} S(\vec{X}, \vec{n}, k) = &\ F(\vec{X}, \vec{n}, k), \Gamma \vdash \Delta, F(\vec{X}, \vec{n}, s(k)) \text{ and} \\ S^*(\vec{X}, \vec{n}, k) = &\ \Gamma \vdash \Delta, F(\vec{X}, \vec{n}, k) \text{ for all } k. \end{aligned}$$

where $F(\vec{X}, \vec{n}, k)$ is a schematic formula in $\vec{X}, \vec{n}, k$ and the (schematic) formulas in $\Gamma, \Delta$ do not contain the parameter $k$. Sequents of this form are called *cf-recursion sequents*. Then any $\mathbf{LK}_{T^+}$ derivation $\rho(\vec{X}, \vec{n}, s(k))$ of the form

$$\dfrac{\dfrac{\dfrac{V_{S^*}(\vec{n}, k)}{S^*(\vec{X}, \vec{n}, k)}\, V_S E \quad \dfrac{\rho'(\vec{X}, \vec{n}, k)}{S(\vec{X}, \vec{n}, k)}}{\Gamma \vdash \Delta, F(\vec{X}, \vec{n}, s(k))}\, cut + c^*}{V_{S^*}(\vec{X}, \vec{n}, s(k))}\, V_S I$$

and with the cut formula $F(\vec{X}, \vec{n}, k)$ (where $\rho'(\vec{X}, \vec{n}, k)$ is an $\mathbf{LK}_{T^+}$ derivation of $S(\vec{X}, \vec{n}, k)$) is said to *admit proof recursion based on* $S(\vec{X}, \vec{n}, k)$ and $S^*(\vec{X}, \vec{n}, k)$. The link expression $V_{S^*}(\vec{n}, k)$ is called the *cf-recursion link expression.*

Now the new definition of proof recursion:

**Definition 7.3.2** (*cf-proof recursion*) Let $S(\vec{X}, \vec{n}, k)$ be a recursion sequent and $\rho(\vec{X}, \vec{n}, k)$ be an $\mathbf{LK}_{T^+}$ derivation admitting cf-proof recursion based on sequents $S(\vec{X}, \vec{n}, k)$ and $S^*(\vec{X}, \vec{n}, k)$ . We abbreviate $V_{S^*}(\vec{X}, \vec{n}, k)$ by $V(k)$ and define

$$\begin{aligned} \{\check{\rho}(\vec{X}, \vec{n}, \bar{0}) =_d &\ \rho_0, \\ \check{\rho}(\vec{X}, \vec{n}, s(k)) =_d &\ \rho(\vec{X}, \vec{n}, s(k))\{V(k) \leftarrow \check{\rho}(\vec{X}, \vec{n}, k)\}\}. \end{aligned}$$

where $\rho_0$ is an $\mathbf{LK}_{T^+}$ derivation of $V_{S^*}(\vec{X}, \vec{n}, \bar{0})$. We say that $\check{\rho}$ is the *cf-inductive closure* of $\rho$, or $\hat{\rho}$ is *obtained via cf-proof recursion* from $\rho$ based on $S(\vec{X}, \vec{n}, k)$ and $S^*(\vec{X}, \vec{n}, k)$ .

Compared to Definition 7.1.12 the new Definition 7.3.2 the if then else-construction is replaced by a set of two defining equations. For any parameter assignment $\sigma$ we either have $\sigma(k) = \mathbf{0}$ or $\sigma(k) > \mathbf{0}$. In the first case the first equation applies and, in the second, the second one (it is clear that the defining equations exclude each other). It is easy to show that proof-recursion and cf-proof recursion are equivalent. As the type of proof objects are the same in both definitions we may apply the same evaluation relation$\downarrow_{T^+}$ relative to an $\omega\iota o$-theory $T^+$ to both of them.

**Proposition 7.3.1** *For all parameter assignments* $\sigma$ *we have*

$$\begin{aligned} \sigma(\check{\rho}(\vec{X}, \vec{n}, \mathbf{0}))\downarrow_{T^+} = &\ \sigma(\hat{\rho}(\vec{X}, \vec{n}, \mathbf{0}))\downarrow_{T^+}, \\ \sigma(\check{\rho}(\vec{X}, \vec{n}, s(k)))\downarrow_{T^+} = &\ \sigma(\hat{\rho}(\vec{X}, \vec{n}, s(k)))\downarrow_{T^+} . \end{aligned}$$

***Proof*** The first equation trivially holds directly by definition. For the second we have (as $s(k){\downarrow} \neq \mathbf{0}$)

$$\begin{aligned}\breve{\rho}(\vec{X}, \vec{n}, s(k)) &= \rho(\vec{X}, \vec{n}, s(k))\{V(k) \leftarrow \breve{\rho}(\vec{X}, \vec{n}, k)\} \text{ and} \\ \hat{\rho}(\vec{X}, \vec{n}, s(k)) &= \rho(\vec{X}, \vec{n}, s(k))\{V(p(s(k))) \leftarrow \hat{\rho}(\vec{X}, \vec{n}, p(s(k)))\}.\end{aligned}$$

As $p(s(k)){\downarrow} = k{\downarrow}$ we get immediately

$$\breve{\rho}(\vec{X}, \vec{n}, s(k)) = \hat{\rho}(\vec{X}, \vec{n}, s(k)).$$

We are now ready to define the general concept of $\mathbf{LK}_{T^+}$-cf-proof schemata, which differ from $\mathbf{LK}_{T^+}$-proof schemata only by the definition of proof recursion. cf-semi-proof schemata are defined analogously to semi-proof schemata and we omit the definition which is straightforward.

**Definition 7.3.3** ($\mathbf{LK}_{T^+}$-*cf-proof schema*) $\mathbf{LK}_{T^+}$-cf-proof schemata are defined inductively: let $A \in \mathcal{A}$ be a schematic sequent in a schematic axiom set $\mathcal{A}$. Then

$$A$$

is an $\mathbf{LK}_{T^+}$-cf-proof schema of $A$ w.r.t $\mathcal{A}$. If $\mathcal{A}$ is the schematic standard axiom set $\mathcal{A}_S$, the above proof schema is an $\mathbf{LK}_{T^+}$-cf-proof schema of $A$ w.r.t $\mathcal{A}_S$, or simply an $\mathbf{LK}_{T^+}$-proof schema.[2]

- If $\rho$ is a cf-proof schema of a schematic sequent $S(\vec{X}, \vec{n}, k)$ then

$$\frac{(\rho)}{V_S(\vec{X}, \vec{n}, k)}\ V_I$$

  is a cf-proof schema of $V_S(\vec{X}, \vec{n}, k)$.
- If $\rho_1$ and $\rho_2$ are $\mathbf{LK}_{T^+}$-cf-proof schemata of $S_1$ and $S_2$ and $\rho =$

$$\frac{\begin{matrix}(\rho_1) & (\rho_2) \\ S_1 & S_2\end{matrix}}{S}\ \xi$$

  for a binary rule $\xi$ then $\rho$ is an $\mathbf{LK}_{T^+}$-cf-proof schema of $S$.
- If $\rho'$ is an $\mathbf{LK}_{T^+}$-cf-proof schema of $S'$ and $\rho =$

$$\begin{matrix}(\rho') \\ \dfrac{S'}{S}\ \xi\end{matrix}$$

[2] From now on we will refer to an $\mathbf{LK}_{T^+}$-cf-proof schema w.r.t $\mathcal{A}$ as an $\mathbf{LK}_{T^+}$-proof schema and assume that the schematic axiom set is clear from the context.

for a unary rule $\xi$ then $\rho$ is an $\mathbf{LK}_{T^+}$-cf-proof schema of $S$.

- Let $\rho$ be an $\mathbf{LK}_{T^+}$-derivation schema of a schematic sequent $S$ (of a link expression $V_S(\vec{r})$) such that all leaves of $\rho$ which are not link expressions are axioms. Let $V_i(\vec{r_i})$ for $i = 1, \ldots, \alpha$ be the link expressions on the leaves of $\rho$ (we assume that all these link expressions are different from $V_S(\vec{r})$ and are different from each other) and $\rho_i$ for $i = 1, \ldots, \alpha$ be $\mathbf{LK}_{T^+}$-cf-proof-schemata of $V_i(\vec{r_i})$. Then $\rho\{V_i(\vec{r_i}) \leftarrow \rho_i \mid i \in \{1, \ldots, \alpha\}\}$ is an $\mathbf{LK}_{T^+}$-cf-proof schema of $S$ (of $V_S(\vec{r})$) .
- Let $S(\vec{X}, \vec{n}, k)$ and $S^*(\vec{X}, \vec{n}, k)$ be cf-recursion sequents and

$$\rho(\vec{X}, \vec{n}, V_S(\vec{X}, \vec{n}, k))$$

for be $\mathbf{LK}_{T^+}$ derivation schema admitting cf-proof recursion based on $S$ and $S^*$. Furthermore assume that $\rho(\vec{X}, \vec{n}, V_S(\vec{X}, \vec{n}, k))$ is a semi-cf-proof schema and

$$D(\check{\rho}) = \{\check{\rho}(\vec{X}, \vec{n}, \mathbf{0}) =_d \rho_0,\\ \check{\rho}(\vec{X}, \vec{n}, s(k)) =_d \rho(\vec{X}, \vec{n}, s(k))\{V(k) \leftarrow \check{\rho}(\vec{X}, \vec{n}, k)\}\}.$$

Assume that $\rho_0$ is an $\mathbf{LK}_{T^+}$-cf-proof schema of $V(0)$ then $\hat{\rho}(\vec{n}, k)$ is an $\mathbf{LK}_{T^+}$-cf-proof schema of $V(k)$.
Note that, as a consequence, $\hat{\rho}(\vec{n}, V(k))\{V(k) \leftarrow S^*(k)\}$ is a cf-proof schema of $S^*(k)$.

**Theorem 7.3.1** *cf-proof schemata and proof-schemata are equivalent.*

***Proof*** Immediate by Proposition 7.3.1.

The **LK**-proof schemata as defined in [12, 31, 32] have no operation of inductive closure but use recursion implicitly in the form of calls. Instead of link variables and proof substitution proof symbols are used for calling other proofs or the proof itself. As an illustration we show the formalization of Example 7.1.1.

**Example 7.3.1** Remember that we had a proof schema in $\mathbf{LK}_{T^+}$ of

$$S(n) = \forall x(P(x) \to P(f(x))) \vdash \forall x(P(x) \to P(\hat{f}(x, n))),$$

where $T^+ = (\emptyset, T^*, \emptyset)$ and

$$D(\hat{f}) = \{\hat{f}(x, n) =_d \underline{\text{if}}\ n = \bar{0}\ \underline{\text{then}}\ x\ \underline{\text{else}}\ f(\hat{f}(x, p(n)))\}.$$

Note, that in [12, 31, 32] $D(\hat{f})$ was defined as follows:

$$D(\hat{f}) = \{\hat{f}(x, \mathbf{0}) = x,\ \ \hat{f}(x, n+1) = f(\hat{f}(x, n))\}.$$

To formalize the proof schema in Example 7.1.1 we use a proof symbol $\delta$ and define

$$S(\delta, n) = \forall x(P(x) \to P(f(x))) \vdash \forall x(P(x) \to P(\hat{f}(x, n)))$$

for the end-sequent corresponding to the proof schema with symbol $\delta$. The proof schema itself is defined as

$$\mathcal{D}_1 = \{(\delta, \rho(\delta, \mathbf{0}), \rho(\delta, n+1))\},$$

where $\rho(\delta, \mathbf{0})$ defines the base case of the recursion and $\rho(\delta, n+1))$ the step case. We define $\rho(\delta, 0)$ as follows:

$$\dfrac{\dfrac{\dfrac{\dfrac{P(\hat{f}(a,\mathbf{0})) \vdash P(\hat{f}(a,\mathbf{0}))}{P(a) \vdash P(\hat{f}(a,\mathbf{0}))}\,T^{+}}{\vdash P(a) \to P(\hat{f}(a,\mathbf{0}))}\,\to_r}{\vdash \forall x(P(x) \to P(\hat{f}(x,\mathbf{0})))}\,\forall : r}{\forall x(P(x) \to P(f(x))) \vdash \forall x(P(x) \to P(\hat{f}(x,\mathbf{0})))}\,w_l$$

$\rho(\delta, n+1)$ as:

$$\dfrac{(\delta, n, \emptyset) \colon S(\delta, n) \qquad (1)}{\forall x(P(x) \to P(f(x))) \vdash \forall x(P(x) \to P(\hat{f}(x, n+1)))}\,cut, c_l$$

Here, $(\delta, n, \emptyset) \colon S(\delta, n)$ expresses that in $\rho(\delta, n+1)$ the proof $\rho(\delta, n)$ is called and the parameter substitution on $n$ is the identity (and thus $\emptyset$). The proof (1) is defined as

$$\dfrac{\dfrac{\dfrac{\dfrac{P(a) \vdash P(a) \qquad \dfrac{\dfrac{P(\hat{f}(a,n)) \vdash P(\hat{f}(a,n)) \qquad \dfrac{P(\hat{f}(a,n+1)) \vdash P(\hat{f}(a,n+1))}{P(f(\hat{f}(a,n))) \vdash P(\hat{f}(a,n+1))}\,T^{+}}{P(\hat{f}(a,n)), P(\hat{f}(a,n)) \to P(f(\hat{f}(a,n))) \vdash P(\hat{f}(a,n+1))}\,\to_l}{P(\hat{f}(a,n)), \forall x(P(x) \to P(f(x))) \vdash P(\hat{f}(a,n+1))}\,\forall_l}{P(a), P(a) \to P(\hat{f}(a,n)), \forall x(P(x) \to P(f(x))) \vdash P(\hat{f}(a,n+1))}\,\to_l}{P(a) \to P(\hat{f}(a,n)), \forall x(P(x) \to P(f(x))) \vdash P(a) \to P(\hat{f}(a,n+1))}\,\to_r}{\forall x(P(x) \to P(\hat{f}(x,n))), \forall x(P(x) \to P(f(x))) \vdash P(a) \to P(\hat{f}(a,n+1))}\,\forall_l}{\forall x(P(x) \to P(\hat{f}(x,n))), \forall x(P(x) \to P(f(x))) \vdash \forall x(P(x) \to P(\hat{f}(x,n+1)))}\,\forall_r$$

We give now a translation of the schema above to a $\mathbf{LK}_{T^{+}}$-cf-proof schema of $S$. We first define, instead of $\rho(\delta, \mathbf{0})$, the proof $\rho' =$

$$\dfrac{\dfrac{\dfrac{\dfrac{P(\hat{f}(a,0)) \vdash P(\hat{f}(a,0))}{P(a) \vdash P(\hat{f}(a,0))}\,T^{+}}{\vdash P(a) \to P(\hat{f}(a,0))}\,\to_r}{\vdash \forall x(P(x) \to P(\hat{f}(x,0)))}\,\forall : r}{\dfrac{S(0)}{V_S(0)}\,V_S I}\,w_l$$

and instead of $\rho(\delta, n+1)$ $\breve{\rho}(s(n))$ where

$$\{\breve{\rho}(\bar{0}) =_d \rho',\ \breve{\rho}(s(n)) =_d \rho(n, V_S(n))\{V_S(n) \leftarrow \breve{\rho}(n)\}.$$

Note that in cf-proof schemata $\breve{\rho}$ takes the role of $\rho$ above as here recursion is expressed explicitly. Here $\rho$ is of the form $\rho(n, V_S(n)) =$

$$\dfrac{\dfrac{\dfrac{V_S(n)}{S(n)}\,V_S E \qquad \dfrac{\dfrac{\dfrac{\dfrac{\dfrac{P(a)\vdash P(a) \qquad \dfrac{\dfrac{P(\hat{f}(a,n))\vdash P(\hat{f}(a,n)) \qquad \dfrac{B\vdash B}{P(f(\hat{f}(a,n)))\vdash B}\,T^{+}}{P(\hat{f}(a,n)), P(\hat{f}(a,n))\to P(f(\hat{f}(a,n)))\vdash B}\,\to_l}{P(\hat{f}(a,n)), A\vdash B}\,\forall_l}{P(a), P(a)\to P(\hat{f}(a,n)), A\vdash B}\,\to_l}{P(a)\to P(\hat{f}(a,n)), A\vdash P(a)\to B}\,\to_r}{\forall x(P(x)\to P(\hat{f}(x,n))), A\vdash P(a)\to B}\,\forall_l}{\forall x(P(x)\to P(\hat{f}(a,n))), A\vdash \forall x(P(x)\to P(\hat{f}(x,s(n))))}\,\forall_r}{S(s(n))}\,cut, c_l}{V_S(s(n))}\,V_S I$$

for $A = \forall x(P(x) \to P(f(x)))$, and $B = P(\hat{f}(a, s(n)))$.

We now give the formal definition of a proof schema in the older formalism:

**Definition 7.3.4** (*schematic deduction and proof schema old*) Let $\mathcal{D}$ be the tuple $(\delta_0, \Delta^*, \Pi)$. $\mathcal{D}$ is called a schematic deduction from a finite set of schematic sequents $\mathcal{S}$ if the following conditions are fulfilled:

- $\Delta^*$ is a finite subset of $\Delta$.
- $\delta_0 \in \Delta^*$, and $\delta_0 > \delta'$ for all $\delta' \in \Delta^*$ such that $\delta' \neq \delta_0$. $\delta_0$ is called the main symbol.
- To every $\delta \in \Delta^*$ we assign a parameter tuple $\vec{n}_\delta$ of pairwise different parameters (called the passive parameters), and a parameter $m_\delta$ (called the active parameter).
- $\Pi$ is a set of pairs $\{(\Pi(\delta, \vec{n}_\delta, m_\delta), S(\delta, \vec{n}_\delta, m_\delta)\}$, where $S(\delta, \vec{n}_\delta, m_\delta)$ is a schematic sequent, and
$$\Pi(\delta, \vec{n}_\delta, m_\delta) = \{(\delta, \vec{n}_\delta, m_\delta) \to \rho(\delta, \vec{n}_\delta, m_\delta)\},$$
where $\rho(\delta, \vec{n}_\delta, 0) = \rho_0(\delta, \vec{n}_\delta)$, and $\rho(\delta, \vec{n}_\delta, s(m_\delta)) = \rho_1(\delta, \vec{n}_\delta, m_\delta)$, and there exists a (possibly empty) finite set of schematic sequents $\mathcal{C}(\delta)$ such that
  1. $\rho_0(\delta, \vec{n}_\delta)$ is a deduction of $S(\delta, \vec{n}_\delta, 0)$ from $\mathcal{S} \cup \mathcal{C}(\delta)$,
  2. $\rho_1(\delta, \vec{n}_\delta, m_\delta)$ is a deduction of $S(\delta, \vec{n}_\delta, m_\delta + 1)$ from
  $$\{(\delta, \Psi) \colon S(\delta, \vec{n}_\delta, m_\delta)\} \cup \mathcal{S} \cup \mathcal{C}(\delta),$$
  where $(\delta, \Psi)$ is a label, and $\Psi$ the empty parameter replacement,
  3. for all $S' \in \mathcal{C}(\delta)$, $S' = (\delta', \Psi) \colon S(\delta', \vec{n}_{\delta'}, m_{\delta'})\Psi$ where $(\delta', \Psi)$ is a label, $\delta' \in \Delta^*$ with $\delta > \delta'$ and $\Psi$ is a parameter replacement on $(\vec{n}_{\delta'}, m_{\delta'})$ w.r.t. $(\vec{n}_\delta, m_\delta)$ such that the conditions 1. and 2. hold for $\delta'$.

If $\mathcal{S} = \mathcal{A}_s$ we call $\mathcal{D}$ a *proof schema* of $S(\delta_0, \vec{n}_{\delta_0}, m_{\delta_0})$.

The ordering of the symbols in $\Delta^*$ reflects the ordinary definition order. In a definition of of a base case $\rho_0(\delta, \vec{n}_\delta)$ only $\delta'$ can be used which has been defined before. Only for a step case $\rho_1(\delta, \vec{n}_\delta, m_\delta + 1)$, a call of $\rho_1(\delta, \vec{n}_\delta, m_\delta)$ is admitted which is reflected by the labeled proof leaf $(\delta, \Psi)\colon S(\delta, \vec{n}_\delta, m_\delta)$. In the formalism introduced in Sect. 7.1 we have no set of labels and the ordering is the implicit definition ordering: only in case of inductive closure $\hat{\rho}(X, n)$ may use $\hat{\rho}(X, p(n))$ or, in the new cf-formalism, $\check{\rho}(X, s(n))$ may use $\check{\rho}(X, n)$. As in the old formalism for proof schemata there is no expression for the inductive closure, a simulation of the old formalism by cf-schemata has to take into account the semantics of the old proof schemata.

**Definition 7.3.5** (*evaluation of proof schema old*) Consider a proof schema $\mathcal{D} = (\delta_0, \Delta^*, \Pi)$, and $\sigma$ a parameter assignment. In defining the evaluation of the proof schema, $\mathcal{D}\sigma{\downarrow}$, we proceed by double induction on the ordering of proof symbols and the assignments $\sigma$.

- Let $\delta_i$ be a minimal element in $\Delta^*$.
  1. $\sigma(m_{\delta_i}) = 0$.
     Then, by definition of a proof schema, $\rho_0(\delta_i, \vec{n}_{\delta_i})$ is a proof with **LK**-inferences and inferences for defined symbols that contain schematic sequents. Let $S_1, \ldots S_n$ be all the schematic sequents in $\rho_0(\delta_i, \vec{n}_{\delta_i})$. Then the evaluation of $\rho_0(\delta_i, \vec{n}_{\delta_i})$ under $\sigma$ is denoted by $\rho_0(\delta_i, \vec{n}_{\delta_i}){\downarrow}$ and obtained by replacing all $S_1, \ldots, S_n$ in $\rho_0(\delta_i, \vec{n}_{\delta_i})$ by $\sigma(S_1){\downarrow}, \ldots, \sigma(S_n){\downarrow}$ and omitting the inferences for defined symbols.
  2. $\sigma(m_{\delta_i}) = \alpha > 0$.
     Evaluate all schematic sequents except the leaves

     $$(\delta_i, \emptyset)\colon S(\delta_i, \vec{n}_{\delta_i}, m_{\delta_i})$$

     under $\sigma$. Let $\sigma[m_{\delta_i}/\alpha - 1]$ be defined as $\sigma[m_{\delta_i}/\alpha - 1](p) = \sigma(p)$ for all $p \neq m_{\delta_i}$ and $\sigma[m_{\delta_i}/\alpha - 1](m_{\delta_i}) = \alpha - 1$. Then we replace the labeled sequent $(\delta_i, \emptyset)\colon S(\delta_i, \vec{n}_{\delta_i}, m_{\delta_i})$ by the proofs $\rho_0(\delta_i, \vec{n}_{\delta_i})\sigma[m_{\delta_i}/\alpha - 1]{\downarrow}$ if $\alpha - 1 = 0$ and by $\rho_1(\delta_i, \vec{n}_{\delta_i}, m_{\delta_i})\sigma[m_{\delta_i}/\alpha - 1]{\downarrow}$ if $\alpha - 1 > 0$. The result is an **LK**-proof.
- $\delta_i \in \Delta^*$ is not minimal.
  1. $\sigma(m_{\delta_i}) = 0$. Evaluate all schematic sequents except the labeled sequents of the form $(\delta', \Psi)\colon S(\delta', \vec{n}_{\delta'}, m_{\delta'})\Psi$ for $\delta_i > \delta'$ and the corresponding parameter replacement $\Psi$ under $\sigma$. Then replace the labeled sequent $(\delta', \Psi)\colon S(\delta', \vec{n}_{\delta'}, m_{\delta'})\Psi$ by the proof $\rho_0(\delta', \vec{n}_{\delta'})\Psi\sigma{\downarrow}$ if $\sigma(m_{\delta'}) = 0$ and by the proof $\rho_1(\delta', \vec{n}_{\delta'}, m_{\delta'})\Psi\sigma{\downarrow}$ otherwise.
  2. $\sigma(m_{\delta_i}) = \alpha > 0$. As above, except for the labeled sequents $(\delta_i, \emptyset)\colon S(\delta_i, \vec{n}_{\delta_i}, m_{\delta_i})$ which are replaced by the proof
     $\rho_0(\delta_i, \vec{n}_{\delta_i})\sigma[m_{\delta_i}/\alpha - 1]$ if $\alpha - 1 = 0$ and by the proof
     $\rho_1(\delta_i, \vec{n}_{\delta_i}, m_{\delta_i})\sigma[m_{\delta_i}/\alpha - 1]$ otherwise.

$\mathcal{D}\sigma\downarrow$ is defined as $\rho_0(\delta_0, \vec{n}_{\delta_0})\sigma\downarrow$ for the $<$-maximal symbol $\delta_0$ if $\sigma(m_{\delta_0}) = 0$, and by $\rho_1(\delta_0, \vec{n}_{\delta_0}, m_{\delta_0})\sigma\downarrow$ if $\sigma(m_{\delta_0}) > 0$.

It remains to prove that cf-schemata simulate the old proof schemata defined above.

**Theorem 7.3.2** *For any old proof schema there exists an equivalent cf-schema.*

***Proof*** We construct a transformation $T_{cf}$ of an old proof schema to a a cf-proof schema.

Consider a minimal symbol $\delta \in \Delta^*$ and the schematic proofs

- $\rho(\delta, \vec{n}, \mathbf{0})\colon \rho_0(\delta, \vec{n}_\delta)$ and
- $\rho(\delta, \vec{n}, m_\delta + 1)\colon \rho_1(\delta, \vec{n}_\delta, m_\delta)$.

As $\delta$ is minimal the leaves in $\rho_0(\delta, \vec{n}_\delta)$ only consist of the axiom sequents $\mathcal{S}$. Assume that $S(\mathbf{0})\colon S(\vec{n}, \mathbf{0})$ is the end-sequent of $\rho_0$. Then

$$T_{cf}(\rho_0(\delta, \vec{n}_\delta)) = \rho'$$

where $\rho'$ differs from $\rho_0$ only by the last inference, i.e. $\rho' =$

$$\cfrac{(\rho_0)}{\cfrac{S}{V_S}\ V_I}$$

Now consider $\rho_1(m_\delta)\colon \rho_1(\delta, \vec{n}, m_\delta)$, which is a derivation of

- $S(m_\delta + 1)\colon S(\vec{n}, m_\delta + 1)$ from $\mathcal{S} \cup \{(\delta, \emptyset)\colon S(m_\delta)\}$.

We may assume w.l.o.g. that $S(m_\delta)$ is of the form

$$\Gamma \vdash \Delta, F(m_\delta)$$

for a formula $F(m_\delta)$ containing the parameter $m_\delta$ and a context $\Gamma$, $\Delta$ not containing $m_\delta$. In fact every sequent can be transformed into such a form by using propositional inferences and cut (and can be retransformed into the original form). Now consider the leaf $(\delta, \emptyset)\colon \Gamma \vdash \Delta, F(m_\delta)$ in $\rho_1(m_\delta)$. We replace it by an axiom leaf $F(m_\delta), \Gamma \vdash \Delta, F(m_\delta)$ and simulate the whole proof $\rho_1(m_\delta)$ with the same inferences. Note that no eigenvariable condition can be violated as the end-sequent in $\rho_1(m_\delta)$ contains all the variables which are also contained in $F(m_\delta)$ (eigenvariables in a proof can never be contained in the end-sequent). Using possibly some contractions we obtain a proof $\rho_1'(m_\delta)$ of the end-sequent

$$S_1(m_\delta)\colon F(m_\delta), \Gamma \vdash \Delta, F(m_\delta + 1).$$

from axiom sequents only. Now let $\rho_2(m_\delta) =$

$$\cfrac{\cfrac{(\delta,\emptyset)\colon \Gamma \vdash \Delta, F(m_\gamma) \quad \rho_1'(m_\delta)}{\Gamma, \Gamma \vdash \Delta, \Delta, F(m_\delta + 1)}\ cut}{\Gamma \vdash \Delta, F(m_\delta + 1)}\ c^*$$

Now it is easy to see that the sequents $S_1(m_\delta)$ and $\Gamma \vdash \Delta, F(m_\delta + 1)$ are cf-recursion sequents and and the proof $\rho_2'(m_\delta)$:

$$\cfrac{\cfrac{\cfrac{\cfrac{V_S(m_\delta)}{\Gamma \vdash \Delta, F(m_\gamma)}\ V_E \quad \rho_1'(m_\delta)}{\Gamma, \Gamma \vdash \Delta, \Delta, F(m_\delta + 1)}\ cut}{\Gamma \vdash \Delta, F(m_\delta + 1)}\ c^*}{V_S(m_\delta + 1)}\ V_I$$

a cf-schema admitting proof recursion (see Definition 7.3.1). Therefore we translate $\rho(\delta, \vec{n}_\delta, m_\delta)$ into the cf-inductive closure

$$\begin{aligned} D(\check{\rho}) = \ & \{\check{\rho}(\vec{n}_\delta, \mathbf{0}) =_d \rho', \\ & \check{\rho}(\vec{n}_\delta, s(m_\delta + 1)) = \rho_2'(\vec{n}_\delta, m_\delta)\{V_S(m_\delta) \leftarrow \check{\rho}(\vec{n}_\delta, m_\delta)\}, \end{aligned}$$

i.e.

$$T_{cf}(\rho(\delta, \vec{n}_\delta, m_\delta)) = \check{\rho}(\vec{n}_\delta, m_\delta).$$

By the definition of the semantics we obtain

$$\sigma(\rho(\delta, \vec{n}, m_\delta){\downarrow} = \sigma(\check{\rho}(\vec{n}, m_\delta)){\downarrow}\ .$$

This concludes the case for minimal $\delta$.

Now consider a nonminimal $\delta \in \Delta^*$ and assume as (IH) that for all $\delta' < \delta$ $T_{cf}(\rho(\delta', \vec{n}_{\delta'}, m_\delta')$ is already defined. Then we have

- $\rho_0(\delta, \vec{n}_\delta)$ is a deduction of $S(\delta, \vec{n}_\delta, 0)$ from $\mathcal{S} \cup \mathcal{C}(\delta)$,
- $\rho_1(\delta, \vec{n}_\delta, m_\delta)$ is a deduction of $S(\delta, \vec{n}_\delta, m_\delta + 1)$ from $\{(\delta, \Psi)\colon S(\delta, \vec{n}_\delta, m_\delta)\} \cup \mathcal{S} \cup \mathcal{C}(\delta)$, where $(\delta, \Psi)$ is a label, and $\Psi$ the empty parameter replacement,
- for all $S' \in \mathcal{C}(\delta)$, $S' = (\delta', \Psi)\colon S(\delta', \vec{n}_{\delta'}, m_{\delta'})\Psi$ where $(\delta', \Psi)$ is a label, $\delta' \in \Delta^*$ with $\delta > \delta'$ and $\Psi$ is a parameter replacement on $(\vec{n}_{\delta'}, m_{\delta'})$ w.r.t. $(\vec{n}_\delta, m_\delta)$ such that the conditions 1. and 2. hold for $\delta'$ .

Like in the minimal case we tranform $\rho_0$ by appending a link variable introduction to the leaf of $\rho_0$. Any leaf which is not an axiom is of the form

$$L\colon (\delta', \Psi)\colon S(\delta', \vec{n}_{\delta'}, m_{\delta'})\Psi$$

for a parameter substitution $\Psi$ and a $\delta' < \delta$. Assume that

- $\Psi(\vec{n}_{\delta'}) = \vec{s}$ and $\Psi(m_{\delta'}) = t$.

By (IH)

$$T_{cf}(\rho(\delta', \vec{n}_{\delta'}, m'_{\delta}) = \breve{\rho}_{\delta'}(\vec{n}_{\delta'}, m'_{\delta}) \text{ for some } \breve{\rho}_{\delta'}.$$

We then replace $L$ by the proof schema $\breve{\rho}_{\delta'}(\vec{s}, t)$. We perform this transformation for any nonaxiomatic leaf and obtain a proof schema $T_{cf}(\rho_0(\delta, \vec{n}_\delta))$.

For transforming $\rho_1(\delta, \vec{n}_\delta, m_\delta)$ we proceed for all leaves of the form $L$ above in the same way as for $\rho_0(\delta, \vec{n}_\delta)$ we still may have the leaf

$$L'\colon (\delta, \emptyset)\colon S(\delta, \vec{n}_\delta, m_\delta).$$

We apply the same proof transformation as for a minimal $\delta$ and finally obtain a cf-schema

- $\rho'_\delta(\vec{n}_\delta, m_\delta)$

which admits cf-proof recursion. We then define (by writing $V_S(m_\delta)$ for $V_S(\vec{n}_\delta, m_\delta)$)

$$\begin{aligned} D(\breve{\rho}_\delta) = \; & \{\breve{\rho}_\delta(\vec{n}_\delta, \mathbf{0}) = T_{cf}(\rho_0(\delta, \vec{n}_\delta)), \\ & \breve{\rho}_\delta(\vec{n}_\delta, m_\delta + 1) = \rho'_\delta(\vec{n}_\delta, m_\delta)\{V_S(m_\delta) \leftarrow \breve{\rho}_\delta(\vec{n}_\delta, m_\delta)\}\}. \end{aligned}$$

and

$$T_{cf}(\rho(\delta, \vec{n}_\delta, m_\delta)) = \breve{\rho}_\delta(\vec{n}_\delta, m_\delta).$$

### 7.3.2 *Variants of Refutation Schemata*

The first formal definition of refutation schema with many parameters can be found in [32]; basically the definition was the same as for input schemata, with the difference that axioms were replaced by sequents of the form $\vdash F$ where $F$ is the schematic formula to be refuted. But to use induction variants for the refutations additional definitions were required and it became clear that refutations of characteristic formula schemata are more complex than the input schemata from which they were extracted. This is also clearly visible when we compare Definitions 7.1.12 and 7.2.5 in this chapter. Indeed, our refutation schemata in Definition 7.2.9 are more expressive than those in [32]. As an alternative to proof schemata in Sect. 7.2 we present the approach defined in [13].

In [13] schematic $\mathrm{RPL}_0{}^\Psi$ derivations where, besides the formulas to be refuted, new kinds of axioms (in the form of labeled sequents) are included. These labeled sequents serve the purpose to establish recursive call structures in the proof. For constructing schematic $\mathrm{RPL}_0{}^\Psi$ derivations we introduce a countably infinite set $\Delta$ of proof symbols which are used to label the individual proofs of a proof schema. A particular proof schema uses a finite set of proof symbols $\Delta^* \subset \Delta$. Like for point transition systems we assign an arity $A(\delta)$ to every $\delta \in \Delta^*$; $A(\delta)$ is just the arity of the input parameters for the proof labeled by $\delta$. Also, we need a concept of *proof labels* which serve the purpose to relate some leafs of the proof tree to recursive calls.

**Definition 7.3.6** (*proof label*) Let $\delta \in \Delta$ and $\vartheta$ be a parameter substitution. Then the pair $(\delta, \vartheta)$ is called a proof label.

We will need to locate specific occurrences of sequents in a proof $\pi$; to this aim we define $\Lambda_\pi$ as the set of all sequent occurrences in $\pi$. By $|\pi|_\lambda$ we denote the sequent occurring at position $\lambda$ for $\lambda \in \Lambda_\pi$. Furthermore, let $\rho$ be a proof with the end-sequent $|\pi|_\lambda$; then by $\pi[\rho]_\lambda$ we denote the proof which results from replacing in $\pi$ the derivation at $\lambda$ by $\rho$.

**Definition 7.3.7** (*labeled sequents and derivations*) Let $S$ be a sequent and $(\delta, \vartheta)$ a proof label, them $(\delta, \vartheta)\colon S$ is a *labeled sequent*. A *labeled* $\mathrm{RPL_0}^\Psi$ derivation is an $\mathrm{RPL_0}^\Psi$ derivation $\pi$ where all leaves are labeled. We distinguish *axiom labels* and *nonaxiom labels*. By $\mathrm{Axiom}(\pi)$ we denote the set of occurrences of leaves labeled by axiom labels, the set of the occurences of leaves with nonaxiom labels is denoted by $\mathrm{Naxiom}(\pi)$. The set of all occurrences of leaves in $\pi$ is denoted by $\mathrm{leaves}(\pi)$; so $\mathrm{leaves}(\pi) = \mathrm{Axiom}(\pi) \cup \mathrm{Naxiom}(\pi)$ and $\mathrm{Axiom}(\pi) \cap \mathrm{Naxiom}(\pi) = \emptyset$.

We will use point transition systems as defined in Chap. 5 as basis for defining proof schemata which we will l-schemata (where l stands for *label*). The following example should serve the purpose to reveal the intuition behind the concept.

**Example 7.3.2** Recall Example 7.2.1 where we have defined a refutation schema for the $\omega\iota o$-theory $T^+\colon (T, T', T'')$ where

$$T = (\hat{p}, \{\hat{p}, \hat{q}\}, D(\{\hat{p}, \hat{q}\}), V, <),$$

$$T' = (\hat{s}, \{\hat{s}\}, D(\{\hat{s}\}), \emptyset)$$

where $X\colon \omega \to \iota$, $Y, Z\colon \iota$ and

$$\begin{aligned} D(\hat{p}) = {} & \{\hat{p}(X, n) =_d \underline{\text{if}}\ n = \bar{0}\ \underline{\text{then}}\ \neg P(X(\bar{0}), \hat{s}(a, \bar{0})) \\ & \underline{\text{else}}\ \hat{p}(X, p(n)) \vee \neg P(X(n), \hat{s}(a, n))\}, \\ D(\hat{q}) = {} & \{\hat{q}(X, Y, Z, n, m) =_d P(\hat{s}(Y, m), Z) \wedge \hat{p}(X, n)\}, \\ D(\hat{s}) = {} & \{\hat{s}(x, n) =_d \underline{\text{if}}\ n = \bar{0}\ \underline{\text{then}}\ x\ \underline{\text{else}}\ h(\hat{s}(x, p(n)))\}. \end{aligned}$$

We create a proof symbol $\delta_0$ for the refutation of $\hat{q}(X, Y, Z, n, m)$. For this refutation we consider the partition $n = 0$ (base case) and $n > 0$ (step case). In order to refute $\hat{q}(X, Y, Z, n, m)$ for $n > 0$ we will need an auxiliary deduction which deduces $\hat{q}(X, Y, Z, 0, m)$ from the axioms $\hat{q}(X, Y, Z, n, m)$. In defining this derivation, which we give the label $\delta_1$, we will need additional parameters $k, l$ which do not occur in the definition of $\hat{q}$ but are needed to control the recursions. For every proof symbol we have a parameter tuple, it is $(n, m)$ for $\delta_0$ and $(n, m, k, l)$ for $\delta_1$. For $\delta_0$ we define the proof schema via the reduction system $\Pi(\delta_0, X, Y, Z, n, m)$ and the specification of the end-sequent $S(\delta_0, X, Y, n, m) = \vdash$. We define

$$\begin{aligned}\Pi(\delta_0, X, Y, Z, n, m) = \; & \{(\delta_0, X, Y, Z, n, m) \to \rho_0(\delta_0, X, Y, Z, n, m) \colon n = 0,\\ & (\delta_0, X, Y, Z, n, m) \to \rho_1(\delta_0, X, Y, Z, n, m) \colon n > 0\}.\end{aligned}$$

which means that for $n = 0$ we select the proof

$$\rho_0(\delta_0, X, Y, Z, n, m)$$

for $n > 0$ the proof

$$\rho_1(\delta_0, X, Z, n, m).$$

By $\rho_0(\delta_0, X, Y, Z, n, m)$ we denote the following derivation:

$$\dfrac{\dfrac{\dfrac{\vdash (\delta_1', (0, m)) \colon \hat{q}(X, Y, Z, 0, m)}{\vdash p(\hat{s}(Y, m), Z) \wedge \hat{p}(X, 0)}\, B\hat{q}r}{\vdash p(\hat{s}(Y, m), Z)}\, \wedge_{r_1} \qquad \dfrac{\dfrac{\dfrac{\dfrac{\vdash (\delta_1', (0, m)) \colon \hat{q}(X, Y, Z, 0, m)}{\vdash P(\hat{s}(Y, m), Z) \wedge \hat{P}(X, 0)}\, B\hat{q}r}{\vdash \hat{p}(X, 0)}\, \wedge_{r_2}}{\vdash \neg P(X(0), \hat{s}(a, 0))}\, B\hat{p}r}{P(X(0), \hat{s}(a, 0)) \vdash}\, \neg : l}{\vdash}\, res(\sigma_1)$$

where $\sigma_1 = \{X(0) \leftarrow \hat{s}(Y, m), Z \leftarrow \hat{s}(a, 0)\}$. Obviously, $\rho_0(\delta_0, X, Y, Z, 0, m)$ is a $\mathrm{RPL_0}^{\Psi}$ refutation of $\hat{q}(X, Y, Z, 0, m)$. The label $(\delta_1', \emptyset)$ means that $\delta_1'$ is an axiom label for the case $(0, m)$. By $\rho_1(\delta_0, X, Y, Z, n, m)$ we denote the following derivation:

$$\dfrac{\dfrac{\dfrac{\begin{array}{c}(\delta_1, (n, m, n, 0))\\ \vdash \hat{q}(X, Y, Z, 0, m)\end{array}}{\vdash p(\hat{s}(Y, m), Z) \wedge \hat{p}(X, 0)}\, B\hat{Q}r}{\vdash p(\hat{s}(Y, m), Z)}\, \wedge_{r_1} \qquad \dfrac{\dfrac{\dfrac{\dfrac{\begin{array}{c}(\delta_1, (n, m, n, 0))\\ \vdash \hat{q}(X, Y, Z, 0, m)\end{array}}{\vdash P(\hat{s}(Y, m), Z) \wedge \hat{p}(X, 0)}\, B\hat{Q}r}{\vdash \hat{p}(X, 0)}\, \wedge_{r_2}}{\vdash \neg P(X(0), \hat{s}(a, 0))}\, B\hat{P}r}{P(X(0), \hat{s}(a, 0)) \vdash}\, \neg : l}{\vdash}\, res\sigma_1$$

In contrast to $\rho_0(\delta_0, X, Y, Z, n, m)$ $\hat{q}(X, Y, Y, 0, m)$ is not an axiom in the proof but rather the end-sequent. The label $(\delta_1, (n, m, n, 0))$ represents a call of the proof $\rho(\delta_1, X, Y, m, n, k, l)$ where $k$ is replaced by $n$ and $l$ by 0. For the proof symbol $\delta_1$ we select the partition

$$\{n = 0,\ n > 0 \wedge l < n,\ n > 0 \wedge l \geq n\},$$

define a reduction system $\Pi(\delta_1, X, Y, Z, m, n, k, l)$ and a sequent

$$S(\delta_1, X, Y, Z, n, m, k, l) = \hat{q}(X, Y, Z, l, m).$$

We define

$$\Pi(\delta_1, X, Y, Z, n, m, k, l) =$$
$$\{(\delta_1, X, Y, Z, n, m, k, l) \to \rho_0(\delta_1, X, Y, Z, n, m, k, l) \colon n = 0,$$
$$(\delta_1, X, Y, Z, n, m, k, l) \to \rho_1(\delta_1, X, Y, Z, n, m, k, l) \colon n > 0 \wedge l < n,$$
$$(\delta_1, X, Y, Z, n, m, k, l) \to \rho_2(\delta_1, X, Y, Z, n, m, k, l) \colon n > 0 \wedge l \geq n\}.$$

where

$$\rho_0(\delta_1, X, Y, Z, n, m, k, l) = (\delta_2', (n, m, k, l)) \colon \hat{q}(X, Y, Z, n, m),$$
$$\rho_2(\delta_1, X, Y, n, m, k, l) = (\delta_3', (n, m, k, l)) \colon \hat{q}(X, Y, Z, n, m).$$

$$(\delta_2', (n, m, k, l)) \text{ and } (\delta_3', (n, m, k, l))$$

are axiom labels, here the computation of the proof stops. The most involved proof is $\rho_1(\delta_1, X, Y, Z, m, n, k, l)$ which performs the real induction. It derives the end sequent $\vdash \hat{q}(X, Y, Z, l, m)$ from the sequent $(\delta_1, (m, n, p(k), s(l)))\colon \hat{q}(X, Y, Z, s(l), m)$. If $l < n$ then $(m, n, p(k), s(l))$ represents a further call of $\rho_1(\delta_1, X, Y, Z, m, n, k, l)$. By this call the leaf $\hat{q}(X, Y, Z, s(l), m)$ is replaced by a derivation of $\hat{q}(X, Y, Z, s(l), m)$ with leaves $\hat{Q}(X, Y, Z, ss(l), m)$ and so on. If the counter reaches $l = n$ the computation stops and we have reached the axioms $\hat{q}(X, Y, Z, n, m)$. Note that, in the whole proof system, we start with $l = 0$ and $k = n$ via the call from $\delta_0$. That is, via the initiation given by the call from $\delta_0$, we obtain a derivation of $\vdash \hat{q}(X, Y, Z, 0, m)$ from axioms $\vdash \hat{q}(X, Y, Z, n, m)$. Putting $\delta_0$ and $\delta_1$ together where $\delta_0$ is the main symbol we eventually obtain a refutation of $\vdash \hat{q}(X, Y, Z, n, m)$.

Below we define $\rho_1(\delta_1, X, Y, Z, m, n, k, l)$. Let $\tau_2 =$

$$\dfrac{\dfrac{\begin{array}{c}(\delta_1, (n, m, p(k), s(l)))\\ \vdash \hat{Q}(X, Y, Z, s(l), m)\end{array}}{\vdash P(\hat{s}(Y, m), Z) \wedge \hat{p}(X, s(l))}\, S\hat{Q}r}{\vdash P(\hat{s}(Y, m), Z)}\, \wedge_{r_1}$$

and $\tau_1 =$

$$\dfrac{\begin{array}{c}(\tau_2)\\ \vdash P(\hat{s}(Y, m), Z)\end{array} \quad \dfrac{\dfrac{\dfrac{\dfrac{\dfrac{\begin{array}{c}(\delta_1, (n, m, p(k), s(l)))\\ \vdash \hat{q}(X, Y, Z, s(l), m)\end{array}}{\vdash P(\hat{s}(Y, m), Z) \wedge \hat{p}(X, s(l))}\, S\hat{q}r}{\vdash \hat{p}(X, s(l))}\, \wedge_{r_2}}{\vdash \neg P(X(l), \hat{s}(a, l)) \vee \hat{p}(X, l)}\, S\hat{p}r}{\vdash \neg P(X(l), \hat{s}(a, l)), \hat{p}(X, l)}\, \vee_r}{P(X(l), \hat{s}(a, l)) \vdash \hat{p}(X, l)}\, \neg_r}{\vdash \hat{p}(X, l)}\, res(\sigma_2)$$

where $\sigma_2 = \{X(l) \leftarrow \hat{f}(Y, m), Z \leftarrow \hat{s}(a, l)\}$.
Finally, $\rho_1(\delta_1, X, Y, Z, m, n, k, l) =$

$$\cfrac{\cfrac{\begin{matrix}(\tau_2) & (\tau_1) \\ \vdash P(\hat{s}(Y, m), Z) & \vdash \hat{p}(X, l)\end{matrix}}{\vdash P(\hat{s}(Y, m), Z) \wedge \hat{p}(X, l)}\ \wedge : Ir}{\vdash \hat{Q}(X, Y, Z, l, m)}\ S\hat{q}r^{+}$$

When we compare the refutation above to the refutation in Example 7.2.1 we see that this formailsm needs additional parameters, while this in Sect. 7.2 does not. Indeed, the additional operation of inductive closure makes additional parameters superfluous. The definition below is a variant of the the corresponding definition in [13]:

**Definition 7.3.8** ($l - \mathrm{RPL_0}^{\Psi}$ *schema*) Let $\mathcal{D}$ be the tuple $(\Psi, \delta_0, \Delta^*, \Delta_a, \vec{X}, \Pi)$. $\mathcal{D}$ is called an $l - \mathrm{RPL_0}^{\Psi}$ *schema* if the following conditions are fulfilled:

- $\Psi$ is an $\omega\iota o$ theory.
- $\Delta^*$ be a finite subset of $\Delta$.
- $\delta_0 \in \Delta^*$, $\delta_0$ is called the main symbol.
- $\Delta_a \subseteq \Delta^*$, the axiom symbols.
- $\vec{X}$ is a tuple of variables.
- To every $\delta \in \Delta^*$ a parameter tuple $\vec{n}_\delta$ of pairwise different parameters is assigned, together with a partition $\mathcal{C}_\delta = \{C_1^\delta, \ldots, C_{k(\delta)}^\delta\}$ of $\mathcal{S}$ where the $C_i^\delta$ contain only parameters in $\vec{n}_\delta$.
- $\Pi$ is a set of pairs

$$\left\{(\Pi(\delta, \vec{X}, \vec{n}_\delta), S(\delta, \vec{X}, \vec{n}_\delta)) \,\middle|\, \delta \in \Delta^*\right\}$$

$S(\delta, \vec{X}, \vec{n}_\delta)$ is a sequent over the variables $\vec{X}$ and the parameters in $\vec{n}_\delta$, and $\Pi(\delta, \vec{X}, \vec{n}_\delta)$ is defined as follows:

$$\{(\delta, \vec{X}, \vec{n}_\delta) \rightarrow \rho_1(\delta, \vec{X}, \vec{n}_\delta) \colon C_1^\delta, \ldots, (\delta, \vec{X}, \vec{n}_\delta) \rightarrow \rho_{k(\delta)}(\delta, \vec{X}, \vec{n}_\delta) \colon C_{k(\delta)}^\delta\}$$

For each $i \in \{1, \ldots, k(\delta)\}$ $\rho_i(\delta, \vec{X}, \vec{n}_\delta)$ is a labeled $\mathrm{RPL_0}^{\Psi}$ derivation of $S(\delta, \vec{X}, \vec{n}_\delta)$ and for each leaf $\lambda \in \mathrm{leaves}(\rho_i(\delta, \vec{X}, \vec{n}_\delta))$ we have

- $|\rho_i(\delta, \vec{X}, \vec{n}_\delta)|_\lambda = (\delta', \vec{n}_\delta) \colon \vdash \hat{P}(\vec{Z}, \vec{r})$ for $\lambda \in \mathrm{Axiom}(\rho_i(\delta, \vec{X}, \vec{n}_\delta))$ where $\hat{P}(\vec{Z}, \vec{r})$ is a variant of the main atom of $\Psi$, $\delta' \in \Delta_a$, $A(\delta') = A(\delta)$ and $\vec{Z}$ is a tuple of variables which is a subtuple of $\vec{X}$.
- $|\rho_i(\delta, \vec{X}, \vec{n}_\delta)|_\lambda = (\delta', \vec{s}) \colon S(\delta', \vec{X}, \vec{s})$ for $\lambda \in \mathrm{Naxiom}(\rho_i(\delta, \vec{X}, \vec{n}_\delta))$ where $\delta' \in \Delta^* \setminus \Delta_a$, and $\vec{s}$ is an $A(\delta')$ tuple of terms in $T_1^\omega$ containing only parameters in $\vec{n}_\delta$.

Furthermore,

- $\vec{n}_\delta = (\vec{r}, \vec{m}_\delta)$, the first part being the parameter tuple $\vec{r}$ for the main symbol of the theory $\Psi$. The arity of $\vec{m}_\delta$ depends on $\delta$.
- $\Pi(\delta_0, \vec{X}, \vec{n}_{\delta_0}) \neq \emptyset$,
- for all $\delta \in \Delta_a$ $\Pi(\delta, \vec{X}, \vec{n}_\delta) = \emptyset$,

$\mathcal{D}$ is called a *refutation schema* of $\Psi$ if $S(\delta_0, \vec{X}, \vec{n}_{\delta_0}) = \vdash$.

In defining $\vec{n} = (\vec{r}, \vec{m}_\delta)$ we distinguish the *static parameters* in $\vec{r}$ (which are used in the recursive formula definition) and the *dynamic parameters* in $\vec{m}_\delta$ which are used for carrying out the inductive steps. In Example 7.3.2 $\delta_0$ has no dynamic parameters, but $\delta_1$ has the dynamic parameters $k$ and $l$. In a recursive call only dynamic parameters are substituted. The tuple $\vec{Z}$ of variables is chosen in a way to guarantee essential disjointness of global variables in the resolution steps. Note that in the refutation schemata defined in Definition 7.2.9 all parameters are potentially dynamic, locating our approach in this book closer to functional programming (of proofs).

Given a parameter assignment $\sigma$, every l-proof schema defines a sequence of proofs in the following way: either we arrive at an axiom and stop or we find a leaf in the proof of the form $(\delta, \vec{p})\colon S(\delta, \vec{X}, \vec{p})$; in the latter case we identify the right condition $C_i^\delta$ and the proof $\rho_i(\delta, \vec{X}, \vec{p})$ and replace the leaf by the derivation $\rho_i(\delta, X, \vec{p})$. If this sequence converges we obtain an $\mathrm{RPL_0}^\Psi$ proof corresponding to $\sigma$. The formal definition is given below.

**Definition 7.3.9** (*semantics of l-proof schemata*) Let $\mathcal{D}$ be a proof schema

$$(\Psi, \delta_0, \Delta^*, \Delta_a, \vec{X}, \Pi).$$

Let $\sigma \in \mathcal{S}$; we define define a sequence $\mathcal{D}(\sigma, \alpha)_{\alpha \in \mathbb{N}}$ and call it the *proof sequence corresponding to* $\sigma$. Let

$$\begin{aligned}\Pi(\delta_0, \vec{X}, \vec{n}_{\delta_0}) = \; & \{(\delta_0, \vec{X}, \vec{n}_{\delta_0}) \to \rho_1(\delta_0, \vec{X}, \vec{n}_{\delta_0})\colon C_1^{\delta_0}, \\ & \ldots, (\delta, \vec{X}, \vec{n}_{\delta_0}) \to \rho_{k(\delta_0)}(\delta_0, \vec{X}, \vec{n}_{\delta_0})\colon C_{k(\delta_0)}^{\delta_0}\}.\end{aligned}$$

As $\{C_1^{\delta_0}, \ldots, C_{k(\delta_0)}^{\delta_0}\}$ is a partition there is exactly one $i \in \{1, \ldots, k(\delta_0)\}$ such that $\sigma[C_i^{\delta_0}] = \top$. Let $\sigma(\vec{n}_{\delta_0}) = \vec{k}$ where $\vec{k} = (\vec{r}_0, \vec{s})$ for $\sigma(\vec{r})\downarrow_\omega = \vec{r}_0$. Then we define

$$\mathcal{D}(\sigma, 0) = \rho_i(\delta_0, \vec{X}, \vec{k}).$$

Note that $\rho_i(\delta_0, \vec{X}, \vec{k})$ is an $\mathrm{RPL_0}^\Psi$ derivation of $S(\delta_0, X, \vec{k})$. The leaves of $\rho_i(\delta_0, \vec{X}, \vec{k})$ are either axioms of the form $(\delta_0', \vec{k})\colon \hat{P}(\vec{X}, \vec{r}_0)$ for $\delta_0' \in \Delta_a$, or they are of the form $(\delta, \vec{l})\colon S(\delta, \vec{X}, \vec{l})$ where $\delta \in \Delta^* \setminus \Delta_a$ and $\vec{l}$ is a ground term tuple obtained in replacing the original one by substituting parameters by numerals.

Assume that $\mathcal{D}(\sigma, \alpha)$ is already defined. We distinguish two cases:

(a) All leaves in $\mathcal{D}(\sigma, \alpha)$ are of the form $(\delta, \vec{p})\colon \hat{P}(\vec{Z}, \vec{s})$ for $\delta \in \Delta_a$ where $\hat{P}$ is the main symbol of $\Psi$. Then we define $\mathcal{D}(\sigma, \alpha + 1) = \mathcal{D}(\sigma, \alpha)$.
(b) There are leaves of the form $(\delta, \vec{k})\colon S(\delta, \vec{X}, \vec{k})$ for an $A(\delta)$ ground term tuple $\vec{k}$ and $\delta \in \Delta^* \backslash \Delta_a$. For each leaf of this form we consider

$$\begin{aligned} \Pi(\delta, \vec{X}, \vec{n}_\delta) = \; & \{(\delta, \vec{X}, \vec{n}_\delta) \to \rho_1(\delta, \vec{X}, \vec{n}_\delta)\colon C_1^\delta, \\ & \ldots, (\delta, \vec{X}, \vec{n}_\delta) \to \rho_{k(\delta)}(\delta, \vec{X}, \vec{n}_\delta)\colon C_{k(\delta)}^\delta\}. \end{aligned}$$

Let $\sigma' \in \mathcal{S}$ such that $\sigma'(\vec{n}_\delta) = \vec{k}$. As $\{C_1^\delta, \ldots, C_{k(\delta)}^\delta\}$ is a partition there is exactly one $i \in \{1, \ldots, k(\delta)\}$ such that $\sigma'[C_i^\delta] = \top$. Now we have to select the production

$$(\delta, \vec{X}, \vec{n}_\delta) \to \rho_i(\delta, \vec{X}, \vec{n}_\delta)\colon C_i^\delta$$

from $\Pi(\delta, \vec{X}, \vec{n}_\delta)$. Finally we replace the leaf $(\delta, \vec{k})\colon S(\delta, \vec{X}, \vec{k})$ by the $\mathrm{RPL}_0{}^\Psi$ derivation $\rho_i(\delta, \vec{X}, \vec{k})$. We perform this transformation for all leaves with $\delta \in \Delta^* \setminus \Delta_a$ and obtain a new $\mathrm{RPL}_0{}^\Psi$ derivation which we call $\mathcal{D}(\sigma, \alpha + 1)$.

If the sequence converges, i.e. $\mathcal{D}(\sigma, \alpha) = \mathcal{D}(\sigma, \alpha + 1)$ for some $\alpha$ we obtain an $\mathrm{RPL}_0{}^\Psi$ proof as a result. Of course it is desirable to obtain such a proof for all parameter assignments $\sigma$.

**Definition 7.3.10** (*convergence of l-proof schemata,* [13]) Let

$$\varphi = \mathcal{D}(\sigma, \alpha)$$

where $\mathcal{D}(\sigma, \alpha) = \mathcal{D}(\sigma, \alpha + 1)$. Then we say that $\mathcal{D}(\sigma, \alpha)_{\alpha \in \mathbb{N}}$ converges. $\mathcal{D}$ is called *convergent* if, for all $\sigma \in \mathcal{S}$, $\mathcal{D}(\sigma, \alpha)_{\alpha \in \mathbb{N}}$ converges.

If $\mathcal{D}(\sigma, \alpha)_{\alpha \in \mathbb{N}}$ converges then it converges to a proof $\varphi \in \mathrm{RPL}_0{}^\Psi$ which is parameter-free. But $\varphi$ can be further reduced to an $\mathrm{RPL}_0$ derivation $\varphi{\downarrow}$ via the method defined in the proof of Theorem 7.2.1. We can now extend the semantics to $\mathrm{RPL}_0$-proofs.

**Definition 7.3.11** Let $\mathcal{D}$ be a proof schema which, for a $\sigma \in \mathcal{S}$, converges to an $\mathrm{RPL}_0{}^\Psi$ derivation $\varphi(\sigma)$. Then $\varphi(\sigma){\downarrow}$ is called the $\mathrm{RPL}_0$ proof defined by $\mathcal{D}$ under $\sigma$. If $\mathcal{D}$ converges for all $\sigma$ then $\{\varphi(\sigma){\downarrow} \mid \sigma \in \mathcal{S}\}$ is called the $\mathrm{RPL}_0$ proof set defined by $\mathcal{D}$.

**Example 7.3.3** Let $\mathcal{D}$ be the proof schema from Example 7.3.2. We compute $\mathcal{D}(\sigma, \alpha)$ for two different $\sigma \in \mathcal{S}$. Let $\sigma_0(n) = \bar{0}$, $\sigma_0(m) = \bar{2}$ and $\sigma_0(k) = \bar{1}$ for all parameters $k \notin \{n, m\}$. We start with $\mathcal{D}(\sigma_0, 0)$. As $\sigma_0(n) = \bar{0}$ we have to replace $(\delta_0, X, Y, (n, m))$ by the proof $\rho_0(\delta_0, X, Y, (\bar{0}, \bar{2}))$. So we define $\mathcal{D}(\sigma_0, 0) =$

$$
\dfrac{
\dfrac{\dfrac{\vdash (\delta_1', (\bar{0}, \bar{2})) : \hat{q}(X, Y, Z, \bar{0}, \bar{2})}{\vdash P(\hat{s}(Y, \bar{2}), Z) \wedge \hat{p}(X, \bar{0})}\, B\hat{q}r}{\vdash P(\hat{s}(Y, \bar{2}), Z)}\, \wedge_{r_1}
\qquad
\dfrac{\dfrac{\dfrac{\dfrac{\vdash (\delta_1', (\bar{0}, \bar{2})) : \hat{q}(X, Y, Z, \bar{0}, \bar{2})}{\vdash P(\hat{s}(Y, \bar{2}), Z) \wedge \hat{p}(X, \bar{0})}\, B\hat{q}r}{\vdash \hat{p}(X, \bar{0})}\, \wedge_{r_2}}{\vdash \neg P(X(0), \hat{s}(a, \bar{0}))}\, B\hat{p}r}{P(X(\bar{0}), \hat{s}(a, \bar{0})) \vdash}\, \neg : l
}{\vdash}\, res(\sigma_1')
$$

for $\sigma_1' = \{X(0) \leftarrow \hat{s}(Y, 2), Z \leftarrow \hat{s}(a, 0)\}$. In $\rho_0(\delta_0, X, Y, Z, (\bar{0}, \bar{2}))$ all leaves are labeled by the axiom labels $\delta_1'$ and there is no leaf to expand; thus $\mathcal{D}(\sigma_0, 1) = \mathcal{D}(\sigma_0, 0)$ and $\mathcal{D}(\sigma_0, \alpha)_{\alpha \in \mathbb{N}}$ converges to $\rho_0(\delta_0, X, Y, Z, (0, 2))$.

Now let $\sigma_1(p) = \sigma_0(p)$ for $p \neq n$ and $\sigma_1(n) = 1$. In this case the condition $n > 0$ holds and we have to choose the proof $\rho_1(\delta_0, X, Y, Z, 1, 2)$. We obtain $\mathcal{D}(\sigma_1, 0) =$

$$
\dfrac{
\dfrac{\dfrac{\begin{array}{c}(\delta_1, (1, 2, 1, 0)) \\ \vdash \hat{q}(X, Y, Z, 0, 2)\end{array}}{\vdash P(\hat{s}(Y, m), Z) \wedge \hat{p}(X, 0)}\, B\hat{q}r}{\vdash P(\hat{s}(Y, 2), Z)}\, \wedge_{r_1}
\qquad
\dfrac{\dfrac{\dfrac{\dfrac{\begin{array}{c}(\delta_1, (1, 2, 1, 0)) \\ \vdash \hat{q}(X, Y, Z, 0, 2)\end{array}}{\vdash P(\hat{s}(Y, 2), Z) \wedge \hat{p}(X, 0)}\, B\hat{q}r}{\vdash \hat{p}(X, 0)}\, \wedge_{r_2}}{\vdash \neg P(X(0), \hat{s}(a, 0))}\, B\hat{p}r}{P(X(0), \hat{s}(a, 0)) \vdash}\, \neg : l
}{\vdash}\, res(\sigma_1')
$$

Now both leaves in $\rho_1(\delta_0, X, Y, 1, 2)$ represent a call to the proof labeled by $\delta_1$. In $\Pi(\delta_1, X, Y, Z, n, m, k, l)$ we have to choose the proof

$$\rho_1(\delta_1, X, Y, Z, 1, 2, 1, 0)$$

via $\sigma'(n) = 1,\ \sigma'(m) = 2,\ \sigma'(k) = 1,\ \sigma'(l) = 0$ (the second condition applies). So we obtain $\mathcal{D}(\sigma_1, 1) =$

$$
\dfrac{
\dfrac{\dfrac{\begin{array}{c}\rho_1(\delta_1, X, Y, Z, (1, 2, 1, 0)) \\ \vdash \hat{q}(X, Y, Z, 0, 2)\end{array}}{\vdash P(\hat{s}(Y, m), Z) \wedge \hat{p}(X, 0)}\, B\hat{q}r}{\vdash P(\hat{s}(Y, m), Z)}\, \wedge_{r_1}
\qquad
\dfrac{\dfrac{\dfrac{\dfrac{\begin{array}{c}\rho_1(\delta_1, X, Y, Z, (1, 2, 1, 0)) \\ \vdash \hat{Q}(X, Y, Z, 0, 2)\end{array}}{\vdash P(\hat{s}(Y, 2), Z) \wedge \hat{p}(X, 0)}\, B\hat{q}r}{\vdash \hat{p}(X, 0)}\, \wedge_{r_2}}{\vdash \neg P(X(0), \hat{s}(a, 0))}\, B\hat{p}r}{P(X(0), \hat{s}(a, 0)) \vdash}\, \neg : l
}{\vdash}\, res\sigma_1
$$

Now $\rho_1(\delta_1, X, Y, (1, 2, 1, 0)) =$

$$
\dfrac{\dfrac{\begin{array}{c}(\delta_1, (1, 2, 0, 1)) \\ \vdash \hat{q}(X, Y, Z, 1, 2)\end{array}}{\vdash P(\hat{s}(Y, 2), Z) \wedge \hat{p}(X, 1)}\, S\hat{Q}r}{\begin{array}{c}\vdash P(\hat{f}(Y, 2), Z) \\ (2)\end{array}}\, \wedge_{r_1}
$$

$$
\dfrac{\begin{array}{c}(2)\\ \vdash P(\hat{f}(Y,2),Z)\end{array} \qquad \dfrac{\dfrac{\dfrac{\dfrac{\dfrac{\begin{array}{c}(\delta_1,(1,2,0,1))\\ \vdash \hat{Q}(X,Y,Z,1,2)\end{array}}{\vdash P(\hat{s}(Y,2),Z) \wedge \hat{p}(X,1)}\ S\hat{q}r}{\vdash \hat{p}(X,1)}\ \wedge_{r2}}{\vdash \neg P(X(0),\hat{s}(a,0)) \vee \hat{p}(X,0)}\ S\hat{p}r}{\vdash \neg P(X(0),\hat{s}(a,0)), \hat{p}(X,0)}\ \vee_r}{P(X(0),\hat{s}(a,0)) \vdash \hat{p}(X,0)}\ \neg_r}{\begin{array}{c}\vdash \hat{p}(X,0)\\ (1)\end{array}}\ res(\sigma_2)
$$

where $\sigma_2 = \{X(0) \leftarrow \hat{f}(Y,2), Z \leftarrow \hat{s}(a,0)\}$.

$$
\dfrac{\dfrac{\begin{array}{cc}(2) & (1)\\ \vdash P(\hat{s}(Y,2),Z) & \vdash \hat{p}(X,0)\end{array}}{P(\hat{s}(Y,2),Z) \wedge \hat{p}(X,0)}\ \wedge : Ir}{\vdash \hat{q}(X,Y,0,2)}\ S\hat{q}r^{+}
$$

In $\mathcal{D}(\sigma, 1)$ we have the leaves $\vdash (\delta_1, (1,2,0,1))\colon \hat{Q}(X,Y,1,2)$. Again we have to call $\delta_1$, this time via $\sigma''$ for $\sigma''(n) = 1, \sigma''(m) = 2, \sigma''(k) = 0, \sigma''(l) = 1$. As now $\sigma''(l) = \sigma''(n)$ we have to call $\rho_2(\delta_1, X, Y, Z, (1,2,0,1))$. But

$$\rho_2(\delta_1, X, Y, 1, 2, 0, 1) = (\delta_3', (1,2,0,1))\colon\ Q(X,Y,Z,1,2)$$

where $\delta_3' \in \Delta_a$. So we obtain $\mathcal{D}(\sigma_1, 2)$ from $\mathcal{D}(\sigma_1, 1)$ by replacing the leaves $\vdash (\delta_1, (1,2,0,1))\colon \hat{Q}(X,Y,Z,1,2)$ by the axiom leaves

$$(\delta_3', (1,2,0,1))\colon\ Q(X,Y,Z,1,2).$$

In $\mathcal{D}(\sigma_1, 2)$ all leaves are axiom leaves and so

$$\mathcal{D}(\sigma_1, 3) = \mathcal{D}(\sigma_1, 2).$$

Therefore also $\mathcal{D}(\sigma_1, \alpha)_{\alpha\in\mathbb{N}}$ converges. It is easy to see that $\mathcal{D}$ itself is convergent. The means to prove this formally will be developed below.

The concept of proof schema $\mathcal{D}$ is defined in a way that the skeleton of $\mathcal{D}$ is, in fact, a point transition system: we just strip the schema of its logical part and obtain the remaining "call part" of the schema.

**Definition 7.3.12** Let $\mathcal{D}\colon (\Psi, \delta_0, \Delta^*, \Delta_a, \vec{X}, \Pi)$ be a proof schema. Assume that for $\delta \in \Delta^*$ we have

$$
\begin{aligned}
\Pi(\delta, \vec{X}, \vec{n}_\delta) = \ & \{(\delta, \vec{X}, \vec{n}_\delta) \to \rho_1(\delta, \vec{X}, \vec{n}_\delta)\colon C_1^\delta, \\
& \ldots, (\delta, \vec{X}, \vec{n}_\delta) \to \rho_{k(\delta)}(\delta, \vec{X}, \vec{n}_\delta)\colon C_{k(\delta)}^\delta\}.
\end{aligned}
$$

Let $t\colon (\delta, \vec{X}, \vec{n}_\delta) \to \rho_i(\delta, \vec{X}, \vec{n}_\delta)\colon C_i^\delta$ be a production in $\Pi(\delta, \vec{X}, \vec{n}_\delta)$. Let $\Lambda(\delta, i)$ the set of all leaves in the proof $\rho_i(\delta, \vec{X}, \vec{n}_\delta)$ . For every $\lambda \in \Lambda(\delta, i)$ of the form $\lambda\colon (\delta', \vec{p})\colon F$ we define $pts(\lambda) = (\delta', \vec{p})$.

Then we replace the production $t$ in $\Pi(\delta, \vec{X}, \vec{n}_\delta)$ by

$$pts(t) = (\delta, \vec{n}_\delta) \to \{pts(\lambda) \mid \lambda \in \Lambda(\delta, i)\}$$

and $\Pi(\delta, \vec{X}, \vec{n}_\delta)$ by

$$pts(\Pi(\delta, \vec{X}, \vec{n}_\delta)) = \{pts(t) \mid t \in \Pi(\delta, \vec{X}, \vec{n}_\delta)\}.$$

Finally we define $\mathbf{P}(\delta) = pts(\Pi(\delta, \vec{X}, \vec{n}_\delta))$ for all $\delta \in \Delta^*$ and set

$$pts(\mathcal{D}) = (\delta_0, \Delta^*, \Delta_a, \mathbf{P}).$$

**Proposition 7.3.2** *Let* $\mathcal{D}\colon (\Psi, \delta_0, \Delta^*, \Delta_a, \vec{X}, \Pi)$ *be a proof schema. Then* $pts(\mathcal{D})$ *is a point transition system.*

***Proof*** Immediate by the Definitions 5.0.10, 7.3.4 and 7.3.12.

**Example 7.3.4** Let $\mathcal{D}$ be the proof schema in Example 7.3.2. Then $pts(\mathcal{D})$ is the point transition system $(\delta_0, \Delta^*, \Delta_a, \mathbf{P})$, where $\Delta^* = \{\delta_0, \delta_1, \delta_1', \delta_2', \delta_3'\}$, $\Delta_a = \{\delta_1', \delta_2', \delta_3'\}$ and $\mathbf{P}$ defined below:

$$\begin{aligned}
\mathbf{P}(\delta_0) = \; & \{(\delta_0, (n, m)) \to \{(\delta_1', (0, m))\}\colon n = 0, \\
& (\delta_0, (n, m)) \to \{(\delta_1, (n, m, n, 0))\}\colon n > 0\} \\
\mathbf{P}(\delta_1) = \; & \{(\delta_1, (n, m, k, l)) \to \{(\delta_2', (n, m, k, l))\}\colon n = 0, \\
& (\delta_1, (n, m, k, l)) \to \{(\delta_1, (n, m, p(k), s(l)))\}\colon n > 0 \wedge l < n, \\
& (\delta_1, (n, m, k, l)) \to \{(\delta_3', (n, m, k, l))\}\colon n > 0 \wedge l \geq n\},
\end{aligned}$$

$\mathbf{P}(\delta_1') = \mathbf{P}(\delta_2') = \mathbf{P}(\delta_3') = \emptyset$.

Note that the right hand sides of the productions in $pts(\mathcal{D})$ are sets - not multisets. Therefore we count different leaves with the same labels just once. It is easy to see that $pts(\mathcal{D})$ is a canonic point transition system and thus terminating. We will prove below that this implies that $\mathcal{D}$ is convergent ([13]).

**Lemma 7.3.1** *Let* $\mathcal{D}(\sigma, \alpha)$ *be as in Definition 7.3.9 and* $T(\mathbf{P}^\star, (\delta, p), \sigma, \alpha)$ *as in Definition 5.0.11. Then, for any* $\alpha \in \mathbb{N}$*:*

*(a) if* $(\delta, \vec{k})\colon S$ *is a leaf node in* $\mathcal{D}(\sigma, \alpha)$ *then* $(\delta, \vec{k})$ *is a leaf node in* $T(pts(\mathcal{D}), (\delta_0, \vec{n}_{\delta_0}), \sigma, \alpha + 1)$*;*
*(b) if* $(\delta, \vec{k})$ *is a leaf node in* $T(pts(\mathcal{D}), (\delta_0, \vec{n}_{\delta_0}), \sigma, \alpha + 1)$ *then there exists a sequent* $S$ *such that* $(\delta, \vec{k})\colon S$ *is a leaf node in* $\mathcal{D}(\sigma, \alpha)$.

***Proof*** We prove (a) by induction on $\alpha$. Let $\alpha = 0$. Let us consider

$$\Pi(\delta_0, \vec{X}, \vec{n}_{\delta_0}) = \{(\delta_0, \vec{X}, \vec{n}_{\delta_0}) \to \rho_1(\delta_0, \vec{X}, \vec{n}_{\delta_0})\colon C_1^{\delta_0}, \ldots, (\delta, \vec{X}, \vec{n}_{\delta_0}) \to \rho_{k(\delta_0)}(\delta_0, \vec{X}, \vec{n}_{\delta_0})\colon C_{k(\delta_0)}^{\delta_0}\}.$$

and let us assume that $\sigma[C_i^{\delta_0}] = \top$. Then $\mathcal{D}(\sigma, 0)$ is defined as

$$\rho_i(\delta_0, \vec{X}, \sigma(\vec{n}_{\delta_0})).$$

Now let $(\delta_j, \vec{p}_j)\colon F_j)$ be the leaves in in $\rho_i(\delta_0, \vec{X}, \vec{n}_{\delta_0})$ for $j = 1, \ldots, \beta$; then $(\delta_j, \sigma(\vec{p}_j))\colon \sigma(F_j))$ are the leaves in $\rho_i(\delta_0, \vec{X}, \sigma(\vec{n}_{\delta_0}))$. By Definition 7.3.12 $pts(\Pi(\delta_0, \vec{X}, \vec{n}_{\delta_0}))$ contains the production

$$(\delta_0, \vec{n}_{\delta_0}) \to \{(\delta_1, \vec{p}_1), \ldots, (\delta_\beta, \vec{p}_\beta)\}.$$

Hence, by Definition 5.0.11 $T(pts(\mathcal{D}), (\delta_0, \vec{n}_{\delta_0}), \sigma, 1)$ has exactly the leaves $(\delta_j, \sigma(\vec{p}_j))$. This concludes the case $\alpha = 0$.

Now assume that, for any leaf $(\delta, \vec{k})\colon F$ in $\mathcal{D}(\sigma, \alpha)$ there exists a leaf $(\delta, \vec{k})$ in $T(pts(\mathcal{D}), (\delta_0, \vec{n}_{\delta_0}), \sigma, \alpha + 1)$. If $\delta \in \Delta_a$ then, by definition, $(\delta, \vec{k})\colon F$ is a leaf in $\mathcal{D}(\sigma, \alpha + 1)$ and $(\delta, \vec{k})$ is a leaf in $T(pts(\mathcal{D}), (\delta_0, \vec{n}_{\delta_0}), \sigma, \alpha + 2)$. Otherwise assume that $\Pi(\delta, \vec{X}, \vec{n}_\delta) \neq \emptyset$ and

$$\Pi(\delta, \vec{X}, \vec{n}_\delta) = \{(\delta, \vec{X}, \vec{n}_\delta) \to \rho_1(\delta, \vec{X}, \vec{n}_\delta)\colon C_1^\delta, \ldots, (\delta, \vec{X}, \vec{n}_\delta) \to \rho_{k(\delta)}(\delta, \vec{X}, \vec{n}_\delta)\colon C_{k(\delta)}^\delta\}.$$

Let $\sigma' \in \mathcal{S}$ such that $\sigma'(\vec{n}_\delta) = \vec{k}$ and $\sigma'[C_i^\delta] = \top$. Now we have to select the production

$$(\delta, \vec{X}, \vec{n}_\delta) \to \rho_i(\delta, \vec{X}, \vec{n}_\delta)\colon C_i^\delta$$

from $\Pi(\delta, \vec{X}, \vec{n}_\delta)$. Again, let us assume that $(\delta_j, \vec{p}_j)\colon F_j$ are the leaves in in $\rho_i(\delta, \vec{X}, \vec{n}_\delta)$ for $j = 1, \ldots, \gamma$, and so $(\delta_j, \sigma'(\vec{p}_j))\colon \sigma'(F_j)$ are the leaves in in $\rho_i(\delta, \vec{X}, \sigma'(\vec{n}_\delta))$. By Definition 7.3.12 $pts(\Pi(\delta, \vec{X}, \vec{n}_\delta))$ contains the production

$$(\delta, \vec{n}_\delta) \to \{(\delta_1, \vec{p}_1), \ldots, (\delta_\beta, \vec{p}_\gamma)\}.$$

By Definition 5.0.11 , $T(pts(\mathcal{D}), (\delta_0, \vec{n}_{\delta_0}), \sigma, \alpha + 2)$ contains the (new) leaves $(\delta_j, \sigma'(\vec{p}_j))$.

The proof of (b) is analogous to that of (a).

**Theorem 7.3.3** ([13]) *Let $\mathcal{D}$ be a proof schema. $\mathcal{D}$ is convergent iff $pts(\mathcal{D})$ is terminating.*

***Proof*** $\mathcal{D}$ is convergent $\Rightarrow$:

For all $\sigma \in \mathcal{S}$ there exists an $\alpha$ such that $\mathcal{D}(\sigma, \alpha) = \mathcal{D}(\sigma, \alpha + 1)$. By definition of $\mathcal{D}$ this happens only if for all leaves $(\delta, \vec{k})\colon S$ in $\mathcal{D}(\sigma, \alpha)$ we have $\delta \in \Delta_a$. By Lemma 7.3.1 all leaves in $T(pts(\mathcal{D}), (\delta_0, \vec{n}_{\delta_0}), \sigma, \alpha + 1)$ are of the form $(\delta, \vec{k})$ for $\delta \in \Delta_a$. As $\Delta_a$ is the set of all end labels in $pts(\mathcal{D})$ we get

$$T(pts(\mathcal{D}), (\delta_0, \vec{n}_{\delta_0}), \sigma, \alpha + 1) = T(pts(\mathcal{D}), (\delta_0, \vec{n}_{\delta_0}), \sigma, \alpha + 2)$$

and, for all $\sigma$, $pts(\mathcal{D})$ terminates on $\sigma$; thus $pts(\mathcal{D})$ is terminating.

$pts(\mathcal{D})$ is terminating $\Rightarrow$:

For all $\sigma \in \mathcal{S}$ there exists an $\alpha$ such that

$$T(pts(\mathcal{D}), (\delta_0, \vec{n}_{\delta_0}), \sigma, \alpha) = T(pts(\mathcal{D}), (\delta_0, \vec{n}_{\delta_0}), \sigma, \alpha + 1).$$

For all $\alpha > 0$ this implies that, for all leaves $(\delta, \vec{k})$ in $T(pts(\mathcal{D}), (\delta_0, \vec{n}_{\delta_0}), \sigma, \alpha)$, $\delta \in \Delta_a$. Now let $(\delta', \vec{p})\colon S$ be a leaf in $\mathcal{D}(\sigma, \alpha - 1)$; by Lemma 7.3.1 $(\delta', \vec{p})$ is a leaf in $T(pts(\mathcal{D}), (\delta_0, \vec{n}_{\delta_0}), \sigma, \alpha)$ and, by assumption, $\delta' \in \Delta_a$. Therefore, by definition of $\mathcal{D}(\sigma, \alpha + 1)$, we get $\mathcal{D}(\sigma, \alpha + 1) = \mathcal{D}(\sigma, \alpha)$ and $\mathcal{D}$ converges on $\sigma$.

We can now characterize the convergence of proof schemata via point transition systems:

**Theorem 7.3.4** *Let $\mathcal{D}$ be a proof schema. Then $\mathcal{D}$ is convergent if $pts(\mathcal{D})$ is canonic.*

***Proof*** Immediate by the Theorems 5.0.1 and 7.3.3.

Now it is easily verified that the proof schema $\mathcal{D}$ from Example 7.3.2 is convergent. The point transition system $pts(\mathcal{D})$ shown in Example 7.3.4 is canonic: for the ordering $<_P$ we use the lexicographic tuple ordering and we partition $\Delta^*$ into $\Delta_1$, $\Delta_2$ by $\Delta_1 = \Delta_a$, $\Delta_2 = \{\delta_0, \delta_1\}$.

Point transition systems are a flexible tool to generalize the form of recursion. So far we have followed the *definition ordering* in defining the proof operations in Definition 7.2.9, i.e. we assume that, in the definition of proof recursion and proof substitutions the proofs we are operating on are already defined. However, there are forms of cyclic recursions which are terminating. Just recall Example 5.0.1, where we gave definitions of computable functions not obeying the definition ordering. In this example $f$ is defined via $g$ and $g$ via $f$; nevertheless the computation terminates as the ordering of the tuples follow the conditions of a canonic point transition system:

$$\begin{aligned}
f(x, y) &= f(p(x), y) + g(x, p(y)) \text{ for } x > 0 \wedge y > 0,\\
f(x, y) &= x \text{ for } x > 0 \wedge y = 0,\\
f(x, y) &= y \text{ for } x = 0 \wedge y = 0,\\
f(x, y) &= 0 \text{ for } x = 0 \wedge y = 0.
\end{aligned}$$

$$g(x, y) = f(x, p(y)) + f(x, p(y)) * x \text{ for } y > 0,$$
$$g(x, y) = x * x \text{ for } y = 0.$$

The corresponding point transition was the following: For $f$ we choose $\delta_0$, $\delta_1$ for $g$ and $\delta_e$ for "end". Then $\Delta^* = \{\delta_0, \delta_1, \delta_e\}$, $\Delta_e = \{\delta_e\}$ and

$$\begin{aligned}
\mathbf{P}(\delta_0) = \ & \{(\delta_0, (x, y)) \to \{(\delta_0, (p(x), y)), (\delta_1, (x, p(y))\} \colon x > 0 \wedge y > 0, \\
& (\delta_0, (x, y)) \to \{(\delta_e, (x, y))\} \colon x > 0 \wedge y = 0, \\
& (\delta_0, (x, y)) \to \{(\delta_e, (x, y))\} \colon x = 0 \wedge y > 0, \\
& (\delta_0, (x, y)) \to \{(\delta_e, (x, y))\} \colon x = 0 \wedge y = 0\}. \\
\mathbf{P}(\delta_1) = \ & \{(\delta_1, (x, y)) \to \{(\delta_0, (x, p(y))\} \colon y > 0, \\
& (\delta_1, (x, y)) \to \{(\delta_e, (x, y))\} \colon y = 0\}. \\
\mathbf{P}(\delta_e) = \ & \emptyset.
\end{aligned}$$

As the point transition system is canonical the definitions of $f$ and $g$ terminate. In [13] Example 23 we used such a cyclic recursion on proofs to specify an l-refutation schema of a characteristic formula schema corresponding to a mathematical proof. Extensions of this type are also possible in refutation schemata as defined in Sect. 7.2. Operations like inductive closure are no longer possible; instead we use a more liberal form of proof substitution beyond the definition ordering. We give a simple example.

**Example 7.3.5** Let $T^*$ be the theory defining the term schemata $\hat{s}, \hat{t}$ below:

$$D(\hat{s}) = \{\hat{s}(x, n) =_d \underline{\text{if}}\ n = \mathbf{0}\ \underline{\text{then}}\ x\ \underline{\text{else}}\ h(\hat{s}(x, p(n)))\},$$
$$D(\hat{t}) = \{\hat{t}(x, n) =_d \underline{\text{if}}\ n = \mathbf{0}\ \underline{\text{then}}\ x\ \underline{\text{else}}\ g(\hat{t}(x, p(n)))\}.$$

We define

- $\rho_1(\mathbf{0}, \mathbf{0}) =$

$$\frac{P(\hat{s}(a, \mathbf{0}), \hat{t}(a, \mathbf{0}))}{V_S(\mathbf{0}, \mathbf{0})}\ V_I$$

- $\rho_2(\mathbf{0}, \mathbf{0}) =$

$$\frac{Q(\hat{s}(a, \mathbf{0}), \hat{t}(a, \mathbf{0}))}{V_{S'}(\mathbf{0}, \mathbf{0})}\ V_I$$

- $S(n, m) = \vdash P(\hat{s}(a, n), \hat{t}(a, m))$,
- $S'(n, m) = \vdash Q(\hat{s}(a, n), \hat{t}(a, m))$.

Futhermore we define

- $\rho_1(s(n), m) =$

$$\cfrac{\cfrac{\cfrac{V_{S'}(n,m)\{V_{S'}(n,m) \leftarrow \rho_2(n,m)\}}{\vdash Q(\hat{s}(a,n), \hat{t}(a,m))}\ V_E \qquad Q(x,y) \vdash P(h(x),y)}{\vdash P(h(\hat{s}(a,n)), \hat{t}(a,m))}\ res}{\cfrac{\vdash P(\hat{s}(a,s(n)), \hat{t}(a,m))}{V_S(s(n),m)}\ V_I}\ T^*$$

- $\rho_1(0, s(m)) = \rho_2(0, m)$ and
- $\rho_2(n, s(m)) =$

$$\cfrac{\cfrac{\cfrac{V_S(n,m)\{V_S(n,m) \leftarrow \rho_1(n,m)\}}{\vdash P(\hat{s}(a,n), \hat{t}(a,m))}\ V_E \qquad P(x,y) \vdash Q(x,g(y))}{\vdash Q(\hat{s}(a,n), g(\hat{t}(a,m)))}\ res}{\cfrac{\vdash Q(\hat{s}(a,n), \hat{t}(a,s(m)))}{V_{S'}(n,s(m))}\ V_I}\ T^*$$

- $\rho_2(s(n), 0) = \rho_1(n, 0)$.

Though there is no definition order among $\rho_1$ and $\rho_2$ possible, both $\rho_1(n, m)$ and $\rho_2(n, m)$ terminate for all parameter assignments $\sigma$. The reason is that

$$(n, m) < (n, s(m)), (0, m) < (0, s(m)), (n, m) < (s(n), m), (n, 0) < (s(n), 0)$$

via the lexicographic tuple ordering. A corresponding point transition system for the computation of $\rho_1$ (where $\delta_e$ is an end state) is given by

$$\begin{aligned}
\mathbf{P}(\delta_0) = \ & \{\delta_0(n,m) \to \{(\delta_e(n,m)\}\colon n = 0 \land m = 0,\\
& \delta_0(n,m) \to \{(\delta_1(p(n),m)\}\colon n > 0,\\
& \delta_0(n,m) \to \{(\delta_1(n,p(m))\}\colon n = 0 \land m > 0\},\\
\mathbf{P}(\delta_1) = \ & \{\delta_1(n,m) \to \{(\delta_e(n,m)\}\colon n = 0 \land m = 0,\\
& \delta_1(n,m) \to \{(\delta_0(n,p(m))\}\colon m > 0\},\\
& \delta_1(n,m) \to \{\delta_0(p(n),m)\}\colon n > 0 \land m = 0\}.
\end{aligned}$$

The admission of cyclic calls based on canonic point transition systems represents a further extension of our formalism of proof recursion which we do not follow further here.

Furthermore we define:

# Chapter 8
# Proof Schemata and PRA

In this chapter we will analyze the expressivity of proof schemata by showing that proof schemata simulate primitive recursive arithmetic (**PRA**). To this aim, we will show the translation of proofs in **PRA** into $\mathbf{LK}_{T^+}$-cf-proof schema as defined in Definition 7.1.15 and vice versa. Note that it was shown in Theorem 7.3.1 that cf-proof schemata are equivalent to proof schemata. In **PRA** induction formulas need to be quantifier-free, but proof schemata allow the use of quantified recursion formulas. If we restrict the proof schemata to those containing only quantifier-free recursion formulas we can show the equivalence between schemata and **PRA**.

**Definition 8.0.1** A proof recursion (see Definition 7.1.12) is called quantifier-free if the recursion formula contains no quantifiers. A proof schema is called quantifier-free with respect to proof recursion if all proof recursions in the schema are quantifier-free.

Other formalisms for schematic proofs have also been compared to proofs in arithmetic. Indeed, it was shown in [31] that proof schemata are equivalent to a specific fragment of arithmetic known as *k-simple induction*. This variant restricts the introduction of new eigenvariables through induction. Recently, the simulation was extended to capture proof schemata that allow an arbitrary number of parameters [30], where a definition of **PRA** was used which does not admit quantifier introductions. This simulation was performed using a different proof schema formalism (see Definition 7.3.4), and recent investigations have shown that it can be extended to a definition of **PRA** where quantifier introduction is allowed. The only restriction is that strong quantification of induction variables is not allowed. In this chapter we will show the same result for cf-proof schemata.

First of all, let us define the calculus for **PRA** as Gentzen's **LK** extended by an equational theory $\mathcal{E}$ and the following induction rule:

$$\frac{\Gamma \vdash \Delta, F(0) \qquad \Gamma, F(y) \vdash \Delta, F(y+1)}{\Gamma \vdash \Delta, F(n)}\ ind$$



A. Leitsch et al., *First-Order Schemata and Inductive Proof Analysis*, Computer Science Foundations and Applied Logic, https://doi.org/10.1007/978-3-032-05741-9_8

where $y$ is a variable of sort $\omega$, $n$ is a variable of sort $\omega$, and $y$ does not occur in $\Gamma, \Delta, F(0)$.

Given the close resemblance of the cf-proof recursion and the *ind* rule in **PRA**, it is easy to show that any cf-proof schema can be translated directly into **PRA**.

**Lemma 8.0.1** *Let $\rho$ be a quantifier-free (with respect to proof recursion)* $\mathbf{LK}_{T^+}$*-cf-proof schema of a schematic sequent $S(\vec{n})$. Then there exists a* **PRA** *proof of $S(\vec{n})$.*

***Proof*** The only non-trivial case occurs when $\rho$ is defined via cf-proof recursion. Indeed, if $\rho$ is defined only via link-substitution then $\rho$ is in **PRA**. Or, if $\rho$ is defined via a condition $C$ then exactly one of $\rho_1$ or $\rho_2$ is chosen and we apply the induction hypothesis.

So let us consider the case where $\rho$ is defined via cf-proof recursion, i.e. $\rho$ is of the form $\breve{\rho}(\vec{n}, k)$. Then, there are cf-recursion sequents $S(\vec{X}, \vec{n}, k)$ and $S^*(\vec{X}, \vec{n}, k)$

$$\begin{aligned} S(\vec{X}, \vec{n}, k) &= F(\vec{X}, \vec{n}, k), \Gamma \vdash \Delta, F(\vec{X}, \vec{n}, s(k)) \text{ and} \\ S^*(\vec{X}, \vec{n}, k) &= \Gamma \vdash \Delta, F(\vec{X}, \vec{n}, k) \text{ for all } k. \end{aligned}$$

such that $\rho(\vec{X}, \vec{n}, V_S(\vec{X}, \vec{n}, k)) =$

$$\dfrac{\dfrac{\dfrac{V_{S^*}(\vec{n}, k)}{S^*(\vec{X}, \vec{n}, k)}\, V_S E \qquad \dfrac{\rho'(\vec{X}, \vec{n}, k)}{S(\vec{X}, \vec{n}, k)}}{\Gamma \vdash \Delta, F(\vec{X}, \vec{n}, s(k))}\, cut + c^*}{V_{S^*}(\vec{X}, \vec{n}, s(k))}\, V_S I$$

is an $\mathbf{LK}_{T^+}$ derivation schema admitting cf-proof recursion based on $S$ and $S^*$. Moreover, $\rho(\vec{X}, \vec{n}, V_S(\vec{X}, \vec{n}, k))$ is a semi-cf-proof schema and

$$\begin{aligned} D(\breve{\rho}) = \{&\breve{\rho}(\vec{X}, \vec{n}, \mathbf{0}) =_d \rho_0, \\ &\breve{\rho}(\vec{X}, \vec{n}, s(k)) =_d \rho(\vec{X}, \vec{n}, s(k))\{V(k) \leftarrow \breve{\rho}(\vec{X}, \vec{n}, k)\}\}. \end{aligned}$$

$\rho_0$ is an $\mathbf{LK}_{T^+}$-cf-proof schema of $V(0)$.

Note that $\rho(\vec{X}, \vec{n}, V_S(\vec{X}, \vec{n}, k))$ resembles the induction rule *ind* in **PRA**. In fact, from the cf-proof schema $\breve{\rho}(\vec{n}, k)$ we directly obtain a proof in **PRA**, by simply taking the derivation of $S(\vec{X}, \vec{n}, k)$ as the right premise of the *ind* rule, and $\rho_0$ as the derivation of the left premise of the *ind* rule. Link introduction and elimination rules are omitted.

Note that general proof schemata (which admit quantified recursion formulas) are more expressive than **PRA**. The translation of **PRA** proofs into cf-proof schemata can be shown in a similar way.

**Lemma 8.0.2** *Let $\pi$ be a* **PRA** *proof of $S$. Then there exists a* $\mathbf{LK}_{T^+}$*-cf-proof schema with end-sequent $S$.*

***Proof*** The proof is by induction on the number of $ind$ inferences in $\pi$, where we start with the uppermost occurrence of $ind$ and proceed to the bottom of the derivation. For each $ind$ inference, a cf-proof schema is constructed. Let

$$\frac{\begin{array}{c}(\pi_1)\\ \Gamma \vdash \Delta, F(0)\end{array} \qquad \begin{array}{c}(\pi_2)\\ \Gamma, F(y) \vdash \Delta, F(y+1)\end{array}}{\Gamma \vdash \Delta, F(n)}\ ind$$

be such an $ind$ inference in $\pi$. We define $\rho_0$ as the derivation obtained from $\pi_1$ after applying a link introduction rule to the end-sequent of $\pi_1$, thus $\rho_0$ is a derivation of $V_S(0)$. Then, we define $\rho'(k)$ from $\pi_2$ by replacing $y$ with $k$ as the derivation of $S(k)$ such that $S(k)$ and $S^*(k)$ are cf-recursion sequents

$$\begin{aligned} S(k) =&\ F(k), \Gamma \vdash \Delta, F(s(k)) \text{ and} \\ S^*(k) =&\ \Gamma \vdash \Delta, F(k) \text{ for all } k. \end{aligned}$$

Note that $S(k)$ is just the right premise of the $ind$ rule above, where the variable $y$ is replaced by a new parameter $k$. We then define $\rho(V_S(k)) =$

$$\cfrac{\cfrac{\cfrac{V_{S^*}(k)}{S^*(k)}\ V_S E \qquad \begin{array}{c}\rho'(k)\\ S(k)\end{array}}{\Gamma \vdash \Delta, F(s(k))}\ cut + c^*}{V_{S^*}(s(k))}\ V_S I$$

as the $\mathbf{LK}_{T^+}$ derivation schema admitting cf-proof recursion based on $S$ and $S^*$. Moreover, $\rho(V_S(k))$ is a semi-cf-proof schema and, finally,

$$\begin{aligned} &D(\check{\rho}) = \{\check{\rho}(\mathbf{0}) =_d \rho_0, \\ &\check{\rho}(s(k)) =_d \rho(s(k))\{V(k) \leftarrow \check{\rho}(k)\}\}. \end{aligned}$$

It is also possible to express interesting mathematical statements and their proofs in the proof schema formalism. As an example, let us illustrate the proof of *commutativity of* $+$ in our formalism, and then translate it into a proof in **PRA**. To this aim, let we extend our calculi by the following two equality axioms:

- $Eq.1$: $s = t, u[s]_\Lambda = v \vdash u[t]_\Lambda = v$, and
- $Eq.2$: $\vdash s = s$.

To construct a proof schema of the commutativity of $+$, we have, of course, to extend our calculus by the Peano axioms.

**Definition 8.0.2** The Peano axioms are defined as

A1 $: \vdash 0 \neq s(x)$
A2 $: \vdash s(x) = s(y) \rightarrow x = y$
A3 $: \vdash x + 0 = x$

$$\frac{\Gamma, x = y \vdash \Delta}{\Gamma, y = x \vdash \Delta}\ sym_l \qquad \frac{\Gamma \vdash \Delta, x = y}{\Gamma \vdash \Delta, y = x}\ sym_r$$

$$\frac{\Gamma' \vdash \Delta', s = t \quad \Gamma'' \vdash \Delta'', t = u}{\Gamma', \Gamma'' \vdash \Delta', \Delta'', s = u}\ trans \qquad \frac{\Gamma \vdash \Delta, x = y}{\Gamma \vdash \Delta, s(x) = s(y)}\ subs$$

$$\frac{\Gamma \vdash \Delta, u[x]_\Lambda = z}{\Gamma \vdash \Delta, u[x+0]_\Lambda = z}\ A3_1 \qquad \frac{\Gamma \vdash \Delta, u[x+0]_\Lambda = z}{\Gamma \vdash \Delta, u[x]_\Lambda = z}\ A3_2$$

$$\frac{\Gamma \vdash \Delta, u[s(x+y)]_\Lambda = z}{\Gamma \vdash \Delta, u[x+s(y)]_\Lambda = z}\ A4_1 \qquad \frac{\Gamma \vdash \Delta, u[x+s(y)]_\Lambda = z}{\Gamma \vdash \Delta, u[s(x+y)]_\Lambda = z}\ A4_2$$

$$\frac{\Gamma \vdash \Delta, u[s(x)]_\Lambda = z}{\Gamma \vdash \Delta, u[x+1]_\Lambda = z}\ A7_1 \qquad \frac{\Gamma \vdash \Delta, u[x+1]_\Lambda = z}{\Gamma \vdash \Delta, u[s(x)]_\Lambda = z}\ A7_2$$

$$\frac{\Gamma \vdash \Delta, u[s(0)]_\Lambda = z}{\Gamma \vdash \Delta, u[1]_\Lambda = z}\ A8_1 \qquad \frac{\Gamma \vdash \Delta, u[1]_\Lambda = z}{\Gamma \vdash \Delta, u[s(0)]_\Lambda = z}\ A8_2$$

**Fig. 8.1** Additional macros. Note that for the inference rules $A3, \ldots, A8_2$, the auxiliary formula takes the form $a = b$, while the principal formula has the form $a' = b$. We use the same label for the corresponding symmetric macro, where the auxiliary formula is $b = a$ and the principal formula is $b = a'$. We omit the formal definition, as the construction via symmetry is straightforward

A4 $: \vdash x + s(y) = s(x + y)$
A5 $: \vdash x \times 0 = 0$
A6 $: \vdash x \times s(y) = x \times y + x$

For convenience we add the axioms

A7 $: \vdash s(x) = x + 1$
A8 $: \vdash s(0) = 1$

First, we will derive some additional axioms and define some macros (see Fig. 8.1).

**Definition 8.0.3**

- **Symmetry**:

  Symmetry is derivable:

$$\frac{\dfrac{Eq.2}{\vdash s = s} \qquad \dfrac{Eq.1}{s = t, s = s \vdash t = s}}{s = t \vdash t = s}\ cut$$

  Symmetry as a macro:

$$\frac{\Gamma \vdash \Delta, x = y}{\Gamma \vdash \Delta, y = x}\ sym_r$$

  is defined by

$$\frac{\dfrac{Assumption}{\Gamma \vdash \Delta, x = y} \qquad \dfrac{Symmetry}{x = y \vdash y = x}}{\Gamma \vdash \Delta, y = x}\ cut$$

and

$$\frac{\Gamma, x = y \vdash \Delta}{\Gamma, y = x \vdash \Delta}\; sym_l$$

is defined by

$$\frac{\dfrac{Assumption}{\Gamma, x = y \vdash \Delta} \qquad \dfrac{Symmetry}{y = x \vdash x = y}}{\Gamma, y = x \vdash \Delta}\; cut$$

- **Transitivity**:

  Transitivity is derivable:

$$\cfrac{\cfrac{Eq.1}{t = s, t = u \vdash s = u}}{s = t, t = u \vdash s = u}\; sym_l$$

  Transitivity as a macro:

$$\frac{\Gamma' \vdash \Delta', s = t \qquad \Gamma'' \vdash \Delta'', t = u}{\Gamma', \Gamma'' \vdash \Delta', \Delta'', s = u}\; trans$$

  is defined by

$$\cfrac{\cfrac{Assumption}{\Gamma'' \vdash \Delta'', t = u} \qquad \cfrac{\cfrac{Assumption}{\Gamma' \vdash \Delta', s = t} \qquad \cfrac{\cfrac{Eq.1}{t = s, t = u \vdash s = u}}{s = t, t = u \vdash s = u}\; sym_l}{\Gamma', t = u \vdash \Delta', s = u}\; cut}{\Gamma', \Gamma'' \vdash \Delta', \Delta'', s = u}\; cut$$

- **Substitutivity of** $s$:

$$\frac{\Gamma \vdash \Delta, x = y}{\Gamma \vdash \Delta, s(x) = s(y)}\; subs$$

  is defined by

$$\cfrac{\cfrac{Assumption}{\Gamma \vdash \Delta, x = y} \qquad \cfrac{\cfrac{\cfrac{Eq.2}{\vdash s(y) = s(y)} \qquad \cfrac{Eq.1}{y = x, s(y) = s(y) \vdash s(x) = s(y)}}{y = x \vdash s(x) = s(y)}\; cut}{x = y \vdash s(x) = s(y)}\; sym_l}{\Gamma \vdash \Delta, s(x) = s(y)}\; cut$$

Note that in all the following macros, the auxiliary formula takes the form $a = b$, while the principal formula has the form $a' = b$. A symmetric macro can be constructed for each of these rules, where the auxiliary formula is $b = a$ and the principal formula is $b = a'$, simply by applying symmetry before and after the macro application. We omit the formal definition and use $Ai_{1/2}$ for $i \in \{3, ..., 10\}$ to denote the application of the macro $Ai$ or its respective symmetric form. Hence, we assign the same label to inferences of the following form:

$$\dfrac{\Gamma \vdash \Delta, a = b}{\Gamma \vdash \Delta, a' = b}\, Ai \qquad \dfrac{\Gamma \vdash \Delta, b = a}{\Gamma \vdash \Delta, b = a'}\, Ai$$

- **A3:**

$$\dfrac{\Gamma \vdash \Delta, u[x]_\Lambda = z}{\Gamma \vdash \Delta, u[x+0]_\Lambda = z}\, A3_1$$

is defined by

$$\dfrac{\dfrac{Assumption}{\Gamma \vdash \Delta, u[x]_\Lambda = z} \qquad \dfrac{\dfrac{\dfrac{A3}{\vdash x+0=x}}{\vdash x = x+0}\, sym_r \qquad \dfrac{Eq.1}{x = x+0, u[x]_\Lambda = z \vdash u[x+0]_\Lambda = z}}{u[x]_\Lambda = z \vdash u[x+0]_\Lambda = z}\, cut}{\Gamma \vdash \Delta, u[x+0]_\Lambda = z}\, cut$$

and

$$\dfrac{\Gamma \vdash \Delta, u[x+0]_\Lambda = z}{\Gamma \vdash \Delta, u[x]_\Lambda = z}\, A3_2$$

is defined by

$$\dfrac{\dfrac{Assumption}{\Gamma \vdash \Delta, u[x+0]_\Lambda = z} \qquad \dfrac{\dfrac{A3}{\vdash x+0=x} \qquad \dfrac{Eq.1}{x+0 = x, u[x+0]_\Lambda = z \vdash u[x]_\Lambda = z}}{u[x+0]_\Lambda = z \vdash u[x]_\Lambda = z}\, cut}{\Gamma \vdash \Delta, u[x]_\Lambda = z}\, cut$$

- **A4**:

$$\dfrac{\Gamma \vdash \Delta, u[s(x+y)]_\Lambda = z}{\Gamma \vdash \Delta, u[x+s(y)]_\Lambda = z}\, A4_1$$

is defined by

$$\dfrac{\dfrac{Assumption}{\Gamma \vdash \Delta, u[s(x+y)]_\Lambda = z} \qquad \overset{(\varphi_1)}{u[s(x+y)]_\Lambda = z \vdash u[x+s(y)]_\Lambda = z}}{\Gamma \vdash \Delta, u[x+s(y)]_\Lambda = z}\, cut$$

where $\varphi_1 =$

$$\dfrac{\dfrac{\dfrac{A4}{\vdash x+s(y)=s(x+y)}}{\vdash s(x+y)=x+s(y)}\,sym_r \qquad \dfrac{Eq.1}{s(x+y)=x+s(y),\, u[s(x+y)]_\Lambda = z \vdash A}}{u[s(x+y)]_\Lambda = z \vdash u[x+s(y)]_\Lambda = z}\,cut$$

where $A = u[x+s(y)]_\Lambda = z$, and

$$\dfrac{\Gamma \vdash \Delta, u[x+s(y)]_\Lambda = z}{\Gamma \vdash \Delta, u[s(x+y)]_\Lambda = z}\,A4_2$$

is defined by

$$\dfrac{\dfrac{Assumption}{\Gamma \vdash \Delta, u[x+s(y)]_\Lambda = z} \qquad \begin{matrix}(\varphi_2)\\ u[x+s(y)]_\Lambda = z \vdash u[s(x+y)]_\Lambda = z\end{matrix}}{\Gamma \vdash \Delta, u[s(x+y)]_\Lambda = z}\,cut$$

where $\varphi_2 =$

$$\dfrac{\dfrac{A4}{\vdash x+s(y)=s(x+y)} \qquad \dfrac{Eq.1}{x+s(y)=s(x+y),\, u[x+s(y)]_\Lambda = z \vdash u[s(x+y)]_\Lambda = z}}{u[x+s(y)]_\Lambda = z \vdash u[s(x+y)]_\Lambda = z}\,cut$$

- **A7**:

$$\dfrac{\Gamma \vdash \Delta, u[s(x)]_\Lambda = z}{\Gamma \vdash \Delta, u[x+1]_\Lambda = z}\,A7_1$$

is defined by

$$\dfrac{\dfrac{Assumption}{\Gamma \vdash \Delta, u[s(x)]_\Lambda = z} \qquad \dfrac{\dfrac{A7}{\vdash s(x)=x+1} \qquad \dfrac{Eq.1}{s(x)=x+1,\, u[s(x)]_\Lambda = z \vdash A}}{u[s(x)]_\Lambda = z \vdash u[x+1]_\Lambda = z}\,cut}{\Gamma \vdash \Delta, u[x+1]_\Lambda = z}\,cut$$

where $A = u[x+1]_\Lambda = z$, and

$$\dfrac{\Gamma \vdash \Delta, u[x+1]_\Lambda = z}{\Gamma \vdash \Delta, u[s(x)]_\Lambda = z}\,A7_2$$

is defined by

$$\dfrac{\dfrac{Assumption}{\Gamma \vdash \Delta, A} \qquad \dfrac{\dfrac{\dfrac{A7}{\vdash s(x)=x+1}}{\vdash x+1=s(x)}\,sym_r \qquad \dfrac{Eq.1}{x+1=s(x),\, B \vdash u[s(x)]_\Lambda = z}}{A \vdash u[s(x)]_\Lambda = z}\,cut}{\Gamma \vdash \Delta, u[s(x)]_\Lambda = z}\,cut$$

where $A = u[x+1]_\Lambda = z$, and $B = u[x+1]_\Lambda = z$.

- **A8**:

$$\frac{\Gamma \vdash \Delta, u[s(0)]_{\Lambda} = z}{\Gamma \vdash \Delta, u[1]_{\Lambda} = z}\ A8_1$$

is defined by

$$\dfrac{\dfrac{\mathit{Assumption}}{\Gamma \vdash \Delta, u[s(0)]_{\Lambda} = z} \qquad \dfrac{\dfrac{A8}{\vdash s(0) = 1} \qquad \dfrac{Eq.1}{s(0) = 1, u[s(0)]_{\Lambda} = z \vdash u[1]_{\Lambda} = z}}{u[s(0)]_{\Lambda} = z \vdash u[1]_{\Lambda} = z}\ \mathit{cut}}{\Gamma \vdash \Delta, u[1]_{\Lambda} = z}\ \mathit{cut}$$

and

$$\frac{\Gamma \vdash \Delta, u[1]_{\Lambda} = z}{\Gamma \vdash \Delta, u[s(0)]_{\Lambda} = z}\ A8_2$$

is defined by

$$\dfrac{\dfrac{\mathit{Assumption}}{\Gamma \vdash \Delta, u[1]_{\Lambda} = z} \qquad \dfrac{\dfrac{\dfrac{A8}{\vdash s(0) = 1}}{\vdash 1 = s(0)}\ \mathit{sym}_r \qquad \dfrac{Eq.1}{1 = s(0), u[1]_{\Lambda} = z \vdash u[s(0)]_{\Lambda} = z}}{u[1]_{\Lambda} = z \vdash u[s(0)]_{\Lambda} = z}\ \mathit{cut}}{\Gamma \vdash \Delta, u[s(0)]_{\Lambda} = z}\ \mathit{cut}$$

Now we are ready to construct the proof schema for the commutativity of $+$.

**Example 8.0.1** Proof of 0 **is a left-identity**: We define a proof schema of

$$S(m): \ \vdash 0 + m = m,$$

where the recursive proof symbol $\breve{\rho}$ is given by:

$$\begin{aligned} D(\breve{\rho}) = \{&\breve{\rho}(\mathbf{0}) =_d \ \rho_0, \\ &\breve{\rho}(s(m)) =_d \rho(s(m))\{V_S(m) \leftarrow \breve{\rho}(m)\}\}. \end{aligned}$$

with $\rho_0 =$

$$\dfrac{\dfrac{A3}{\vdash 0 + 0 = 0}}{V_S(0)}\ V_S I$$

and $\rho(s(m)) =$

$$\dfrac{\dfrac{\dfrac{V_S(m)}{\vdash 0+m=m}\,V_SE \qquad \dfrac{\dfrac{0+m=m\vdash 0+m=m}{0+m=m\vdash s(0+m)=s(m)}\,subs}{0+m=m\vdash 0+s(m)=s(m)}\,A4_1}{\vdash 0+s(m)=s(m)}\,cut}{V_S(s(m))}\,V_SI$$

Proof of **associativity**: We define a proof schema of

$$S'(a,b,m') : \ \vdash (a+b)+m' = a+(b+m'),$$

where the recursive proof symbol $\breve{\rho}'$ is given by:

$$D(\breve{\rho}') = \{\breve{\rho}'(a,b,\mathbf{0}) =_d \ \rho'_0,$$
$$\breve{\rho}'(a,b,s(m')) =_d \rho'(a,b,s(m'))\{V_{S'}(a,b,m') \leftarrow \breve{\rho}'(a,b,m')\}\}.$$

with $\rho'_0 =$

$$\dfrac{\dfrac{\dfrac{\dfrac{Eq.2}{\vdash a+b=a+b}}{\vdash a+b=a+(b+0)}\,A3_1}{\vdash (a+b)+0=a+(b+0)}\,A3_1}{V_{S'}(a,b,0)}\,V_{S'}I$$

and $\rho'(a,b,s(m')) =$

$$\dfrac{\dfrac{\dfrac{V_{S'}(a,b,m')}{\vdash (a+b)+m'=a+(b+m')}\,V_{S'}E \qquad \dfrac{\dfrac{\dfrac{\dfrac{A\vdash A}{A\vdash s((a+b)+m')=s(a+(b+m'))}\,subs}{A\vdash s((a+b)+m')=a+s(b+m')}\,A4_1}{A\vdash s((a+b)+m')=a+(b+s(m'))}\,A4_1}{A\vdash (a+b)+s(m')=a+(b+s(m'))}\,A4_1}{\vdash (a+b)+s(m')=a+(b+s(m'))}\,cut}{V_{S'}(a,b,s(m'))}\,V_{S'}I$$

where $A = (a+b)+m' = a+(b+m')$. To prove **commutativity**, we first need to define a proof schema of

$$S''(m'') : \ \vdash m''+1 = 1+m''.$$

Let $\breve{\rho}''$ be

$$D(\breve{\rho}'') = \{\breve{\rho}''(\mathbf{0}) =_d \ \rho''_0,$$
$$\breve{\rho}''(s(m'')) =_d \rho''(s(m''))\{V_{S''}(m'') \leftarrow \breve{\rho}''(m'')\}\}.$$

with $\rho''_0 =$

$$\dfrac{\dfrac{\dfrac{V_S(1)}{\vdash 0+1=1}\,V_S E}{\vdash 0+1=1+0}\,A3_1}{V_{S''}(0)}\,V_{S''} I$$

and $\rho''(s(m''))=$

$$\dfrac{\dfrac{V_{S''}(m'')}{\vdash m''+1=1+m''}\,V_{S''}E \qquad \dfrac{\dfrac{\dfrac{\dfrac{\dfrac{\dfrac{A\vdash A}{A\vdash B}\,subs \qquad \dfrac{\dfrac{Eq.2}{\vdash 1+s(m'')=1+s(m'')}}{\vdash s(1+m'')=1+s(m'')}\,A4_2}{A\vdash s(m''+1)=1+s(m'')}\,trans}{A\vdash s((m''+1)+0)=1+s(m'')}\,A3_1}{A\vdash s(s(m'')+0)=1+s(m'')}\,A7_2}{A\vdash s(m'')+s(0)=1+s(m'')}\,A4_1}{A\vdash s(m'')+1=1+s(m'')}\,A9\ 2x,\ A8_1}{\dfrac{\vdash s(m'')+1=1+s(m'')}{V_{S''}(s(m''))}\,V_{S''}I}\,cut$$

where $A = 1+m'' = 1+m''$, and $B = s(m''+1) = s(1+m'')$. Now we define a proof schema of

$$S^k(n,k):\ \vdash n+k=k+n.$$

Let $\check{\rho}^k$ be

$$D(\check{\rho}^k) = \{\check{\rho}^k(n,\mathbf{0}) =_d \rho^k_0,$$
$$\check{\rho}^k(n,s(k)) =_d \rho^k(n,s(k))\{V(k) \leftarrow \check{\rho}^k(n,k)\}\}.$$

with $\rho^k_0=$

$$\dfrac{\dfrac{\dfrac{\dfrac{V_S(n)}{\vdash 0+n=n}\,V_S E}{\vdash n=0+n}\,sym_r}{\vdash n+0=0+n}\,A3_1}{V_{S^k}(0)}\,V_{S^k} I$$

and $\rho^k(n,s(k))=$

$$\dfrac{\dfrac{V^k_S(k)}{\vdash n+k=k+n}\,V_{S^k}E \qquad \begin{matrix}(\varphi)\\ n+k=k+n\vdash n+s(k)=s(k)+n\end{matrix}}{\dfrac{n+s(k)=s(k)+n}{V_{S^k}(s(k))}\,V_{S^k}I}\,cut$$

where $\varphi=$

$$\dfrac{\dfrac{\dfrac{V_{S'}(n,k,1)}{\vdash n+(k+1)=(n+k)+1}\,V_{S'}E \qquad \begin{matrix}(\varphi_1)\\ n+k=k+n\vdash (n+k)+1=(k+1)+n\end{matrix}}{n+k=k+n\vdash n+(k+1)=(k+1)+n}\,trans}{n+k=k+n\vdash n+s(k)=s(k)+n}\,A7_2\ 2x$$

where $\varphi_1 =$

$$
\dfrac{\dfrac{\dfrac{\dfrac{\dfrac{n+k=k+n \vdash n+k=k+n}{n+k=k+n \vdash s(n+k)=s(k+n)}\,\mathit{subs} \qquad \dfrac{\dfrac{\begin{array}{c}(\varphi_2)\\ \vdash k+(n+1)=(k+1)+n\end{array}}{\vdash k+s(n)=(k+1)+n}\,A7_2}{\vdash s(k+n)=(k+1)+n}\,A4_2}{n+k=k+n \vdash s(n+k)=(k+1)+n}\,\mathit{trans}}{n+k=k+n \vdash s((n+k)+0)=(k+1)+n}\,A3_1}{n+k=k+n \vdash (n+k)+s(0)=(k+1)+n}\,A4_1}{n+k=k+n \vdash (n+k)+1=(k+1)+n}\,A8_1
$$

$\varphi_2$ is

$$
\dfrac{\dfrac{V_{S'}(k,1,n)}{\vdash k+(1+n)=(k+1)+n}\,V_{S'}E \qquad \begin{array}{c}(\varphi_3)\\ k+(1+n)=(k+1)+n \vdash k+(n+1)=(k+1)+n\end{array}}{\vdash k+(n+1)=(k+1)+n}\,\mathit{cut}
$$

and $\varphi_3$ is

$$
\dfrac{\dfrac{\dfrac{V_{S''}(n)}{\vdash n+1=1+n}\,V_{S''}E}{\vdash 1+n=n+1}\,\mathit{sym}_r \qquad \dfrac{Eq.1}{1+n=n+1, k+(1+n)=(k+1)+n \vdash k+(n+1)=(k+1)+n}}{k+(1+n)=(k+1)+n \vdash k+(n+1)=(k+1)+n}\,\mathit{cut}
$$

In the example below we will use the schematic standard axiom set extended by the Peano axioms to provide a translation of the proof schema of commutativity from Example 8.0.1 to **PRA**.

**Example 8.0.2** We translate the proof schema $\breve{\rho}^k$ into:

$$
\dfrac{\dfrac{\dfrac{\dfrac{\dfrac{A3}{\vdash 0+0=0} \qquad \dfrac{\dfrac{0+y=y \vdash 0+y=y}{0+y=y \vdash s(0+y)=s(y)}\,\mathit{subs}}{0+y=y \vdash 0+s(y)=s(y)}\,A4_1}{\vdash 0+n=n}\,\mathit{ind}}{\vdash n=0+n}\,\mathit{sym}_r}{\vdash n+0=0+n}\,A3_1 \qquad \begin{array}{c}(\varphi(n,y))\\ S'\end{array}}{\vdash n+m=m+n}\,\mathit{ind}
$$

where $S' = n+y=y+n \vdash n+s(y)=s(y)+n$, and $\varphi(n,y) =$

$$
\dfrac{\dfrac{\begin{array}{c}(\varphi_{S'}(n,y,1))\\ \vdash n+(y+1)=(n+y)+1\end{array} \qquad \begin{array}{c}(\varphi_1(n,y))\\ n+y=y+n \vdash (n+y)+1=(y+1)+n\end{array}}{n+y=y+n \vdash n+(y+1)=(y+1)+n}\,\mathit{trans}}{n+y=y+n \vdash n+s(y)=s(y)+n}\,A7_2\,2x
$$

where $\varphi_1(n,y) =$

$$
\dfrac{\dfrac{\dfrac{\dfrac{\dfrac{n+y=y+n \vdash n+y=y+n}{n+y=y+n \vdash s(n+y)=s(y+n)}\,\mathit{subs} \qquad \dfrac{\dfrac{\begin{array}{c}(\varphi_2(n,y))\\ \vdash y+(n+1)=(y+1)+n\end{array}}{\vdash y+s(n)=(y+1)+n}\,A7_2}{\vdash s(y+n)=(y+1)+n}\,A4_2}{n+y=y+n \vdash s(n+y)=(y+1)+n}\,\mathit{trans}}{n+y=y+n \vdash s((n+y)+0)=(y+1)+n}\,A3_1}{n+y=y+n \vdash (n+y)+s(0)=(y+1)+n}\,A4_1}{n+y=y+n \vdash (n+y)+1=(y+1)+n}\,A8_1
$$

$\varphi_2(n, y)$ is

$$\dfrac{\overset{(\varphi_{S'}(y,1,n))}{\vdash y+(1+n)=(y+1)+n} \qquad \overset{(\varphi_3(n,y))}{y+(1+n)=(y+1)+n \vdash y+(n+1)=(y+1)+n}}{\vdash y+(n+1)=(y+1)+n}\ cut$$

and $\varphi_3(n, y)$ is

$$\dfrac{\dfrac{\overset{(\varphi_{S''}(n))}{\vdash n+1=1+n}}{\vdash 1+n=n+1}\ sym_r \qquad \dfrac{Eq.1}{1+n=n+1,\ y+(1+n)=(y+1)+n \vdash y+(n+1)=(y+1)+n}}{y+(1+n)=(y+1)+n \vdash y+(n+1)=(y+1)+n}\ cut$$

where $\varphi_{S''}(n) =$

$$\dfrac{\dfrac{\dfrac{\dfrac{A3}{\vdash 0+0=0} \qquad \dfrac{\dfrac{0+y=y \vdash 0+y=y}{0+y=y \vdash s(0+y)=s(y)}\ subs}{0+y=y \vdash 0+s(y)=s(y)}\ A4_1}{\vdash 0+1=1}\ ind}{\vdash 0+1=1+0}\ A3_1 \qquad \overset{(\varphi^1_{S''}(n))}{S''}}{\vdash n+1=1+n}\ ind$$

for $S'' = 1+y = 1+y \vdash s(y)+1 = 1+s(y)$, and $\varphi^1_{S''}(n) =$

$$\dfrac{\dfrac{\dfrac{\dfrac{\dfrac{\dfrac{1+y=1+y \vdash 1+y=1+y}{1+y=1+y \vdash s(y+1)=s(1+y)}\ subs \qquad \dfrac{\dfrac{Eq.2}{\vdash 1+s(y)=1+s(y)}}{\vdash s(1+y)=1+s(y)}\ A4_2}{1+y=1+y \vdash s(y+1)=1+s(y)}\ trans}{1+y=1+y \vdash s((y+1)+0)=1+s(y)}\ A3_1}{1+y=1+y \vdash s(s(y)+0)=1+s(y)}\ A7_2}{1+y=1+y \vdash s(y)+s(0)=1+s(y)}\ A4_1}{1+y=1+y \vdash s(y)+1=1+s(y)}\ A8_1$$

and $\varphi_{S'}(a, b, n) =$

$$\dfrac{\dfrac{\dfrac{\dfrac{Eq.2}{\vdash a+b=a+b}}{\vdash a+b=a+(b+0)}\ A3_1}{\vdash (a+b)+0=a+(b+0)}\ A3_1 \qquad \dfrac{\dfrac{\dfrac{\dfrac{A \vdash A}{A \vdash s((a+b)+y)=s(a+(b+y))}\ subs}{A \vdash s((a+b)+y)=a+s(b+y)}\ A4_1}{A \vdash s((a+b)+y)=a+(b+s(y))}\ A4_1}{A \vdash (a+b)+s(y)=a+(b+s(y))}\ A4_1}{\vdash (a+b)+s(n)=a+(b+s(n))}\ ind$$

for $A = (a+b)+y = a+(b+y)$.

# Chapter 9
# The Schematic CERES Method

In this chapter we will introduce the schematic CERES method.

## 9.1 Schematic CERES

The schematic CERES method for proof analysis is very similar to the first-order CERES method. We are working in a schematic setting, and thus need the concepts of characteristic formula schemata and projection schemata. The usual first-order definition of the characteristic formula and projection is based on the cut-status of the formula occurrences in a proof, i.e. whether a given formula occurrence is a cut-ancestor or not. However, a formula occurrence in a proof schema corresponds to many formula occurrences in its evaluation, some of which will be cut-ancestors and some will be not. We will illustrate this problem on some examples below. Let us first have a look on a simple example, where the cut-status of the formulas does not change with different evaluations.

**Example 9.1.1** Let $\hat{\rho}$ be the regular proof schema of the schematic sequent $S(n) = \forall x(P(x) \to P(f(x))) \vdash \forall x(P(x) \to P(\hat{f}(x,n)))$ defined in Example 7.1.8. The schematic proof $\hat{\rho}(n, V_S(n))$ is given by the recursive definition

$$D(\hat{\rho}) = \{\hat{\rho}(n, V_S(n)) =_d \underline{\text{if}}\ n = \bar{0}\ \underline{\text{then}}\ \rho'$$
$$\underline{\text{else}}\ \rho(n, V_S(n))\{V_S(p(n)) \leftarrow \hat{\rho}(p(n), V_S(p(n)))\}\},$$

where $\rho'$ is



A. Leitsch et al., *First-Order Schemata and Inductive Proof Analysis*, Computer Science Foundations and Applied Logic, https://doi.org/10.1007/978-3-032-05741-9_9

$$\dfrac{\dfrac{\dfrac{\dfrac{\dfrac{P(\hat{f}(X^{\rho'}(0),0)) \vdash P(\hat{f}(X^{\rho'}(0),0))}{P(X^{\rho'}(0)) \vdash P(\hat{f}(X^{\rho'}(0),0))}\ T^{+}}{\vdash P(X^{\rho'}(0)) \to P(\hat{f}(X^{\rho'}(0),0))}\ \to_r}{\vdash \forall x(P(x) \to P(\hat{f}(x,0)))}\ \forall_r}{S(0)}\ w_l}{V_S(0)}\ V_S I$$

and $\rho(n, V_S(n))$ is

$$\dfrac{\dfrac{\dfrac{V_S(p(n))}{S(p(n))}\ V_S E \qquad \begin{array}{c}(\rho_1(n))\\ \forall x(P(x) \to P(\hat{f}(x,p(n)))), A \vdash \forall x(P(x) \to P(\hat{f}(x,n)))\end{array}}{S(n)}\ cut, c_l}{V_S(n)}\ V_S I$$

for $A = \forall x(P(x) \to P(f(x)))$, and $\rho_1(n) =$

$$\dfrac{\dfrac{\dfrac{\dfrac{P(X^{\rho}(n)) \vdash P(X^{\rho}(n)) \qquad \begin{array}{c}(\varphi(n))\\ P(\hat{f}(X^{\rho}(n),p(n))), \forall x(P(x) \to P(f(x))) \vdash P(\hat{f}(X^{\rho}(n),n))\end{array}}{P(X^{\rho}(n)), P(X^{\rho}(n)) \to P(\hat{f}(X^{\rho}(n),p(n))), \forall x(P(x) \to P(f(x))) \vdash P(\hat{f}(X^{\rho}(n),n))}\ \to_l}{P(X^{\rho}(n)) \to P(\hat{f}(X^{\rho}(n),p(n))), \forall x(P(x) \to P(f(x))) \vdash P(X^{\rho}(n)) \to P(\hat{f}(X^{\rho}(n),n))}\ \to_r}{\forall x(P(x) \to P(\hat{f}(x,p(n)))), \forall x(P(x) \to P(f(x))) \vdash P(X^{\rho}(n)) \to P(\hat{f}(X^{\rho}(n),n))}\ \forall_l}{\forall x(P(x) \to P(\hat{f}(x,p(n)))), \forall x(P(x) \to P(f(x))) \vdash \forall x(P(x) \to P(\hat{f}(x,n)))}\ \forall_r$$

where $\varphi(n) =$

$$\dfrac{\dfrac{P(\hat{f}(X^{\rho}(n),p(n))) \vdash P(\hat{f}(X^{\rho}(n),p(n))) \qquad \dfrac{P(\hat{f}(X^{\rho}(n),n)) \vdash P(\hat{f}(X^{\rho}(n),n))}{P(f(\hat{f}(X^{\rho}(n),p(n)))) \vdash P(\hat{f}(X^{\rho}(n),n))}\ T^{+}}{P(\hat{f}(X^{\rho}(n),p(n))), P(\hat{f}(X^{\rho}(n),p(n))) \to P(f(\hat{f}(X^{\rho}(n),p(n)))) \vdash P(\hat{f}(X^{\rho}(n),n))}\ \to_l}{P(\hat{f}(X^{\rho}(n),p(n))), \forall x(P(x) \to P(f(x))) \vdash P(\hat{f}(X^{\rho}(n),n))}\ \forall_l$$

Note that this proof schema cannot be analyzed by CERES. In fact, we would need to skolemize it first, as the formula $\forall x(P(x) \to P(\hat{f}(x,n)))$ in $S(n)$ is not in Skolem form. However, as demonstrated in Example 7.1.8, $\hat{\rho}$ is used in defining the final proof schema of

$$\forall x(P(x) \to P(f(x))) \vdash (P(\hat{f}(c,n)) \to P(g(c,n))) \to (P(c) \to P(g(c,n))).$$

And indeed, $\forall x(P(x) \to P(\hat{f}(x,n)))$ in $S(n)$ goes into a cut in the final proof schema. In this example we will still only analyze $\hat{\rho}$, keeping in mind that the formula $\forall x(P(x) \to P(\hat{f}(x,n)))$ goes into a cut in the final proof schema. $\forall x(P(x) \to P(\hat{f}(x,n)))$ is therefore a so-called global cut-ancestor in $\hat{\rho}$, and thus defined via a so-called *global configuration* for $\hat{\rho}$, a notion that will be formally introduced in Definition 9.1.1.

Let us demonstrate how the characteristic formula schema of $\hat{\rho}$ is computed for different parameters $n$.

In the case where $n = \bar{0}$, $\hat{\rho}(0, V_S(0))$ is a schematic proof of $V_S(0)$. Let $\rho^*$ be $\rho'$ without the last inference, i.e. $\rho' =$

$$\frac{\rho*}{V_S(0)}\ V_S I$$

Then $\rho^* =$

$$\cfrac{\cfrac{\cfrac{\cfrac{\mathbf{P}(\hat{\mathbf{f}}(\mathbf{X}^{\rho'}(\mathbf{0}),\mathbf{0})) \vdash \mathbf{P}(\hat{\mathbf{f}}(\mathbf{X}^{\rho'}(\mathbf{0}),\mathbf{0}))}{\mathbf{P}(\mathbf{X}^{\rho'}(\mathbf{0})) \vdash \mathbf{P}(\hat{\mathbf{f}}(\mathbf{X}^{\rho'}(\mathbf{0}),\mathbf{0}))}\ T^+}{\vdash \mathbf{P}(\mathbf{X}^{\rho'}(\mathbf{0})) \to \mathbf{P}(\hat{\mathbf{f}}(\mathbf{X}^{\rho'}(\mathbf{0}),\mathbf{0}))}\ \to_r}{\vdash \forall \mathbf{x}(\mathbf{P}(\mathbf{x}) \to \mathbf{P}(\hat{\mathbf{f}}(\mathbf{x},\mathbf{0})))}\ \forall_r}{\forall x(P(x) \to P(f(x))) \vdash \forall \mathbf{x}(\mathbf{P}(\mathbf{x}) \to \mathbf{P}(\hat{\mathbf{f}}(\mathbf{x},\mathbf{0})))}\ w_l$$

Even though there are no cut-inferences in $\rho'$, and therefore no *local cut-ancestors* in $\hat{\rho}(0, V_S(0))$, there are still formulas that are ancestors of cuts (denoted in bold letters): The occurrence of the formula

$$\forall x(P(x) \to P(\hat{f}(x, 0)))$$

is the global configuration of $\rho'$, and therefore $\hat{\rho}(0, V_S(0))$ does contain cut-ancestors. Following the general method for the construction of the characteristic formula as given in Definition 3.2.1 (the formal definition of the characteristic formula schema will be given in Definition 9.1.2), the characteristic formula schema of $\hat{\rho}(0, V_S(0))$ is $\mathrm{CF}(\rho', \{(1, r)\}) = \mathrm{CF}(\rho^*, \{(1, r)\})$, where

$$\mathrm{CF}(\rho^*, \{(1, r)\}) = \neg P(\hat{f}(X^{\rho'}(0), 0)) \vee P(\hat{f}(X^{\rho'}(0), 0)),$$

where the set $\{(1, r)\}$, the configuration for $\rho^*$, denotes that the first formula on the right side of the end-sequent of $\rho^*$ is a global cut-ancestor.

Now let us consider the case $n = \bar{1}$. $\hat{\rho}(1, V_S(1))$ is a schematic proof of

$$S(1) = \forall x(P(x) \to P(f(x))) \vdash \forall x(P(x) \to P(\hat{f}(1, x)))$$

and is given by

$$\cfrac{\begin{matrix}(\rho^*) \\ \forall x(P(x) \to P(f(x))) \vdash \forall \mathbf{x}(\mathbf{P}(\mathbf{x}) \to \mathbf{P}(\hat{\mathbf{f}}(\mathbf{0},\mathbf{x})))\end{matrix} \qquad \begin{matrix}(\rho'') \\ S_{\rho''}\end{matrix}}{\forall x(P(x) \to P(f(x))) \vdash \forall \mathbf{x}(\mathbf{P}(\mathbf{x}) \to \mathbf{P}(\hat{\mathbf{f}}(\mathbf{1},\mathbf{x})))}\ cut, c_l$$

where $S_{\rho''} =$

$$\forall \mathbf{x}(\mathbf{P}(\mathbf{x}) \to \mathbf{P}(\hat{\mathbf{f}}(\mathbf{x},\mathbf{0}))), \forall x(P(x) \to P(f(x))) \vdash \forall \mathbf{x}(\mathbf{P}(\mathbf{x}) \to \mathbf{P}(\hat{\mathbf{f}}(\mathbf{x},\mathbf{1}))),$$

and $\rho^*$ is defined above and where $\rho''$ is

$$
\dfrac{\dfrac{\dfrac{\dfrac{\mathbf{P(X^{\rho}(1))} \vdash \mathbf{P(X^{\rho}(1))} \qquad \overset{(\rho'''(1))}{\mathbf{P(\hat{f}(X^{\rho}(1),0))}, \forall x(P(x) \to P(f(x))) \vdash \mathbf{P(\hat{f}(X^{\rho}(1),1))}}}{\mathbf{P(X^{\rho}(1))}, \mathbf{P(X^{\rho}(1)) \to P(\hat{f}(X^{\rho}(1),0))}, \forall x(P(x) \to P(f(x))) \vdash \mathbf{P(\hat{f}(X^{\rho}(1),1))}}\ {\to_l}}{\mathbf{P(X^{\rho}(1)) \to P(\hat{f}(X^{\rho}(1),0))}, \forall x(P(x) \to P(f(x))) \vdash \mathbf{P(X^{\rho}(1)) \to P(\hat{f}(X^{\rho}(1),1))}}\ {\to_r}}{\mathbf{\forall x(P(x) \to P(\hat{f}(x,0)))}, \forall x(P(x) \to P(f(x))) \vdash \mathbf{P(X^{\rho}(1)) \to P(\hat{f}(X^{\rho}(1),1))}}\ {\forall_l}}{\mathbf{\forall x(P(x) \to P(\hat{f}(x,0)))}, \forall x(P(x) \to P(f(x))) \vdash \mathbf{\forall x(P(x) \to P(\hat{f}(x,1)))}}\ {\forall_r}
$$

for $\rho'''(1) =$

$$
\dfrac{\dfrac{\mathbf{P(\hat{f}(X^{\rho}(1),0))} \vdash P(\hat{f}(X^{\rho}(1),0)) \qquad \dfrac{P(\hat{f}(X^{\rho}(1),1)) \vdash \mathbf{P(\hat{f}(X^{\rho}(1),1))}}{P(f(\hat{f}(X^{\rho}(1),0))) \vdash \mathbf{P(\hat{f}(X^{\rho}(1),1))}}\ {T^{+}}}{\mathbf{P(\hat{f}(X^{\rho}(1),0))}, P(\hat{f}(X^{\rho}(1),0)) \to P(f(\hat{f}(X^{\rho}(1),0))) \vdash \mathbf{P(\hat{f}(X^{\rho}(1),1))}}\ {\to_l}}{\mathbf{P(\hat{f}(X^{\rho}(1),0))}, \forall x(P(x) \to P(f(x))) \vdash \mathbf{P(\hat{f}(X^{\rho}(1),1))}}\ {\forall_l}
$$

Again, we assume a global configuration for $\hat{\rho}(1, V_S(1))$, where the formula schema $\forall x(P(x) \to P(\hat{f}(1,x)))$ is a global cut-ancestor. There is one local cut inference in $\hat{\rho}(1, V_S(1))$, with cut-formula $\forall x(P(x) \to P(\hat{f}(x,0)))$. All the formulas that are cut-ancestors are denoted in bold letters, and can be found in $\rho'$ and in $\rho''$.

Let us have a look at the characteristic formula schema of $\hat{\rho}(1, V_S(1))$. It is computed as

$$\mathrm{CF}(\hat{\rho}(1, V_S(1)), \{(1,r)\}) = \mathrm{CF}(\rho^*) \wedge \mathrm{CF}(\rho'', \{(1,l), (1,r)\}),$$

where $\mathrm{CF}(\rho')$ is the characteristic formula schema of $\rho'$ w.r.t. the local configuration $\forall x(P(x) \to P(\hat{f}(x,0)))$, and $\mathrm{CF}(\rho'', \{(1,l), (1,r)\})$ the characteristic formula schema of $\rho''$ w.r.t. the local configuration $\forall x(P(x) \to P(\hat{f}(x,0)))$, and the global configuration $\forall x(P(x) \to P(\hat{f}(1,x)))$. We have to combine the two characteristic formula schemata with a conjunction, as the two branches $\rho'$ and $\rho''$ are combined via a cut, and this inference obviously operates on cut-ancestors (only if the two branches are combined via an inference operating on ancestors of formulas in the end-sequent, we combine the formulas with a disjunction, see Definition 3.2.1 for the characteristic formula construction).

We start with $\mathrm{CF}(\rho^*)$, which is computed from the axiom $P(\hat{f}(X^{\rho'}(0),0)) \vdash P(\hat{f}(X^{\rho'}(0),0))$, where both formulas are ancestors of cuts, i.e.

$$\mathrm{CF}(\rho') = \neg P(\hat{f}(X^{\rho'}(0),0)) \vee P(\hat{f}(X^{\rho'}(0),0)).$$

For $\mathrm{CF}(\rho'', \{(1,l), (1,r)\})$, the formulas extracted from initial sequents containing cut-ancestors are

$$\neg P(X^{\rho}(1)) \vee P(X^{\rho}(1)), \quad \neg P(\hat{f}(X^{\rho}(1),0)), \quad \text{and} \quad P(\hat{f}(X^{\rho}(1),1)).$$

The implication-left inference combining the branches of $\neg P(\hat{f}(X^{\rho}(1),0))$ and $P(\hat{f}(X^{\rho}(1),1))$ operates on ancestors of the end-sequent, thus the two formulas are combined with a disjunction, $\neg P(\hat{f}(X^{\rho}(1),0)) \vee P(\hat{f}(X^{\rho}(1),1))$, and the $\to_l$

inference combining this branch with the one of $\neg P(X^{\rho}(1)) \vee P(X^{\rho}(1))$ operates on cut-ancestors, therefore
$\mathrm{CF}(\rho'', \{(1,l),(1,r)\}) =$

$$(\neg P(X^{\rho}(1)) \vee P(X^{\rho}(1))) \wedge (\neg P(\hat{f}(X^{\rho}(1),0)) \vee P(\hat{f}(X^{\rho}(1),1))).$$

Putting things together, the characteristic formula schema of $\hat{\rho}(1, V_S(1))$ is
$\mathrm{CF}(\hat{\rho}(1, V_S(1)), \{(1,r)\}) =$

$$\begin{array}{l}(\neg P(\hat{f}(X^{\rho'}(0),0)) \vee P(\hat{f}(X^{\rho'}(0),0)))\wedge \\ (\neg P(X^{\rho}(1)) \vee P(X^{\rho}(1)) \wedge \neg P(\hat{f}(X^{\rho}(1),0)) \vee P(\hat{f}(X^{\rho}(1),1))).\end{array}$$

The last parameter instantiation we will consider in this example is $n = \bar{2}$. We construct $\hat{\rho}(2, V_S(2))$, which is a proof schema of

$$S(2) = \forall x(P(x) \to P(f(x))) \vdash \forall x(P(x) \to P(\hat{f}(2,x)))$$

$$\cfrac{\begin{array}{c}(\rho)\\ S(1)\end{array} \qquad \cfrac{\cfrac{\cfrac{\cfrac{\mathbf{P(X^{\rho}(2))} \vdash \mathbf{P(X^{\rho}(2))} \qquad \begin{array}{c}(\rho'''(2))\\ \mathbf{P(\hat{f}(X^{\rho}(2),1))}, A \vdash \mathbf{P(\hat{f}(X^{\rho}(2),2))}\end{array}}{\mathbf{P(X^{\rho}(2))}, \mathbf{P(X^{\rho}(2))} \to \mathbf{P(\hat{f}(X^{\rho}(2),1))}, A \vdash \mathbf{P(\hat{f}(X^{\rho}(2),2))}}\,\to_l}{\mathbf{P(X^{\rho}(2))} \to \mathbf{P(\hat{f}(X^{\rho}(2),1))}, A \vdash \mathbf{P(X^{\rho}(2))} \to \mathbf{P(\hat{f}(X^{\rho}(2),2))}}\,\to_r}{\forall \mathbf{x}(\mathbf{P(x)} \to \mathbf{P(\hat{f}(x,1))}), A \vdash \mathbf{P(X^{\rho}(2))} \to \mathbf{P(\hat{f}(X^{\rho}(2),2))}}\,\forall_l}{\forall \mathbf{x}(\mathbf{P(x)} \to \mathbf{P(\hat{f}(x,1))}), A \vdash \forall \mathbf{x}(\mathbf{P(x)} \to \mathbf{P(\hat{f}(x,2))})}\,\forall_r}{A \vdash \forall \mathbf{x}(\mathbf{P(x)} \to \mathbf{P(\hat{f}(2,x))})}\,cut, c_l$$

for $A = \forall x(P(x) \to P(f(x)))$, and $\rho'''(2) =$

$$\cfrac{\cfrac{\mathbf{P(\hat{f}(X^{\rho}(2),1))} \vdash P(\hat{f}(X^{\rho}(2),1)) \qquad \cfrac{P(\hat{f}(X^{\rho}(2),2)) \vdash \mathbf{P(\hat{f}(X^{\rho}(2),2))}}{P(f(\hat{f}(X^{\rho}(2),1))) \vdash \mathbf{P(\hat{f}(X^{\rho}(2),2))}}\,T^{+}}{\mathbf{P(\hat{f}(X^{\rho}(2),1))}, P(\hat{f}(X^{\rho}(2),1)) \to P(f(\hat{f}(X^{\rho}(2),1))) \vdash \mathbf{P(\hat{f}(X^{\rho}(2),2))}}\,\to_l}{\mathbf{P(\hat{f}(X^{\rho}(2),1))}, \forall x(P(x) \to P(f(x))) \vdash \mathbf{P(\hat{f}(X^{\rho}(2),2))}}\,\forall_l$$

and where $S(1) = \forall x(P(x) \to P(f(x))) \vdash \forall \mathbf{x}(\mathbf{P(x)} \to \mathbf{P(\hat{f}(1,x))})$ and $\rho$ is

$$\cfrac{\begin{array}{c}(\rho^{*})\\ \forall x(P(x) \to P(f(x))) \vdash \forall \mathbf{x}(\mathbf{P(x)} \to \mathbf{P(\hat{f}(0,x))})\end{array} \qquad \begin{array}{c}(\rho'')\\ S_{\rho''}\end{array}}{\forall x(P(x) \to P(f(x))) \vdash \forall \mathbf{x}(\mathbf{P(x)} \to \mathbf{P(\hat{f}(1,x))})}\,cut, c_l$$

where $S_{\rho''} =$

$$\forall \mathbf{x}(\mathbf{P(x)} \to \mathbf{P(\hat{f}(x,0))}), \forall x(P(x) \to P(f(x))) \vdash \forall \mathbf{x}(\mathbf{P(x)} \to \mathbf{P(\hat{f}(x,1))}),$$

and for $\rho^{*}$ and $\rho''$ defined above. All formulas that are ancestors of cut-formulas are indicated in bold letters. The characteristic formula schema of $\hat{\rho}(2, V_S(2))$ is computed as

$\mathrm{CF}(\hat{\rho}(2, V_S(2)), \{(1, r)\}) =$

$$\mathrm{CF}(\rho)\wedge$$
$$(\neg P(X^{\rho}(2)) \vee P(X^{\rho}(2))) \wedge (\neg P(\hat{f}(X^{\rho}(2), 1)) \vee P(\hat{f}(X^{\rho}(2), 2))),$$

where $\mathrm{CF}(\rho)$ is the characteristic formula schema from the left branch of the lowermost cut in $\hat{\rho}(2, V_S(2))$ (i.e. the characteristic formula schema of $\rho$ w.r.t. the local configuration $\forall x(P(x) \to P(\hat{f}(x, 1))))$, and the other part of the formula is just the characteristic formula schema computed from the right branch of the lowermost cut in $\hat{\rho}(2, V_S(2))$ w.r.t. the local configuration $\forall x(P(x) \to P(\hat{f}(x, 1))))$ and the global cut configuration $\forall x(P(x) \to P(\hat{f}(x, 1)))$.

We have already computed $\mathrm{CF}(\rho)$ by $\rho = \hat{\rho}(1, V_S(1))$, thus

$$\begin{aligned}&\mathrm{CF}(\rho, \{(1, r)\}) = \mathrm{CF}(\hat{\rho}(1, V_S(1)), \{(1, r)\}) =\\&(\neg P(\hat{f}(X^{\rho'}(0), 0)) \vee P(\hat{f}(X^{\rho'}(0), 0)))\wedge\\&\ (\neg P(X^{\rho}(1)) \vee P(X^{\rho}(1)) \wedge \neg P(\hat{f}(X^{\rho}(1), 0)) \vee P(\hat{f}(X^{\rho}(1), 1))).\end{aligned}$$

Putting things together we finally obtain

$$\begin{aligned}&\mathrm{CF}(\hat{\rho}(2, V_S(2)), \{1, r\}) = (\neg P(\hat{f}(X^{\rho'}(0), 0)) \vee P(\hat{f}(X^{\rho'}(0), 0)))\wedge\\&\ (\neg P(X^{\rho}(1)) \vee P(X^{\rho}(1)) \wedge (\neg P(\hat{f}(X^{\rho}(1), 0)) \vee P(\hat{f}(X^{\rho}(1), 1)))\wedge\\&\ (\neg P(X^{\rho}(2)) \vee P(X^{\rho}(2))) \wedge (\neg P(\hat{f}(X^{\rho}(2), 1)) \vee P(\hat{f}(X^{\rho}(2), 2))).\end{aligned}$$

In the next example the cut-status of a formula will change for different evaluations.

**Example 9.1.2** Let

$$S(n) = \forall x(Px \to Pf(x)) \vdash \exists x(Px \to P\hat{f}(x, n)).$$

For improving readability we write $Pt$ instead of $P(t)$ throughout the whole example. In contrast to the last example this sequent is skolemized and could be an end-sequent for schematic CERES. We consider the following proof $\rho'$ of $S(\bar{0})$:

$\rho' =$

$$\dfrac{\dfrac{\dfrac{\dfrac{Pa \vdash Pa}{Pa \vdash P\hat{f}(a, \bar{0})}\ T^{+}}{\vdash Pa \to P\hat{f}(a, \bar{0})}\ \to: r}{\vdash \exists x(Px \to P\hat{f}(x, \bar{0}))}\ \exists: r}{\dfrac{\forall x(Px \to Pf(x)) \vdash \exists x(Px \to P\hat{f}(x, \bar{0}))}{S(\bar{0})}\ V_S I}\ w: l$$

If we consider the last sequent of $\rho'$ as an end-sequent we get $\mathrm{CF}(\rho') = \bot$. Assume now that the formula $\exists x(Px \to P\hat{f}(x, \bar{0}))$ is an ancestor of a cut. As in Example 9.1.1 we write $\mathrm{CF}(\rho', \{(1, r)\})$ to express that now the first formula on the right of $S(\bar{0})$ is an ancestor of a cut. Then, obviously

$$\mathrm{CF}(\rho', \{(1, r)\}) = Pa \vee \neg Pa.$$

Now consider $\rho(X, n)$ for $n > \bar{0}$ where $\rho(X, n) = \rho_1(y, n)\{y \leftarrow X(n)\}$ and $\rho_1(y, n) =$

$$
\dfrac{\dfrac{V_S(p(n))}{S(p(n))} \qquad \dfrac{\dfrac{\dfrac{\dfrac{\dfrac{\dfrac{Py \vdash Py \quad \dfrac{P\hat{f}(y, p(n)) \vdash P\hat{f}(y, p(n)) \quad \dfrac{P\hat{f}(y, n) \vdash P\hat{f}(y, n)}{Pf(\hat{f}(y, p(n))) \vdash P\hat{f}(y, n)}\,T^{+}}{P\hat{f}(y, p(n)), P\hat{f}(y, p(n)) \to Pf(\hat{f}(y, p(n))) \vdash P\hat{f}(y, n)}\,{\to}\colon l}{Py, Py \to P\hat{f}(y, p(n)), P\hat{f}(y, p(n)) \to Pf(\hat{f}(y, p(n))) \vdash P\hat{f}(y, n)}\,{\to}\colon l}{Py \to P\hat{f}(y, p(n)), P\hat{f}(y, p(n)) \to Pf(\hat{f}(y, p(n))) \vdash Py \to P\hat{f}(y, n)}\,{\to}\colon r}{Py \to P\hat{f}(y, p(n)), P\hat{f}(y, p(n)) \to Pf(\hat{f}(y, p(n))) \vdash \exists x(Px \to P\hat{f}(x, n))}\,\exists\colon r}{Py \to P\hat{f}(y, p(n)), \forall x(Px \to Pf(x)) \vdash \exists x(Px \to P\hat{f}(x, n))}\,\forall\colon l}{\exists x(Px \to P\hat{f}(x, p(n))), \forall x(Px \to Pf(x)) \vdash \exists x(Px \to P\hat{f}(x, n))}\,\exists\colon l}{S(n)\colon \forall x(Px \to Pf(x)) \vdash \exists x(Px \to P\hat{f}(x, n))}\,cut + c\colon l
$$

In $\rho(X, n)$ we have a cut with the cut formula $\exists x(Px \to P\hat{f}(x, p(n)))$, where on the left-hand-side of the cut we have a link variable but no derivation. In computing the characteristic formula we simply use the link variable $V_S(p(n))$ which stands for a proof to be inserted. If the last sequent $S(n)$ of $\rho(X, n)$ is an end-sequent then the characteristic formula for the empty cut-configuration of $S(n)$ (i.e. $S(n)$ has no cut ancestors) is

$$\mathrm{CF}(\rho(X, n), \emptyset) = V_S(p(n)) \wedge PX(n) \wedge \neg P\hat{f}(X(n), p(n)).$$

If, however, the last formula $\exists x(Px \to P\hat{f}(x, n))$ in $S(n)$ is a cut-ancestor we get the cut configuration $\{(1, r)\}$ of $S(n)$ and $\mathrm{CF}(\rho(X, n), \{(1, r)\}) =$

$$V_S(p(n)) \wedge PX(n) \wedge (\neg P\hat{f}(X(n), p(n)) \vee P(\hat{f}(X(n), n))).$$

We consider now the inductive closure $\hat{\rho}$ of $\rho$:

$$\hat{\rho}(X, n) =_d \underline{\text{if}}\ n = \bar{0}\ \underline{\text{then}}\ \rho'\ \underline{\text{else}}\ \rho(X, n)\{V(p(n)) \leftarrow \hat{\rho}(X, p(n))\}.$$

When we compute $\mathrm{CF}(\hat{\rho}(X, \bar{0}))$ we have no cut-ancestors and so

$$\mathrm{CF}(\hat{\rho}(X, \bar{0}), \emptyset) = \bot.$$

But in computing $\mathrm{CF}(\hat{\rho}(X, \bar{1}))$ (with empty cut configuration in $S(\bar{1})$) we use $\rho'$ but now with the cut ancestor $(1, r)$ in $\rho'$ and therefore $\mathrm{CF}(\hat{\rho}(X, \bar{1})) =$

$$(Pa \vee \neg Pa) \wedge PX(\bar{1}) \wedge \neg P\hat{f}(X(\bar{1}), \bar{0}).$$

If, on the other hand, the last formula $\exists x(Px \to P\hat{f}(x, \bar{1}))$ in $S(\bar{1})$ is a cut-ancestor (which happens if we compute $\hat{\rho}(X, \bar{2})$) we obtain
$\mathrm{CF}(\hat{\rho}(X, \bar{1}), \{(1, r)\}) =$

$$(Pa \vee \neg Pa) \wedge PX(\bar{1}) \wedge (\neg P\hat{f}(X(\bar{1}), \bar{0}) \vee P(\hat{f}(X(\bar{1}), \bar{1}))).$$

So when we compute $\mathrm{CF}(\hat{\rho}(X, \bar{2}))$ we obtain the formula

$$(Pa \vee \neg Pa) \wedge PX(\bar{1}) \wedge (\neg P\hat{f}(X(\bar{1}), \bar{0}) \vee P(\hat{f}(X(\bar{1}), \bar{1})))\wedge$$
$$PX(\bar{2}) \wedge \neg P\hat{f}(X(\bar{2}), \bar{1}).$$

It is easy to see that for $n \geq \bar{1}$ $\mathrm{CF}(\hat{\rho}(X, n), \{(1, r)\}) =$

$$(Pa \vee \neg Pa) \wedge \bigwedge_{i=1}^{n} PX(i) \wedge (\neg P\hat{f}(X(i), p(i)) \vee P(\hat{f}(X(i), i)))$$

and

$$\mathrm{CF}(\hat{\rho}(X, n)) = \mathrm{CF}(\hat{\rho}^*(X, p(n)), \{(1, r)\}) \wedge PX(n) \wedge \neg P\hat{f}(X(n), p(n)).$$

The two last examples have shown that we need some machinery to track the cut-status of formula occurrences in proof schemata, which leads to a first difference between the first-order and the schematic CERES method: we add a new parameter to the characteristic formula and projection of a proof symbol which will denote the set of formula occurrences in the end-sequent of that proof symbol that are cut-ancestors.

**Definition 9.1.1** (*configuration*) The set $\Omega$ of formula occurrences in the end-sequent of a subproof $\pi$ of a proof $\varphi$ which are ancestors of a cut in $\varphi$ is called a *global configuration* for $\pi$ w.r.t. $\varphi$. If $\nu$ is a node in $\pi$ we define by $S(\nu, \Omega)$ the subsequent of $S(\nu)$ consisting of all formulas which are ancestors of $\Omega$ or of a cut in $\pi$. Let $V_S$ be a link expression corresponding to $S$ then we define $V_S(\nu, \Omega)$ as corresponding to $S(\nu, \Omega)$.

Note that the configuration at a node $\nu$ might differ from the configuration of another node $\nu'$, which occurs above $\nu$ in a proof schema. Indeed, by the application of cut inferences or contractions, the set of formula occurrences which are ancestors of cuts might grow, and by the application of weakening it might decrease. Furthermore, the position of these formulas that are ancestors of cuts can be changed.

Below we will give a formal definition of the characteristic formula schema and projection schema. Recall that in the first-order CERES method, the projection of a proof $\varphi$ of end-sequent $\Gamma \vdash \Delta$ is constructed by eliminating all inferences operating on cut-ancestors from $\varphi$, resulting in a cut-free proof of $\Gamma \vdash \Delta, F$, where $F$ is the characteristic formula of $\varphi$. This is also the case in the schematic setting: in fact the projection schema is a cut-free proof schema of the same end-sequent as the original proof schema, concatenated with the characteristic formula schema.

**Definition 9.1.2** Let $\rho(X, \mathbf{n}, k)$ be a regular $LK_{T^+}$-proof schema of

$$S(X, \mathbf{n}, k),$$

where $\mathbf{n}$ and $k$ are all the parameters and $X$ all the variables introduced in $\rho$ during regularization. For each node $\nu$ in $\rho(X, \mathbf{n}, k)$ we define the characteristic formula schema

$$\mathrm{CF}(\rho, \nu, \Omega)$$

and the projection schema

$$\mathrm{Proj}(\rho, \nu, \Omega)$$

w.r.t. a configuration $\Omega$. Then, the characteristic formula schema

$$\mathrm{CF}(\rho(X, \mathbf{n}, k), \Omega)$$

and the projection schema

$$\mathrm{Proj}(\rho(X, \mathbf{n}, k), \Omega)$$

of $\rho(X, \mathbf{n}, k)$ w.r.t. a configuration $\Omega$ of $S(X, \mathbf{n}, k)$ are defined inductively as $\mathrm{CF}(\rho, \nu_0, \Omega)$ and $\mathrm{Proj}(\rho, \nu_0, \Omega)$, where $\nu_0$ is the node at $S(X, \mathbf{n}, k)$:

- If $\rho(X, \mathbf{n}, k)$ at node $\nu$ is a schematic axiom $A \vdash A$ we distinguish the cases:

  1. $S(\nu, \Omega) = \vdash$. Then $\mathrm{CF}(\rho, \nu, \Omega) = \bot$, and $\mathrm{Proj}(\rho, \nu, \Omega) = A \vdash A$.
  2. $S(\nu, \Omega) = \vdash A$. Then $\mathrm{CF}(\rho, \nu, \Omega) = A$, and $\mathrm{Proj}(\rho, \nu, \Omega) = A \vdash \mathrm{CF}(\rho, \nu, \Omega)$.
  3. $S(\nu, \Omega) = A \vdash$. Then $\mathrm{CF}(\rho, \nu, \Omega) = \neg A$, and $\mathrm{Proj}(\rho, \nu, \Omega) =$

  $$\frac{A \vdash A}{\vdash A, \mathrm{CF}(\rho, \nu, \Omega)} \neg : r$$

  4. $S(\nu, \Omega) = A \vdash A$. Then $\mathrm{CF}(\rho, \nu, \Omega) = A \vee \neg A$, and $\mathrm{Proj}(\rho, \nu, \Omega) =$

  $$\frac{\dfrac{A \vdash A}{\vdash A, \neg A} \neg : r}{\vdash \mathrm{CF}(\rho, \nu, \Omega)} \vee : r$$

- Let $\rho(X, \mathbf{n}, k)$ be a proof such that the sequents at all leaf nodes $\nu$ of the form $S(\nu)$ or $V_S(\nu)$ such that all nodes $S(\nu)$ alre labeled by axioms. Let $M$ be the set of leaves $\nu$ in $\rho$ labeled by link expressions $V_S(\nu)$ and assume that there exists schematic proofs $\rho(\nu)$ of $V_S(\nu)$ for all $\nu \in M$. For every node $\nu$ in $M$ let $\Omega_\nu$ be the local configurations of $S(\nu)$ in $\rho$ and $\Omega$ be the global configurations in $\rho$. Then

$$\begin{aligned}&\mathrm{CF}(\rho(X, \mathbf{n}, k, \Omega)\{V_S(\nu) \leftarrow \rho(\nu)\}) = \\ &\mathrm{CF}(\rho, \nu, \Omega)\{V_S(\nu, \Omega \cup \Omega_\nu) \leftarrow \mathrm{CF}(\rho(\nu, \Omega \cup \Omega_u))\}.\end{aligned}$$

and

$$\begin{aligned}&\mathrm{Proj}(\rho(X, \mathbf{n}, k, \Omega)\{V_S(\nu) \leftarrow \rho(\nu)\}) = \\ &\mathrm{Proj}(\rho, \nu, \Omega)\{V_S(\nu, \Omega \cup \Omega_\nu) \leftarrow \mathrm{Proj}(\nu, \Omega \cup \Omega_u)\}.\end{aligned}$$

- Assume that $\xi$ is a unary inference with premise $\nu'$ and conclusion $\nu$ in $\rho(X, \mathbf{n}, k)$. Then $\mathrm{CF}(\rho, \nu, \Omega) = \mathrm{CF}(\rho, \nu', \Omega')$, where $\Omega'$ contains the ancestors of $\Omega$ in $\nu'$. For defining $\mathrm{Proj}(\rho, \nu, \Omega)$ we distinguish three cases:

1. Let $\nu'$ be the premise and $\nu$ the conclusion of an inference $\xi$ of the form

$$\frac{S(X, \mathbf{n}, k)}{V_S(X, \mathbf{n}, k)}\ V_S I$$

then $\mathrm{Proj}(\rho, \nu, \Omega) = \mathrm{Proj}(\rho, \nu', \Omega)$,
2. the principal formula of $\xi$ is an ancestor of $S(\nu, \Omega)$, then $\mathrm{Proj}(\rho, \nu, \Omega) = \mathrm{Proj}(\rho, \nu', \Omega')$,
3. the principal formula of $\xi$ is not an ancestor of $S(\nu, \Omega)$, then $\mathrm{Proj}(\rho, \nu, \Omega)$ is

$$\frac{\begin{array}{c}(\mathrm{Proj}(\rho, \nu', \Omega'))\\ \Gamma \vdash \Delta, \mathrm{CF}(\rho, \nu', \Omega')\end{array}}{\Gamma' \vdash \Delta', \mathrm{CF}(\rho, \nu, \Omega)}\ \xi$$

where $\Gamma \vdash \Delta$ are not ancestors of $S(\nu', \Omega')$, and $\Gamma' \vdash \Delta'$ are not ancestors of $S(\nu, \Omega)$,

where $\Omega'$ contains the ancestors of $\Omega$ in $\nu'$.

- Assume $\xi$ is a binary inference with premises $\nu_1, \nu_2$ and conclusion $\nu$ in $\rho(X, \mathbf{n}, k)$. We distinguish two cases:

1. the auxiliary formulas in $\xi$ are ancestors of $S(\nu, \Omega)$. Then we define

$$\mathrm{CF}(\rho, \nu, \Omega) = \mathrm{CF}(\rho, \nu_1, \Omega_1) \wedge \mathrm{CF}(\rho, \nu_2, \Omega_2),$$

where $\Omega_1$ contains the ancestors of $\Omega$ in $\nu_1$, and $\Omega_2$ contains the ancestors of $\Omega$ in $\nu_2$. Moreover, $\mathrm{Proj}(\rho, \nu, \Omega)$ is

$$\frac{\begin{array}{c}(\mathrm{Proj}(\rho, \nu_1, \Omega_1))\\ \Gamma_1 \vdash \Delta_1, \mathrm{CF}(\rho, \nu_1, \Omega_1)\end{array} \qquad \begin{array}{c}(\mathrm{Proj}(\rho, \nu_2, \Omega_2))\\ \Gamma_2 \vdash \Delta_2, \mathrm{CF}(\rho, \nu_2, \Omega_2)\end{array}}{\Gamma_1, \Gamma_2 \vdash \Delta_1, \Delta_2, \mathrm{CF}(\rho, \nu, \Omega)}\ \wedge : r$$

where $\Gamma_1 \vdash \Delta_1$ are not ancestors of $S(\nu_1, \Omega_1)$ and $\Gamma_2 \vdash \Delta_2$ are not ancestors of $S(\nu_2, \Omega_2)$.
2. the auxiliary formulas in $\xi$ are not ancestors of $S(\nu, \Omega)$. Then we define

$$\mathrm{CF}(\rho, \nu, \Omega) = \mathrm{CF}(\rho, \nu_1, \Omega_1) \vee \mathrm{CF}(\rho, \nu_2, \Omega_2),$$

where $\Omega_1$ contains the ancestors of $\Omega$ in $\nu_1$, and $\Omega_2$ contains the ancestors of $\Omega$ in $\nu_2$. Moreover, $\mathrm{Proj}(\rho, \nu, \Omega)$ is

$$\dfrac{\dfrac{\begin{array}{c}(\mathrm{Proj}(\rho, \nu_1, \Omega_1)) \\ \Gamma_1 \vdash \Delta_1, \mathrm{CF}(\rho, \nu_1, \Omega_1)\end{array} \qquad \begin{array}{c}(\mathrm{Proj}(\rho, \nu_2, \Omega_2)) \\ \Gamma_2 \vdash \Delta_2, \mathrm{CF}(\rho, \nu_2, \Omega_2)\end{array}}{\Gamma_1', \Gamma_2' \vdash \Delta_1', \Delta_2', \mathrm{CF}(\rho, \nu_1, \Omega_1), \mathrm{CF}(\rho, \nu_2, \Omega_2)}\ \xi}{\Gamma_1', \Gamma_2' \vdash \Delta_1', \Delta_2', \mathrm{CF}(\rho, \nu, \Omega)}\ \vee : r$$

where $\Gamma_1 \vdash \Delta_1$ are not ancestors of $S(\nu_1, \Omega_1)$ and $\Gamma_2 \vdash \Delta_2$ are not ancestors of $S(\nu_2, \Omega_2)$.

- If $\rho(X, \mathbf{n}, k)$ is defined via proof recursion by $\rho'$, then $\rho$ is some $\hat{\rho}$, where

$$\begin{aligned} D(\hat{\rho}) = \{\hat{\rho}(X, \mathbf{n}, k) =_d\ & \underline{\text{if}}\ k = \bar{0}\ \underline{\text{then}}\ \rho_0 \\ & \underline{\text{else}}\ \rho'(X, (\mathbf{n}, k), V(k))\{V(p(k)) \leftarrow \hat{\rho}(X, (\mathbf{n}, p(k)))\}\}, \end{aligned}$$

and $\hat{\rho}(X, \mathbf{n}, k)$ is a subproof of the whole proof schema $\varphi(X, \mathbf{n}, k)$. It is important that the end-sequents of $\rho'(X, (\mathbf{n}, k, V(k))$ are of the form $S(X, \mathbf{n}, k)$ and the global configurations of all these sequents and of $\rho_0$ are equal. Let $\Omega$ be the global configuration of $\rho'(X, \mathbf{n}, k)$ and of $\rho_0$. Then

$$\mathrm{CF}(\hat{\rho}(X, \mathbf{n}, \bar{0}), \Omega) = \mathrm{CF}(\rho_0, \Omega),$$

and

$$\mathrm{Proj}(\hat{\rho}(X, \mathbf{n}, \bar{0}), \Omega) = \mathrm{Proj}(\rho_0, \Omega).$$

Now we define $\mathrm{CF}(\hat{\rho}(X, \mathbf{n}, m), \Omega)$ and $\mathrm{Proj}(\hat{\rho}(X, \mathbf{n}, m), \Omega)$ for $m > 0$. If $1 \leq k < m$ the proof $\hat{\rho}(X, \mathbf{n}, k)$ is of the form

$$\rho'(X, \mathbf{n}, k)\{V(p(k)) \leftarrow \hat{\rho}(X, (\mathbf{n}, p(k)))\}$$

and the end-sequent of $\hat{\rho}(X, \mathbf{n}, k)$ has cut-ancestors $\Omega'$, namely the formula occurrences in the end-sequent $S(X, \mathbf{n}, k)$ being ancestors of cuts in $\hat{\rho}(X, \mathbf{n}, s(k))$. Note that $\Omega'$ is the same configuration for all $1 \leq k < m$. For all these $k$ we define

$$\mathrm{CF}(\hat{\rho}(X, \mathbf{n}, k), \Omega) = \mathrm{CF}(\hat{\rho}(X, \mathbf{n}, p(k)), \Omega) \wedge \mathrm{CF}(\rho'(X, \mathbf{n}, k), \Omega \cup \Omega'),$$

and $\mathrm{Proj}(\hat{\rho}(X, \mathbf{n}, k), \Omega) =$

$$\mathrm{Proj}(\rho'(X, (\mathbf{n}, k), V(k), \Omega \cup \Omega')\{V(p(k)) \leftarrow \mathrm{Proj}(\hat{\rho}(X, (\mathbf{n}, p(k))), \Omega)\}.$$

For $m$ itself we obtain

$$\mathrm{CF}(\hat{\rho}(X, \mathbf{n}, m), \Omega) = \mathrm{CF}(\hat{\rho}(X, \mathbf{n}, p(m)), \Omega) \wedge \mathrm{CF}(\rho'(X, \mathbf{n}, m), \Omega),$$

and $\mathrm{Proj}(\hat{\rho}(X, \mathbf{n}, m), \Omega) =$

$$\mathrm{Proj}(\rho'(X, \mathbf{n}, m, V(m), \Omega)\{V(p(m)) \leftarrow \mathrm{Proj}(\hat{\rho}(X, (\mathbf{n}, p(m))), \Omega)\}.$$

For a regular skolemized proof schema $\rho(X, \mathbf{n}, k)$ that is not used in the definition of another proof schema and thus can be used as input proof schema for CERES, the configuration of $\rho(X, \mathbf{n}, k)$ is empty, and we define $\mathrm{CF}(\rho(X, \mathbf{n}, k)) = \mathrm{CF}(\rho(X, \mathbf{n}, k), \emptyset)$ as the characteristic formula schema and $\mathrm{Proj}(\rho(X, \mathbf{n}, k)) = \mathrm{Proj}(\rho(X, \mathbf{n}, k), \emptyset)$ as the projection schema of $\rho(X, \mathbf{n}, k)$.

**Example 9.1.3** In this example we compute the characteristic formula schema $\mathrm{CF}(\rho_0(n))$ and the projection schema $\mathrm{Proj}(\rho_0(n))$ of the regular proof schema $\rho_0(n)$ from Example 7.1.8 (it is called $\rho_1^R$ there):

$$\dfrac{\begin{matrix}(\hat{\rho}(n, V_S(n)) \\ \{V_S(0) \leftarrow \rho'\}) \\ S(n)\end{matrix} \qquad \dfrac{\dfrac{\dfrac{\dfrac{P(c) \vdash P(c) \qquad \dfrac{P(\hat{f}(c,n)) \vdash P(\hat{f}(c,n)) \qquad P(g(c,n)) \vdash P(g(c,n))}{P(\hat{f}(c,n)) \to P(g(c,n)), P(\hat{f}(c,n)) \vdash P(g(c,n))}\,\to_l}{P(c), P(\hat{f}(c,n)) \to P(g(c,n)), P(c) \to P(\hat{f}(c,n)) \vdash P(g(c,n))}\,\to_l}{P(\hat{f}(c,n)) \to P(g(c,n)), P(c) \to P(\hat{f}(c,n)) \vdash P(c) \to P(g(c,n))}\,\to_r}{P(c) \to P(\hat{f}(c,n)) \vdash (P(\hat{f}(c,n)) \to P(g(c,n))) \to (P(c) \to P(g(c,n)))}\,\to_r}{\forall x(P(x) \to P(\hat{f}(x,n))) \vdash (P(\hat{f}(c,n)) \to P(g(c,n))) \to (P(c) \to P(g(c,n)))}\,\forall_l}{\forall x(P(x) \to P(f(x))) \vdash (P(\hat{f}(c,n)) \to P(g(c,n))) \to (P(c) \to P(g(c,n)))}\,cut$$

where $S(n) = \forall x(P(x) \to P(f(x))) \vdash \forall x(P(x) \to P(\hat{f}(x,n)))$. $\hat{\rho}(n, V_S(n))$ is given by

$$D(\hat{\rho}) = \{\hat{\rho}(n, V_S(n)) =_d \underline{\text{if}}\ n = \bar{0}\ \underline{\text{then}}\ \rho' \\ \underline{\text{else}}\ \rho(n, V_S(n))\{V_S(p(n)) \leftarrow \hat{\rho}(p(n), V_S(p(n)))\}\}.$$

as in Example 9.1.1. $\mathrm{CF}(\rho_0(n))$ is computed inductively, starting at the end-sequent of $\rho_0(n)$. The first inference (from bottom to top) is a cut, therefore

$$\mathrm{CF}(\rho_0(n)) = \mathrm{CF}(\hat{\rho}(n, V_S(n)), \{(1, r)\}) \wedge \mathrm{CF}(\varphi, \{(1, l)\}),$$

where with $\varphi$ we denote the sub-derivation on the right branch above the cut. The characteristic formula schema of $\varphi$ w.r.t. the cut-ancestor at $\{(1, l)\})$ is

$$\mathrm{CF}(\varphi, \{(1, l)\}) = P(c) \wedge \neg P(\hat{f}(c, n)),$$

and the projection schema of $\varphi$ w.r.t. the cut-ancestor at $\{(1, l)\})$ is

$$\dfrac{\dfrac{\dfrac{P(c) \vdash P(c) \qquad \dfrac{\dfrac{P(\hat{f}(c,n)) \vdash P(\hat{f}(c,n))}{\vdash P(\hat{f}(c,n)), \neg P(\hat{f}(c,n))}\,\neg_r \qquad P(g(c,n)) \vdash P(g(c,n))}{P(\hat{f}(c,n)) \to P(g(c,n)) \vdash P(g(c,n)), \neg P(\hat{f}(c,n))}\,\to_l}{P(c), P(\hat{f}(c,n)) \to P(g(c,n)) \vdash P(g(c,n)), P(c) \wedge \neg P(\hat{f}(c,n))}\,\wedge_r}{P(\hat{f}(c,n)) \to P(g(c,n)) \vdash P(c) \to P(g(c,n)), P(c) \wedge \neg P(\hat{f}(c,n))}\,\to_r}{\vdash (P(\hat{f}(c,n)) \to P(g(c,n))) \to (P(c) \to P(g(c,n))), P(c) \wedge \neg P(\hat{f}(c,n))}\,\to_r$$

The projection schema of $\rho_0(n)$, $\mathrm{Proj}(\rho_0(n))$, is then

$$
\dfrac{\dfrac{(\mathrm{Proj}(\hat{\rho}(n, V_S(n)), \{(1,r)\}))}{S(n)} \qquad \dfrac{\dfrac{\dfrac{P(c) \vdash P(c) \qquad \dfrac{\dfrac{P(\hat{f}(c,n)) \vdash P(\hat{f}(c,n))}{\vdash P(\hat{f}(c,n)), \neg P(\hat{f}(c,n))}\neg_r \qquad P(g(c,n)) \vdash P(g(c,n))}{P(\hat{f}(c,n)) \to P(g(c,n)) \vdash P(g(c,n)), \neg P(\hat{f}(c,n))}\to_l}{P(c), P(\hat{f}(c,n)) \to P(g(c,n)) \vdash P(g(c,n)), P(c) \wedge \neg P(\hat{f}(c,n))}\wedge_r}{P(\hat{f}(c,n)) \to P(g(c,n)) \vdash P(c) \to P(g(c,n)), P(c) \wedge \neg P(\hat{f}(c,n))}\to_r}{\vdash (P(\hat{f}(c,n)) \to P(g(c,n))) \to (P(c) \to P(g(c,n))), P(c) \wedge \neg P(\hat{f}(c,n))}\to_r}{\forall x(P(x) \to P(f(x))) \vdash (P(\hat{f}(c,n)) \to P(g(c,n))) \to (P(c) \to P(g(c,n))), \mathrm{CF}(\mathrm{Proj}(\rho_0(n)))}\wedge_r
$$

where $S(n)$ is the end-sequent of $\mathrm{Proj}(\hat{\rho}(n, V_S(n)), \{(1,r)\})$. Now let us compute $\mathrm{CF}(\hat{\rho}(n, V_S(n)), \{(1,r)\})$ and $\mathrm{Proj}(\hat{\rho}(n, V_S(n)), \{(1,r)\})$. From Example 9.1.1 we know that

$$
\begin{aligned}
\mathrm{CF}(\hat{\rho}(0, V_S(0)), \{(1,r)\}) &= \mathrm{CF}(\rho', \{(1,r)\}) \\
&= \neg P(\hat{f}(X^{\rho'}(0), 0)) \vee P(\hat{f}(X^{\rho'}(0), 0)).
\end{aligned}
$$

Now consider $\mathrm{CF}(\hat{\rho}(n, V_S(n)), \{(1,r)\})$ for $n > 0$. If $1 \leq k \leq n$ the proof $\hat{\rho}(k, V_S(k))$ is of the form

$$
\rho(k, V_S(k))\{V_S(p(k)) \leftarrow \hat{\rho}(p(k), V_S(p(k)))\}
$$

where $\rho(k, V_S(k))$ is

$$
\dfrac{\dfrac{V_S(p(k))}{S(p(k))}V_S E \qquad \begin{array}{c}(\rho_1(k)) \\ \forall x(\mathbf{P(x)} \to \mathbf{P(\hat{f}(x, p(k))))}, \forall x(P(x) \to P(f(x))) \vdash \forall x(P(x) \to P(\hat{f}(x,k)))\end{array}}{\dfrac{S(k)}{V_S(k)}V_S I}\mathit{cut}, c_l
$$

for $\rho_1(k) =$

$$
\dfrac{\dfrac{\dfrac{\dfrac{P(X^\rho(k)) \vdash \mathbf{P(X^\rho(k))} \qquad \begin{array}{c}(\rho_2(k) \\ \mathbf{P(\hat{f}(X^\rho(k), p(k)))}, \forall x(P(x) \to P(f(x))) \vdash P(\hat{f}(X^\rho(k), k))\end{array}}{P(X^\rho(k)), \mathbf{P(X^\rho(k))} \to \mathbf{P(\hat{f}(X^\rho(k), p(k)))}, \forall x(P(x) \to P(f(x))) \vdash P(\hat{f}(X^\rho(k), k))}\to_l}{\mathbf{P(X^\rho(k))} \to \mathbf{P(\hat{f}(X^\rho(k), p(k)))}, \forall x(P(x) \to P(f(x))) \vdash P(X^\rho(k)) \to P(\hat{f}(X^\rho(k), k))}\to_r}{\forall \mathbf{x}(\mathbf{P(x)} \to \mathbf{P(\hat{f}(x, p(k)))}), \forall x(P(x) \to P(f(x))) \vdash P(X^\rho(k)) \to P(\hat{f}(X^\rho(k), k))}\forall_l}{\forall \mathbf{x}(\mathbf{P(x)} \to \mathbf{P(\hat{f}(x, p(k)))}), \forall x(P(x) \to P(f(x))) \vdash \forall x(P(x) \to P(\hat{f}(x,k)))}\forall_r
$$

and $\rho_2(k) =$

$$
\dfrac{\dfrac{\mathbf{P(\hat{f}(X^\rho(k), p(k)))} \vdash P(\hat{f}(X^\rho(k), p(k))) \qquad \dfrac{P(\hat{f}(X^\rho(k), k)) \vdash P(\hat{f}(X^\rho(k), k))}{P(f(\hat{f}(X^\rho(k), p(k)))) \vdash P(\hat{f}(X^\rho(k), k))}T^+}{\mathbf{P(\hat{f}(X^\rho(k), p(k)))}, P(\hat{f}(X^\rho(k), p(k))) \to P(f(\hat{f}(X^\rho(k), p(k)))) \vdash P(\hat{f}(X^\rho(k), k))}\to_l}{\mathbf{P(\hat{f}(X^\rho(k), p(k)))}, \forall x(P(x) \to P(f(x))) \vdash P(\hat{f}(X^\rho(k), k))}\forall_l
$$

The local cut-ancestors are indicated in bold letters (occurring at $\{(1,l)\}$). The end-sequent of $\hat{\rho}(k, V_S(k))$ has cut-ancestors $\Omega'$, which are formula occurrences in the end-sequent that are ancestors of cuts in the proof schema $\hat{\rho}(s(k), V_S(s(k)))$. For all these $k$ we have
$\mathrm{CF}(\hat{\rho}(k, V_S(k)), \{(1,r)\}) =$

$$\begin{aligned}&\mathrm{CF}(\hat{\rho}(p(k), V_S(p(k))), \{(1,r)\}) \wedge \mathrm{CF}(\rho_1(k), \{(1,l),(1,r)\}\\ =\ &\mathrm{CF}(\hat{\rho}(p(k), V_S(p(k))), \{(1,r)\}) \wedge\\ &(\neg P(X^{\rho}(k)) \vee P(X^{\rho}(k))) \wedge (\neg P(\hat{f}(X^{\rho}(k), p(k))) \vee P(\hat{f}(X^{\rho}(k), k))).\end{aligned}$$

For $n$ itself we obtain $\mathrm{CF}(\hat{\rho}(n, V_S(n)), \{(1,r)\}) =$

$$\begin{aligned}&\mathrm{CF}(\hat{\rho}(p(n), V_S(p(n))), \{(1,r)\}) \wedge \mathrm{CF}(\rho(n, V_S(n)), \{(1,r)\})\\ =\ &\mathrm{CF}(\hat{\rho}(p(n), V_S(p(n))), \{(1,r)\}) \wedge\\ &(\neg P(X^{\rho}(n)) \vee P(X^{\rho}(n))) \wedge (\neg P(\hat{f}(X^{\rho}(n), p(n))) \vee P(\hat{f}(X^{\rho}(n), n))).\end{aligned}$$

For the projection schema, $\mathrm{Proj}(\hat{\rho}(n, V_S(n)), \{(1,r)\})$ is given by

$$\begin{aligned}&D(\mathrm{Proj}(\hat{\rho})) = \{\mathrm{Proj}(\hat{\rho}(n, V_S(n))) =_d \underline{\text{if}}\ n = \bar{0}\ \underline{\text{then}}\ \mathrm{Proj}(\rho', \{(1,r)\})\\ &\underline{\text{else}}\ \mathrm{Proj}(\rho(n, V_S(n))\{V_S(p(n)) \leftarrow \mathrm{Proj}(\hat{\rho}(p(n), V_S(p(n))))\}\},\end{aligned}$$

where $\mathrm{Proj}(\hat{\rho}(0, V_S(0)), \{(1,r)\}) = \mathrm{Proj}(\rho', \{(1,r)\}) =$

$$\dfrac{\dfrac{\dfrac{\dfrac{P(\hat{f}(X^{\rho'}(0),0)) \vdash P(\hat{f}(X^{\rho'}(0),0))}{\vdash \neg P(\hat{f}(X^{\rho'}(0),0)), P(\hat{f}(X^{\rho'}(0),0))}\,\neg_r}{\vdash \neg P(\hat{f}(X^{\rho'}(0),0)) \vee P(\hat{f}(X^{\rho'}(0),0))}\,\vee_r}{\forall x(P(x) \to P(f(x))) \vdash \neg P(\hat{f}(X^{\rho'}(0),0)) \vee P(\hat{f}(X^{\rho'}(0),0))}\,w_l}{V_S(0)}\,V_S I$$

and for a parameter $k < n$, we have we have $\mathrm{Proj}(\hat{\rho}(k, V_S(k)), \{(1,r)\}) =$

$$\mathrm{Proj}(\rho(k, V_S(k)), \{(1,r)\})\{V_S(p(k)) \leftarrow \mathrm{Proj}(\hat{\rho}(p(k), V_S(p(k))), \{(1,r)\})\},$$

where $\mathrm{Proj}(\rho(k, V_S(k)), \{(1,r)\})$ is

$$\dfrac{\dfrac{\dfrac{V_S(p(k))}{S(p(k))}\,V_S E \qquad \begin{matrix}(\mathrm{Proj}(\rho_1(k), \{(1,l),(1,r)\}))\\ \forall x(P(x) \to P(f(x))) \vdash F(k)\end{matrix}}{\forall x(P(x) \to P(f(x))) \vdash \mathrm{CF}(\hat{\rho}(k, V_S(k)), \{(1,r)\})}\,\wedge_r}{V_S(k)}\,V_S I$$

$F(k) = (\neg P(X^{\rho}(k)) \vee P(X^{\rho}(k))) \wedge (\neg P(\hat{f}(X^{\rho}(k), p(k)))) \vee P(\hat{f}(X^{\rho}(k), k))$ and $\mathrm{Proj}(\rho_1(k), \{(1,l),(1,r)\}) =$

$$\dfrac{\dfrac{\dfrac{P(X^{\rho}(k)) \vdash P(X^{\rho}(k))}{\vdash \neg P(X^{\rho}(k)), P(X^{\rho}(k))}\,\neg_r}{\vdash \neg P(X^{\rho}(k)) \vee P(X^{\rho}(k))}\,\vee_r \qquad \begin{matrix}(\varphi(k)\\ \forall x(P(x) \to P(f(x))) \vdash \neg P(\hat{f}(X^{\rho}(k), p(k))) \vee P(\hat{f}(X^{\rho}(k), k))\end{matrix}}{\forall x(P(x) \to P(f(x))) \vdash (\neg P(X^{\rho}(k)) \vee P(X^{\rho}(k))) \wedge (\neg P(\hat{f}(X^{\rho}(k), p(k))) \vee P(\hat{f}(X^{\rho}(k), k)))}\,\wedge_r$$

where $\varphi(k) =$

$$
\dfrac{\dfrac{\dfrac{\dfrac{P(\hat{f}(X^{\rho}(k), p(k))) \vdash P(\hat{f}(X^{\rho}(k), p(k)))}{\vdash P(\hat{f}(X^{\rho}(k), p(k))), \neg P(\hat{f}(X^{\rho}(k), p(k)))}\neg_r \quad \dfrac{P(\hat{f}(X^{\rho}(k), k)) \vdash P(\hat{f}(X^{\rho}(k), k))}{P(f(\hat{f}(X^{\rho}(k), p(k)))) \vdash P(\hat{f}(X^{\rho}(k), k))}T^{+}}{P(\hat{f}(X^{\rho}(k), p(k))) \to P(f(\hat{f}(X^{\rho}(k), p(k)))) \vdash \neg P(\hat{f}(X^{\rho}(k), p(k))), P(\hat{f}(X^{\rho}(k), k))}\to_l}{P(\hat{f}(X^{\rho}(k), p(k))) \to P(f(\hat{f}(X^{\rho}(k), p(k)))) \vdash \neg P(\hat{f}(X^{\rho}(k), p(k))) \vee P(\hat{f}(X^{\rho}(k), k))}\vee_r}{\forall x(P(x) \to P(f(x))) \vdash \neg P(\hat{f}(X^{\rho}(k), p(k))) \vee P(\hat{f}(X^{\rho}(k), k))}\forall_l
$$

For $n$ itself $\mathrm{Proj}(\hat{\rho}(n, V_S(n)), \{(1, r)\})$ is as above, only that $k = n$.

Below we show the soundness of characteristic formula schemata and projection schemata.

**Theorem 9.1.1** *Let $\sigma$ be a parameter assignment and $\rho$ a regular proof schema. Then*

1. *the characteristic formula of $\sigma(\rho)$ is $\sigma(\mathrm{CF}(\rho))$, and*
2. *the projection of $\sigma(\rho)$ is $\sigma(\mathrm{Proj}(\rho))$.*

***Proof*** 1. We prove by induction on the number of links and for each link by induction on $\sigma$ that for any configuration $\Omega$ we have that $\mathrm{CF}(\sigma(\rho), \Omega) = \sigma(\mathrm{CF}(\rho, \Omega))$. By a straightforward induction of formulas, one can show that $\sigma(\mathrm{CF}(\rho, \Omega))$ is obtained from $\mathrm{CF}(\sigma(\rho), \Omega)$ by replacing each formula $\mathrm{CF}(\sigma(\rho), \nu, \Omega)$ by $\sigma(\mathrm{CF}(\rho', \Omega'))$ where $\nu$ is a leaf in $\rho$ containing a link-variable $V_{S'}$ and $S'$ the end-sequent of $\rho'$, and $\Omega'$ are all occurrences in $S'$ which are ancestors of $\Omega$ and cuts in $\rho$. By induction hypothesis, we have $\sigma(\mathrm{CF}(\rho', \Omega')) = \mathrm{CF}(\sigma(\rho'), \Omega')$. Thus, $\sigma(\mathrm{CF}(\rho, \Omega)) = \mathrm{CF}(\sigma(\rho), \Omega)$.

2. We prove by induction on the number of links and for each link by induction on $\sigma$ that for any configuration $\Omega$ we have that $\mathrm{Proj}(\sigma(\rho), \Omega) = \sigma(\mathrm{Proj}(\rho, \Omega))$. By a straightforward induction on derivations, one can show that $\sigma(\mathrm{Proj}(\rho, \Omega))$ is obtained from $\mathrm{Proj}(\sigma(\rho), \Omega)$ by replacing each $\mathrm{Proj}(\sigma(\rho), \nu, \Omega)$ by $\sigma(\mathrm{Proj}(\rho', \Omega'))$, where $\nu$ is a leaf in $\rho$ that contains a link variable $V_{S'}$ and $S'$ the end-sequent of $\rho'$, and $\Omega'$ are all occurrences in $S'$ that are ancestors of $\Omega$ and cuts in $\rho$. By induction hypothesis we have $\sigma(\mathrm{Proj}(\rho', \Omega')) = \mathrm{Proj}(\sigma(\rho'), \Omega')$. Thus, $\sigma(\mathrm{Proj}(\rho, \Omega)) = \mathrm{Proj}(\sigma(\rho), \Omega)$.

## 9.2 Herbrand Systems

Just like in the first-order CERES method, the schematic Herbrand sequent (the so-called Herbrand system) can be constructed from the Herbrand system of the projection schema and the Herbrand system of the refutation schema.

### *9.2.1 Herbrand Systems from Projection Schemata*

When we consider an **LK**-proof $\varphi$ of a prenex skolemized end-sequent and its proof projection $\pi$ then $\pi$ is a cut-free **LK**-proof of a skolemized prenex end sequent $S$, thus $\pi$ defines a Herbrand sequent and, for every formula $F$ occurring in $S$, a set of substitutions $S(F, \pi)$ defining the Herbrand substitutions for the formula $F$ in $\pi$. However, when we consider proof schemata the situation changes. Proof schemata are based on $\mathbf{LK}_{T^+}$-derivation schemata, and the initial sequents of $\mathbf{LK}V$-derivations may correspond to (recursive) calls and may not be axioms.

We use the notion of thread, trace and yield as in Sect. 3.1, the only difference being that our initial sequents might not be axioms. Note that when we admit initial sequents that are not axioms we do not obtain a full but merely a partial Herbrand instance.

**Definition 9.2.1** Let $\pi$ be a cut-free $\mathbf{LK}_{T^+}$-derivation of a skolemized prenex end-sequent $S$ and let $\mu_0$ be an occurrence of a formula

$$F\colon Qx_1 \ldots Qx_\beta . F'(x_1, \ldots, x_\beta)$$

in $S$ such that $Q \in \{\forall, \exists\}$ and $F'$ is quantifier-free. Let $\tau : \mu_0, \ldots, \mu_\alpha$ be a trace of $\mu_0$ in $\pi$. Then either

1. $\mathit{formula}(\mu_\alpha) = F$ or
2. there exists an $i \in \{1, \ldots, \beta - 1\}$ and terms $t_1, \ldots, t_i$ such that

$$\mathit{formula}(\mu_\alpha) = Qx_{i+1} \ldots Qx_\beta . F'(t_1, \ldots, t_i, x_{i+1}, \ldots, x_\beta) \text{ or}$$

3. there exist terms $t_1, \ldots, t_\beta$ such that $\mathit{formula}(\mu_\alpha)$ is a subformula of $F'(t_1, \ldots, t_\beta)$ and $F'(t_1, \ldots, t_\beta)$ occurs in the yield of $\tau$.

Definition 9.2.1 gives us the means to assign a substitution $hi(\tau)$ to every trace $\tau$ in a cut-free proof $\pi$ of a skolemized prenex end-sequent such that either $hi(\tau)$ is a real Herbrand substitution or just a "partial" one.

**Definition 9.2.2** Let $\pi$ be a cut-free $\mathbf{LK}_{T^+}$-derivation of a skolemized prenex end-sequent $S$ and let $\mu_0$ be an occurrence of a formula

$$F\colon Qx_1 \ldots Qx_\beta . F'(x_1, \ldots, x_\beta)$$

in $S$ such that $Q \in \{\forall, \exists\}$ and $F'$ is quantifier-free. Let $\tau : \mu_0, \ldots, \mu_\alpha$ be a trace of $\mu_0$ in $\pi$ and consider the cases 1., 2. and 3. in Definition 9.2.1. We define

- case 1.: $hi(\tau) = \emptyset$,
- case 2.: $hi(\tau) = \{x_1 \leftarrow t_1, \ldots, x_i \leftarrow t_i\}$,
- case 3.: $hi(\tau) = \{x_1 \leftarrow t_1, \ldots, x_\beta \leftarrow t_\beta\}$.

**Example 9.2.1** Consider the following derivation $\varphi$ where

$$\hat{f}(x, n) \simeq \underline{\text{if}}\ n = \bar{0}\ \underline{\text{then}}\ x\ \underline{\text{else}}\ h(\hat{f}(x, p(n))) \text{ is a definition in } T^+.$$

$$\dfrac{\dfrac{\begin{array}{c}\psi\\ \nu_2\colon P(h(\hat{f}(t,n))),\Gamma\vdash\Delta\end{array}}{\nu_1\colon P(\hat{f}(t,s(n))),\Gamma\vdash\Delta}\;T^+}{\nu_0\colon \forall x P(x),\Gamma\vdash\Delta}\;\forall\colon l$$

And assume $P(h(\hat{f}(t,n)))$ goes into an axiom in $\psi$, i.e. $P(h(\hat{f}(t,n)))$ does not change in $\psi$. Then $\theta : \nu_0, \nu_1, \nu_2, \ldots$ is a thread (which continues in $\psi$). Let $\mu_0$ be the formula occurrence of $\forall x P(x)$ in $Seq(\nu_0)$, and $\mu_{i+1}$ the occurrence in $Seq(\nu_{i+1})$ which is an ancestor occurrence of $\mu_i$ in $Seq(\nu_i)$. Then $\tau\colon \mu_0, \mu_1, \mu_2, \ldots$ is a trace of $\mu_0$ in $\varphi$. The yield of $\tau$ is the sequence of formulas

$$\forall x P(x),\;\; P(\hat{f}(t,s(n))),\;\; P(h(\hat{f}(t,n))),\ldots$$

$hi(\tau) = \{x \leftarrow h(\hat{f}(t,n))\}$.

We are now ready to define the schematic set of Herbrand substitutions defined by a projection schema. To this aim we will extend Definition 3.1.6 to proof recursions.

**Definition 9.2.3** Let $\hat{\rho}$ be given by proof recursion

$$\begin{array}{l} D(\hat{\rho}) = \{\hat{\rho}(\mathbf{X},\mathbf{n},k) =_d \underline{\text{if}}\ k=\bar{0}\ \underline{\text{then}}\ \rho_0\\ \qquad\qquad \underline{\text{else}}\ \rho^*(\mathbf{X},\mathbf{n},k)\{V(p(k)) \leftarrow \hat{\rho}(\mathbf{X},\mathbf{n},p(k))\}\}.\end{array}$$

The set of all traces of a formula occurrence $\mu_0(\mathbf{n},k)$ in the end-sequent of $\hat{\rho}(\mathbf{X},\mathbf{n},k)$, $T(\hat{\rho},\mu_0(\mathbf{n},k))$, is defined depending on $k$. If $k=\bar{0}$, the traces in $\rho_0$ are considered, otherwise the traces are computed from $\rho^*(\mathbf{X},\mathbf{n},k)$, where the link variables $V(p(k))$ are replaced by the traces of $\hat{\rho}(\mathbf{X},\mathbf{n},p(k))$:

$$\begin{array}{l} D(T) =\\ \{T(\hat{\rho},\mu_0(\mathbf{n},k)) =_d \underline{\text{if}}\ k=\bar{0}\ \underline{\text{then}}\ T(\rho_0,\mu_0(\mathbf{n},\bar{0}))\\ \ \underline{\text{else}}\ T(\rho^*(\mathbf{X},\mathbf{n},k),\mu_0(\mathbf{n},k))\{V(p(k)) \leftarrow T(\hat{\rho}(\mathbf{X},\mathbf{n},p(k))\},\mu_0(\mathbf{n},p(k)))\}.\end{array}$$

**Example 9.2.2** Consider the projection schema $\hat{\rho}(n, V_S(n))$ as in Example 9.1.3 (it is called $\mathrm{Proj}(\hat{\rho}(n, V_S(n)), \{(1,r)\})$ there), which is given by

$$\begin{array}{l} D(\hat{\rho}) = \{\hat{\rho}(n,V_S(n)) =_d \underline{\text{if}}\ n=\bar{0}\ \underline{\text{then}}\ \rho'\\ \underline{\text{else}}\ \rho(n,V_S(n))\{V_S(p(n)) \leftarrow \hat{\rho}(p(n),V_S(p(n)))\}\},\end{array}$$

where $\hat{\rho}(0, V_S(0)) = \rho' =$

$$\dfrac{\dfrac{\dfrac{\dfrac{P(\hat{f}(X^{\rho'}(0),0)) \vdash P(\hat{f}(X^{\rho'}(0),0))}{\vdash \neg P(\hat{f}(X^{\rho'}(0),0)), P(\hat{f}(X^{\rho'}(0),0))}\;\neg_r}{\vdash \neg P(\hat{f}(X^{\rho'}(0),0)) \vee P(\hat{f}(X^{\rho'}(0),0))}\;\vee_r}{\forall \mathbf{x}(\mathbf{P}(\mathbf{x}) \to \mathbf{P}(\mathbf{f}(\mathbf{x}))) \vdash \neg P(\hat{f}(X^{\rho'}(0),0)) \vee P(\hat{f}(X^{\rho'}(0),0))}\;w_l}{V_S(0)}\;V_S I$$

and $\hat{\rho}(n, V_S(n)) =$

$$\rho(n, V_S(n))\{V_S(p(n)) \leftarrow \hat{\rho}(p(n), V_S(p(n)))\},$$

where $\rho(n, V_S(n))$ is

$$\dfrac{\dfrac{\dfrac{\mathbf{V_S(p(n))}}{\mathbf{S(p(n))}}\, V_S E \qquad \begin{array}{c}(\rho_1(n))\\ \forall \mathbf{x}(\mathbf{P(x)} \to \mathbf{P(f(x))}) \vdash F(n)\end{array}}{\forall \mathbf{x}(\mathbf{P(x)} \to \mathbf{P(f(x))}) \vdash \mathrm{CF}(\hat{\rho}(n, V_S(n)), \{(1, r)\})}\, \wedge_r}{V_S(n)}\, V_S I$$

$F(n) = (\neg P(X^\rho(n)) \vee P(X^\rho(n))) \wedge (\neg P(\hat{f}(X^\rho(n), p(n)))) \vee P(\hat{f}(X^\rho(n), n))$
and $\rho_1(n)) =$

$$\dfrac{\dfrac{\dfrac{P(X^\rho(n)) \vdash P(X^\rho(n))}{\vdash \neg P(X^\rho(n)), P(X^\rho(n))}\, \neg_r}{\vdash \neg P(X^\rho(n)) \vee P(X^\rho(n))}\, \vee_r \qquad \begin{array}{c}(\varphi(n))\\ \forall \mathbf{x}(\mathbf{P(x)} \to \mathbf{P(f(x))}) \vdash \neg P(\hat{f}(X^\rho(n), p(n))) \vee P(\hat{f}(X^\rho(n), n))\end{array}}{\forall \mathbf{x}(\mathbf{P(x)} \to \mathbf{P(f(x))}) \vdash (\neg P(X^\rho(n)) \vee P(X^\rho(n))) \wedge (\neg P(\hat{f}(X^\rho(n), p(n))) \vee P(\hat{f}(X^\rho(n), n)))}\, \wedge_r$$

where $\varphi(n) =$

$$\dfrac{\dfrac{\dfrac{\dfrac{P(\hat{f}(X^\rho(n), p(n))) \vdash \mathbf{P}(\hat{\mathbf{f}}(\mathbf{X}^\rho(\mathbf{n}), \mathbf{p(n)}))}{\vdash \mathbf{P}(\hat{\mathbf{f}}(\mathbf{X}^\rho(\mathbf{n}), \mathbf{p(n)})), \neg P(\hat{f}(X^\rho(n), p(n)))}\, \neg_r \qquad \dfrac{\mathbf{P}(\hat{\mathbf{f}}(\mathbf{X}^\rho(\mathbf{n}), \mathbf{n})) \vdash P(\hat{f}(X^\rho(n), n))}{\mathbf{P}(\mathbf{f}(\hat{\mathbf{f}}(\mathbf{X}^\rho(\mathbf{n}), \mathbf{p(n)}))) \vdash P(\hat{f}(X^\rho(n), n))}\, T^+}{\mathbf{P}(\hat{\mathbf{f}}(\mathbf{X}^\rho(\mathbf{n}), \mathbf{p(n)})) \to \mathbf{P}(\mathbf{f}(\hat{\mathbf{f}}(\mathbf{X}^\rho(\mathbf{n}), \mathbf{p(n)}))) \vdash \neg P(\hat{f}(X^\rho(n), p(n))), P(\hat{f}(X^\rho(n), n))}\, \to_l}{\mathbf{P}(\hat{\mathbf{f}}(\mathbf{X}^\rho(\mathbf{n}), \mathbf{p(n)})) \to \mathbf{P}(\mathbf{f}(\hat{\mathbf{f}}(\mathbf{X}^\rho(\mathbf{n}), \mathbf{p(n)}))) \vdash \neg P(\hat{f}(X^\rho(n), p(n))) \vee P(\hat{f}(X^\rho(n), n))}\, \vee_r}{\forall \mathbf{x}(\mathbf{P(x)} \to \mathbf{P(f(x))}) \vdash \neg P(\hat{f}(X^\rho(n), p(n))) \vee P(\hat{f}(X^\rho(n), n))}\, \forall_l$$

Consider the formula occurrence $\mu_0(n)$ of $\forall x P(x) \to P(f(x))$ in the end-sequent of $\hat{\rho}$. We indicate all occurrences of this formula and its ancestor occurrences in bold letters in the derivations above. The set of all traces for $\mu_0(n)$ is $D(T) =$

$$\{T(\hat{\rho}, \mu_0(n)) =_d \underline{\text{if}}\ n = \bar{0}\ \underline{\text{then}}\ T(\rho', \mu_0(\bar{0}))$$
$$\underline{\text{else}}\ T(\rho(n, V_S(n)), \mu_0(n))\{V_S(p(n)) \leftarrow T(\hat{\rho}(p(n), V_S(p(n))), \mu_0(p(n)))\}\},$$

$T(\rho', \mu_0(\bar{0})) = \emptyset$, as there is no thread in $\rho'$ for $\mu_0(\bar{0})$. $T(\rho(n, V_S(n)), \mu_0(n))$ defines three traces for $\mu_0(n)$: $\tau_0(n) =$

$$\forall x(P(x) \to P(f(x))),$$
$$S(p(n))(= \forall \mathbf{x}(\mathbf{P(x)} \to \mathbf{P(f(x))}) \vdash \mathrm{CF}(\hat{\rho}(p(n), V_S(p(n))), \{(1, r)\})),$$
$$V_S(p(n))$$

$\tau_1(n) =$

$$\forall x(P(x) \to P(f(x))),$$
$$\forall x(P(x) \to P(f(x))),$$
$$\forall x(P(x) \to P(f(x)),$$
$$P(\hat{f}(X^\rho(n), p(n))) \to P(f(\hat{f}(X^\rho(n), p(n)))),$$

$$
\begin{array}{l}
P(\hat{f}(X^{\rho}(n), p(n))) \to P(f(\hat{f}(X^{\rho}(n), p(n)))),\\
P(f(\hat{f}(X^{\rho}(n), p(n)))),\\
P(\hat{f}(X^{\rho}(n), n))
\end{array}
$$

and $\tau_2(n) =$

$$
\begin{array}{l}
\forall x(P(x) \to P(f(x))),\\
\forall x(P(x) \to P(f(x))),\\
\forall x(P(x) \to P(f(x)),\\
P(\hat{f}(X^{\rho}(n), p(n))) \to P(f(\hat{f}(X^{\rho}(n), p(n)))),\\
P(\hat{f}(X^{\rho}(n), p(n))) \to P(f(\hat{f}(X^{\rho}(n), p(n)))),\\
P(\hat{f}(X^{\rho}(n), p(n))),\\
P(\hat{f}(X^{\rho}(n), p(n)))
\end{array}
$$

(the threads are indicated in bold letters in $\rho(n, V_S(n))$.

$$T(\rho(n, V_S(n)), \mu_0(n))\{V_S(p(n)) \leftarrow T(\hat{\rho}(p(n), V_S(p(n))), \mu_0(p(n)))\}$$

is then obtained from $T(\rho(n, V_S(n)), \mu_0(n))$ by replacing the link variables $V_S(p(n))$ by the traces $T(\hat{\rho}(p(n), V_S(p(n))), \mu_0(p(n)))$.

Just like formula schemata, traces can also be evaluated for any parameter assignment $\sigma$. Let $\sigma(n) = \bar{1}$. Then the evaluation of $T(\hat{\rho}, \mu_0(n))$ under the parameter assignment $\sigma$ is

$$T(\rho(\bar{1}, V_S(\bar{1})), \mu_0(\bar{1}))\{V_S(p(\bar{1})) \leftarrow T(\hat{\rho}(p(\bar{1}), V_S(p(\bar{1}))), \mu_0(p(\bar{1})))\}$$

and contains the two traces $\tau_1(\bar{1}) =$

$$
\begin{array}{l}
\forall x(P(x) \to P(f(x))),\\
\forall x(P(x) \to P(f(x))),\\
\forall x(P(x) \to P(f(x)),\\
P(\hat{f}(X^{\rho}(\bar{1}), p(\bar{1}))) \to P(f(\hat{f}(X^{\rho}(\bar{1}), p(\bar{1})))),\\
P(\hat{f}(X^{\rho}(\bar{1}), p(\bar{1}))) \to P(f(\hat{f}(X^{\rho}(\bar{1}), p(\bar{1})))),\\
P(f(\hat{f}(X^{\rho}(\bar{1}), p(\bar{1})))),\\
P(\hat{f}(X^{\rho}(\bar{1}), \bar{1}))
\end{array}
$$

and $\tau_2(\bar{1}) =$

$$
\begin{array}{l}
\forall x(P(x) \to P(f(x))),\\
\forall x(P(x) \to P(f(x))),\\
\forall x(P(x) \to P(f(x)),
\end{array}
$$

$$P(\hat{f}(X^{\rho}(\bar{1}), p(\bar{1}))) \rightarrow P(f(\hat{f}(X^{\rho}(\bar{1}), p(\bar{1})))),$$
$$P(\hat{f}(X^{\rho}(\bar{1}), p(\bar{1}))) \rightarrow P(f(\hat{f}(X^{\rho}(\bar{1}), p(\bar{1})))),$$
$$P(\hat{f}(X^{\rho}(\bar{1}), p(\bar{1}))),$$
$$P(\hat{f}(X^{\rho}(\bar{1}), p(\bar{1})))$$

$\tau_0(\bar{1}) =$

$$\forall x(P(x) \rightarrow P(f(x))),$$
$$S(p(\bar{1}))(= \forall \mathbf{x}(\mathbf{P}(\mathbf{x}) \rightarrow \mathbf{P}(\mathbf{f}(\mathbf{x}))) \vdash \mathrm{CF}(\hat{\rho}(p(\bar{1}), V_S(p(\bar{1}))), \{(1, r)\})),$$
$$V_S(p(\bar{1})) \leftarrow T(\hat{\rho}(p(\bar{1}), V_S(p(\bar{1}))), \mu_0(p(\bar{1})))(= \emptyset)$$

is not a trace as the occurrence of the formula $\forall x(P(x) \rightarrow P(f(x))$ disappears.

Now let $\sigma(n) = \bar{2}$. Then the evaluation of $T(\hat{\rho}, \mu_0(n))$ is

$$T(\rho(\bar{2}, V_S(\bar{2})), \mu_0(\bar{2}))\{V_S(p(\bar{2})) \leftarrow T(\hat{\rho}(p(\bar{2}), V_S(p(\bar{2}))), \mu_0(p(\bar{2})))\}$$

and contains the traces $\tau_1(\bar{2}) =$

$$\forall x(P(x) \rightarrow P(f(x))),$$
$$\forall x(P(x) \rightarrow P(f(x))),$$
$$\forall x(P(x) \rightarrow P(f(x)),$$
$$P(\hat{f}(X^{\rho}(\bar{2}), p(\bar{2}))) \rightarrow P(f(\hat{f}(X^{\rho}(\bar{2}), p(\bar{2})))),$$
$$P(\hat{f}(X^{\rho}(\bar{2}), p(\bar{2}))) \rightarrow P(f(\hat{f}(X^{\rho}(\bar{2}), p(\bar{2})))),$$
$$P(f(\hat{f}(X^{\rho}(\bar{2}), p(\bar{2})))),$$
$$P(\hat{f}(X^{\rho}(\bar{2}), \bar{2}))$$

$\tau_2(\bar{2}) =$

$$\forall x(P(x) \rightarrow P(f(x))),$$
$$\forall x(P(x) \rightarrow P(f(x))),$$
$$\forall x(P(x) \rightarrow P(f(x)),$$
$$P(\hat{f}(X^{\rho}(\bar{2}), p(\bar{2}))) \rightarrow P(f(\hat{f}(X^{\rho}(\bar{2}), p(\bar{2})))),$$
$$P(\hat{f}(X^{\rho}(\bar{2}), p(\bar{2}))) \rightarrow P(f(\hat{f}(X^{\rho}(\bar{2}), p(\bar{2})))),$$
$$P(\hat{f}(X^{\rho}(\bar{2}), p(\bar{2}))),$$
$$P(\hat{f}(X^{\rho}(\bar{2}), p(\bar{2})))$$

and

$$\forall x(P(x) \rightarrow P(f(x))),$$
$$S(p(\bar{2}))(= \forall \mathbf{x}(\mathbf{P}(\mathbf{x}) \rightarrow \mathbf{P}(\mathbf{f}(\mathbf{x}))) \vdash \mathrm{CF}(\hat{\rho}(p(\bar{2}), V_S(p(\bar{2}))), \{(1, r)\})),$$

$$V_S(p(\bar{2})) \leftarrow T(\hat{\rho}(p(\bar{2}), V_S(p(\bar{2}))), \mu_0(p(\bar{2})))$$

which are the two traces $\tau_{0,1}(\bar{2}) =$

$$\begin{aligned}
&\forall x(P(x) \to P(f(x))),\\
&S(p(\bar{2}))(= \mathbf{\forall x(P(x) \to P(f(x)))} \vdash \mathrm{CF}(\hat{\rho}(p(\bar{2}), V_S(p(\bar{2}))), \{(1,r)\})),\\
&\forall x(P(x) \to P(f(x))),\\
&\forall x(P(x) \to P(f(x))),\\
&\forall x(P(x) \to P(f(x)),\\
&P(\hat{f}(X^{\rho}(\bar{1}), p(\bar{1}))) \to P(f(\hat{f}(X^{\rho}(\bar{1}), p(\bar{1})))),\\
&P(\hat{f}(X^{\rho}(\bar{1}), p(\bar{1}))) \to P(f(\hat{f}(X^{\rho}(\bar{1}), p(\bar{1})))),\\
&P(f(\hat{f}(X^{\rho}(\bar{1}), p(\bar{1})))),\\
&P(\hat{f}(X^{\rho}(\bar{1}), \bar{1}))
\end{aligned}$$

and $\tau_{0,2}(\bar{2}) =$

$$\begin{aligned}
&\forall x(P(x) \to P(f(x))),\\
&S(p(\bar{2}))(= \mathbf{\forall x(P(x) \to P(f(x)))} \vdash \mathrm{CF}(\hat{\rho}(p(\bar{2}), V_S(p(\bar{2}))), \{(1,r)\})),\\
&\forall x(P(x) \to P(f(x))),\\
&\forall x(P(x) \to P(f(x))),\\
&\forall x(P(x) \to P(f(x)),\\
&P(\hat{f}(X^{\rho}(\bar{1}), p(\bar{1}))) \to P(f(\hat{f}(X^{\rho}(\bar{1}), p(\bar{1})))),\\
&P(\hat{f}(X^{\rho}(\bar{1}), p(\bar{1}))) \to P(f(\hat{f}(X^{\rho}(\bar{1}), p(\bar{1})))),\\
&P(\hat{f}(X^{\rho}(\bar{1}), p(\bar{1}))),\\
&P(\hat{f}(X^{\rho}(\bar{1}), p(\bar{1})))
\end{aligned}$$

**Definition 9.2.4** (*Herbrand schema of projection schema*) Let

$$\rho(X, \mathbf{n})$$

be a regular proof schema and $\pi\colon \mathrm{Proj}(\rho(X, \mathbf{n}))$ its projection schema, with end-sequent $S$. Let $\mu_0$ be the occurrence of a formula

$$F(\mathbf{n})\colon Qx_1, \ldots, Qx_\beta.F'(x_1, \ldots, x_\beta)$$

in $S$ such that $Q \in \{\forall, \exists\}$ and $\tau\colon \mu_0, \ldots, \mu_\alpha$ a trace of $\mu_0$ in $\pi$. We define the set of Herbrand substitutions $S_S(\pi, \tau)$ for the trace $\tau$ and distinguish the cases:

- *formula*$(\mu_\alpha)$ is quantifier-free. We define $S_S(\pi, \tau) = \{hi(\tau)\}$.
- *formula*$(\mu_{\alpha-1}) = F(\mathbf{n})$. This can only be the case if the sequent occurring at node $\nu_\alpha$ (for the corresponding thread $\nu_0, \ldots, \nu_\alpha$) is a link variable $V_i(\mathbf{n}, p(k))$. Then

the proof schema is defined via a link substitution: there is an **LK**$V$-derivation $\rho_i$ with link-variable $V_i(\mathbf{n}, p(k))$ occurring at its root. We define

$$S_S(\pi, \tau) = \bigcup\{S_S(\pi', \tau') \mid \tau' \in T(\pi', \mu_\alpha)\}$$

for $\pi' = \rho_i$.

- *formula*$(\mu_{\alpha-1}) = Qx_{i+1} \dots Qx_\beta.F'(t_1, \dots, t_i, x_{i+1}, \dots, x_\beta)$ for $i < \beta$. This can only be the case if the sequent occurring at the node $\nu_\alpha$ is a link variable $V_i(\mathbf{n}, p(k))$. Then the proof schema is defined via a link substitution, i.e. there is an **LK**$V$-derivation $\rho_i$ with link-variable $V_i(\mathbf{n}, p(k))$ occurring at its root. We define

$$S_S(\pi, \tau) = \{x_1 \leftarrow t_1, \dots, x_i \leftarrow t_i\} \bigcup\{S_S(\pi', \tau') \mid \tau' \in T(\pi', \mu_\alpha)\}$$

for $\pi' = \rho_i$.

In the end we define the Herbrand system for $\mu_0$ as

$$S_S(\pi, \mu_0) = \bigcup\{S_S(\pi, \tau) \mid \tau \in T(\pi, \mu_0)\}.$$

**Example 9.2.3** Consider the projection schema $\mathrm{Proj}(\rho_0(n))$ from Example 9.1.3:

$$\dfrac{\begin{array}{c}(\mathrm{Proj}(\hat{\rho}(n, V_S(n))\\ \{V_S(0) \leftarrow \rho'\}, \{(1, r)\}))\\ S(n)\end{array} \qquad \begin{array}{c}(\varphi(n))\\ \vdash F\end{array}}{\forall x(P(x) \to P(f(x))) \vdash (P(\hat{f}(c,n)) \to P(g(c,n))) \to (P(c) \to P(g(c,n))), \mathrm{CF}(\mathrm{Proj}(\rho_0(n)))}\ \wedge_r$$

where $F =$

$$(P(\hat{f}(c,n)) \to P(g(c,n))) \to (P(c) \to P(g(c,n))), P(c) \wedge \neg P(\hat{f}(c,n))$$

and $\varphi(n) =$

$$\dfrac{\dfrac{P(c) \vdash P(c) \qquad \dfrac{\dfrac{P(\hat{f}(c,n)) \vdash P(\hat{f}(c,n))}{\vdash P(\hat{f}(c,n)), \neg P(\hat{f}(c,n))}\ \neg_r \qquad P(g(c,n)) \vdash P(g(c,n))}{P(\hat{f}(c,n)) \to P(g(c,n)) \vdash P(g(c,n)), \neg P(\hat{f}(c,n))}\ \to_l}{\dfrac{P(c), P(\hat{f}(c,n)) \to P(g(c,n)) \vdash P(g(c,n)), P(c) \wedge \neg P(\hat{f}(c,n))}{P(\hat{f}(c,n)) \to P(g(c,n)) \vdash P(c) \to P(g(c,n)), P(c) \wedge \neg P(\hat{f}(c,n))}\ \to_r}\ \wedge_r}{\vdash (P(\hat{f}(c,n)) \to P(g(c,n))) \to (P(c) \to P(g(c,n))), P(c) \wedge \neg P(\hat{f}(c,n))}\ \to_r$$

Let $\mu_0(n)$ be the occurrence of the formula $\forall x(P(x) \to P(f(x)))$ in the end-sequent of $\mathrm{Proj}(\rho_0(n))$. Then $S_S(\mathrm{Proj}(\rho_0(n)), \mu_0(n)) =$

$$\bigcup\{S_S(\mathrm{Proj}(\rho_0(n)), \tau) \mid \tau \in T(\mathrm{Proj}(\rho_0(n)), \mu_0(n))\}.$$

The traces of $\mu_0$ are as in Example 9.2.2. Therefore,

$$\begin{aligned} S_S(\text{Proj}(\rho_0(n)), \mu_0(n)) = &\{x \leftarrow \hat{f}(X^\rho(n), p(n))\} \cup \\ &\{x \leftarrow \hat{f}(X^\rho(n-1), p(n-1))\} \cup \\ &\ldots \cup \\ &\{x \leftarrow \hat{f}(X^\rho(\bar{1}), p(\bar{1}))\}. \end{aligned}$$

Given a parameter assignment $\sigma$ the Herbrand system $S_S(\rho(\mathbf{n}, k), \mu_i)$ can be evaluated by computing $S_S(\rho(\boldsymbol{\alpha}, \beta), \mu_i)$ for the chosen numerals $\boldsymbol{\alpha}, \beta$. It can be shown that for a projection schema $\pi$, $\sigma(S_S(\pi, \mu_i)) = S(\sigma(\pi), \mu_i)$, where $S(\sigma(\pi), \mu_i)$ is the set of Herbrand substitutions for the quantified formula occurrence $\mu_i$ in the end-sequent of $\sigma(\pi)$, see Definition 3.1.7.

**Proposition 9.2.1** *Let* $\pi = \text{Proj}(\rho(X, \mathbf{n}, k), \emptyset)$ *be a projection schema,* $S_S(\pi, \mu_i(k))$ *a corresponding Herbrand schema for the formula occurrence* $\mu_i(k)$ *in the end-sequent of* $\pi$*, and* $\sigma$ *a parameter assignment. Then*

$$\sigma(S_S(\pi, \mu_i(k))) = S(\sigma(\pi), \mu_i),$$

*where* $\sigma(\pi)$ *is the evaluated projection schema* $\pi$ *under* $\sigma$*, and* $\mu$ *is the evaluated formula occurrence* $\mu_i(k)$ *under* $\sigma$*, i.e. the corresponding formula occurrence in the end-sequent of* $\sigma(\pi)$*.*

***Proof*** $S_S(\pi, \mu_i(k))$ is defined as the union of all $S_S(\pi, \tau)$ for all the traces $\tau$ of $\mu_i(k)$ in $\pi$. By construction, $S_S(\pi, \mu_i(k))$ is a set of s-substitutions, as, if the formula occurrence does not disappear, eventually case 3. in Definition 9.2.2 applies. Depending on whether for a given trace $\tau : \mu_1(k), \ldots, \mu_\alpha(k)$ of $\mu_i(k)$, the formula occurrence at $\mu_\alpha(k)$ is quantifier-free or not, we take the set of substitutions $hi(\tau)$ or further expand $\pi$ by evaluating the derivations that are linked to through link-variables. Computing $\sigma(S_S(\pi, \mu_i(k)))$ means that we evaluate a set of s-substitutions under $\sigma$. As $\pi$ and $\sigma(\pi)$ have the same end-sequent (modulo $\sigma$), quantified variables in quantified formulas in the end-sequent remain the same, only the terms that are used in the proof to eliminate the quantifiers change (modulo $\sigma$). Therefore, evaluating $\pi$ under $\sigma$ and then computing $S(\sigma(\pi), \mu_i)$ results in the same set of substitutions as when we compute $\sigma(S_S(\pi, \mu_i))$.

### 9.2.2 Herbrand Systems from Refutation Schemata

Using the formalism for refutation schemata in Chap. 7.2, it is possible to extract a schematic structure representing the Herbrand sequent of the refutation.

**Example 9.2.4** Consider the refutation schema $\rho_0$, as defined in Example 7.2.1. $\rho_0$ defines the s-substitution

$$\Theta(\bar{0}, m) = \{X(\bar{0}) \leftarrow \hat{s}(Y(0), m), Z(0) \leftarrow \hat{s}(a, \bar{0})\}.$$

Moreover, as $\rho_0$ contains the recursive proof symbol $\hat{\rho}$ as axiom, we have to take the substitutions coming from $\hat{\rho}$ into account as well! In fact, $\Theta(\bar{0}, m)$ will be applied to all the other s-substitutions coming from the rule applications "above", i.e. to all the substitutions in derivations corresponding to the proof variables in the leaves. $\hat{\rho}$ is recursively defined over the derivation $\rho$, which defines the s-substitution

$$\Theta(n, m) = \{X(n) \leftarrow \hat{s}(Y(n), m),\ Z(n) \leftarrow \hat{s}(a, n)\}.$$

By construction, as $\hat{\rho}$ is recursive, $\Theta(p(n), m)$ is applied to $\Theta(n, m)$; to the result we apply $\Theta(p(p(n)), m)$ and so on. Intuitively, $\Theta^*(n, m)$ is is the sequence

$$\Theta(n, m)\Theta(p(n), m)\Theta(p(p(n)), m) \cdots \Theta(\bar{0}, m).$$

or expressed recursively $D(\Theta) =$

$$\{\Theta^*(n, m) =_d\ n = 0\ \underline{\text{then}}\ \Theta(\bar{0}, m)\ \underline{\text{else}}\ \Theta(n, m) \circ \Theta^*(p(n), m)\}.$$

**Definition 9.2.5** (*Herbrand system of a schematic $F$-proof*) The Herbrand system of a schematic $F$-proof is defined inductively as:

- For $\vdash F$ the Herbrand system is $\{\emptyset\}$, the set containing the identical substitution.
- Let $\rho$ be a schematic $F$-proof of the form

$$\frac{\begin{array}{cc}(\rho_1) & (\rho_2)\\ S_1 & S_2\end{array}}{S}\ \xi$$

  and assume the Herbrand system of $\rho_1$ and $\rho_2$ are $\Theta_1$ and $\Theta_2$. If $\xi$ is a binary rule different to the resolution rule, then the Herbrand system of $\rho$ is the global s-unifier of $\Theta_1 \cup \Theta_2$, which can be computed after regularization of the proof. If $\xi$ is a resolution rule of the form $R(\Theta)$, then prior regularization is mandatory and the Herbrand system is $(\Theta_1 \cup \Theta_2) \circ \Theta$.
- Let $\rho$ be a schematic $F$-proof of the form

$$\frac{\begin{array}{c}(\rho')\\ S'\end{array}}{S}\ \xi$$

  and assume that the Herbrand system of $\rho'$ is $\Theta$, then the Herbrand system of $\rho$ is $\Theta$.
- Let $\rho$ be given by link-substitution, so $\rho$ is of the form

$$\rho\{V_{S_i}(\mathbf{X}, \mathbf{r_i}) \leftarrow \rho_i \mid i \in \{1, \ldots, \alpha\}\}.$$

  Let $\Theta_i$ be the Herbrand system of $\rho_i$, then we define $\Theta'$ as the global s-unifier of

$$\bigcup_{i=1,\ldots\alpha} \Theta_i.$$

Then the Herbrand system of $\rho$ is

$$\Theta' \circ \Theta,$$

where $\Theta$ is the global s-unifier of $\rho$.

- Let $\hat{\rho}(\bar{X}, \mathbf{n}, k)$ be given by standard $\mathrm{RPL}_0^\Psi V$-recursion, so $\hat{\rho}$ is defined as

$$\hat{\rho}(\bar{X}, \mathbf{n}, k) =_d \underline{\text{if}}\ k = \bar{0}\ \underline{\text{then}}\ \rho(\bar{0})$$
$$\underline{\text{else}}\ \hat{\rho}(\bar{X}, \mathbf{n}, p(k))\{V(k) \leftarrow \rho(k)\}.$$

with the abbreviations $V(k)$ for $V_S(\mathbf{X}, (\mathbf{n}, k))$, and $\rho(k)$ for $\rho(\bar{X}, (\mathbf{n}, k), V_S(\mathbf{X}, (\mathbf{n}, k)))$.
Let $\Theta(\mathbf{n}, k)$ be the Herbrand system of $\rho(k)$. Then the Herbrand system of $\hat{\rho}$ is defined as $D(\Theta^*) =$

$$\{\Theta^*(\mathbf{n}, k) =_d \underline{\text{if}}\ k = \bar{0}\ \underline{\text{then}}\ \Theta(\mathbf{n}, \bar{0})\ \ \underline{\text{else}}\ \Theta(\mathbf{n}, k) \circ \Theta^*(\mathbf{n}, p(k))\}.$$

- Let $\hat{\rho}(\bar{X}, \mathbf{n}, k)$ be given by $\mathrm{RPL}_0^\Psi V$-recursion, so $\hat{\rho}$ is defined as

$$\hat{\rho}(\bar{X}, \mathbf{n}, k)) =_d \underline{\text{if}}\ k = \bar{0}\ \underline{\text{then}}\ \rho(\bar{X}, (\mathbf{n}, \bar{0}), V_S(\mathbf{X}, (\mathbf{n}, \bar{0}))$$
$$\underline{\text{else}}\ \hat{\rho}((\bar{X}, \mathbf{n}, p(k))\{V_i(k) \leftarrow \rho_i(k) \mid i \in \{1, \ldots, m\}\},$$

where $V_i(k)$ is an abbreviation for $U_S^i(\mathbf{X}, \eta_i((n_1, \ldots, n_\alpha, p(k))))$ and $\rho_i(k)$ for

$$\rho(\bar{X}, \eta_i(\mathbf{n}, p(k)), V_S(\mathbf{X}, \eta_i(\mathbf{n}, p(k))).$$

Let $\Theta(\mathbf{n}, 0)$ be the Herbrand system of

$$\rho(\bar{X}, (\mathbf{n}, \bar{0}), V_S(\mathbf{X}, (\mathbf{n}, \bar{0}))$$

and for $k > 0$ let $\Theta_i(\mathbf{n}, k)$ be the Herbrand system of

$$\rho(\bar{X}, \eta_i(\mathbf{n}, p(k)), V_S(\mathbf{X}, \eta_i(\mathbf{n}, p(k))).$$

Then the Herbrand system of $\hat{\rho}$ is defined as $D(\Theta^*) =$

$$\{\Theta^*(\mathbf{n}, k) =_d \underline{\text{if}}\ k = \bar{0}\ \underline{\text{then}}\ \Theta(\mathbf{n}, \bar{0})\ \ \underline{\text{else}}\ \Theta(\mathbf{n}, k) \circ \Theta^*(\mathbf{n}, p(k))\}.$$

where $\Theta(\mathbf{n}, k)$ is the global s-unifier of

$$\bigcup_{i=1,\ldots m} \Theta_i(\mathbf{n}, k).$$

- Let $\rho(\mathbf{X_1}, \mathbf{X_2}, \mathbf{n}) =$

$$\underline{\text{if}}\ C\ \underline{\text{then}}\ \rho_1(\mathbf{X_1}, \mathbf{n})\ \underline{\text{else}}\ \rho_2(\mathbf{X_2}, \mathbf{n}).$$

  be a schematic $F$-proof, and assume that the Herband systems of $\rho_1$ and $\rho_2$ are $\Theta_1$ and $\Theta_2$. Then the Herbrand system of $\rho$ is

$$\underline{\text{if}}\ C\ \underline{\text{then}}\ \Theta_1\ \underline{\text{else}}\ \Theta_2.$$

**Example 9.2.5** Consider the Herbrand system from Example 9.2.4 for the parameter assignments $\sigma(n) = \bar{1}$ and $\sigma(m) = \bar{0}$. Then,

$$\begin{aligned}
\Theta^*(\bar{1}, \bar{0}) &= \Theta(\bar{1}, \bar{0})\Theta^*(p(\bar{1}), \bar{0}) \\
&= \{\{X(\bar{1}) \leftarrow \hat{s}(Y(\bar{1}), \bar{0}), Z(\bar{1}) \leftarrow \hat{s}(a, \bar{1})\}\}\{\{X(\bar{0}) \leftarrow \hat{s}(Y(\bar{0}), \bar{0}), \\
&\quad Z(\bar{0}) \leftarrow \hat{s}(a, \bar{0})\}\} \\
&= \{\{X(\bar{1}) \leftarrow Y(\bar{1}),\ Z(\bar{1}) \leftarrow h(a),\ X(\bar{0}) \leftarrow Y(\bar{0}),\ Z(\bar{0}) \leftarrow a\}\}.
\end{aligned}$$

Applying $\Theta^*(\bar{1}, \bar{0})$ to the initial sequents results in an unsatisfiable set of sequents.

**Theorem 9.2.1** *Let $\Theta$ be a Herbrand system of a refutation schema $\rho$ of $\hat{q}(\mathbf{X}, n_1, \ldots, n_\alpha)$. Then, for all numerals $\nu_1, \ldots, \nu_\alpha$, the set of sequents*

$$\{\vdash \hat{q}(\mathbf{X}, n_1, \ldots, n_\alpha)\Theta\}\{n_1 \leftarrow \nu_1, \ldots, n_\alpha \leftarrow \nu_\alpha\}$$

*is unsatisfiable.*

***Proof*** For all numerals, first apply the s-substitutions defined in $\Theta$ to $\rho$. The initial sequents are then instances of $\vdash \hat{q}(\mathbf{X}, n_1, \ldots, n_\alpha)$, and the resolution rules turn into applications of cut rules. The thus obtained derivation is in $\text{RPL}_0^\Psi$. As this calculus is sound the set of sequents $\{\vdash \hat{q}(\mathbf{X}, n_1, \ldots, n_\alpha)\Theta\}\{n_1 \leftarrow \nu_1, \ldots, n_\alpha \leftarrow \nu_\alpha\}$ occurring at the leaves is unsatisfiable.

### 9.2.3 *Herbrand Systems with* CERES

Analogous to the first-order CERES method as described in Sect. 3.2.2, the construction of a schematic CERES normal form is superfluous for the extraction of Herbrand systems. In fact, we can combine the Herbrand system of the projection schema with the Herbrand system of the refutation schema to obtain the final Herbrand system for a schematic end-sequent.

**Definition 9.2.6** Let $\rho$ be a regular proof schema of a skolemized prenex schematic end-sequent, $\mathrm{CF}(\rho)$ its characteristic formula schema and $\mathrm{Proj}(\rho)$ its projection schema. Let $\rho'$ be the schematic $F$-proof from $\mathrm{CF}(\rho)$ (i.e. the refutation schema of $\mathrm{CF}(\rho)$).

For each quantified formula occurrence $\mu_i$ in the end-sequent of $\mathrm{Proj}(\rho)$ let $S_S(\mathrm{Proj}(\rho), \mu_i)$ be the Herbrand schema of $\mathrm{Proj}(\rho)$ for the formula occurrence $\mu_i$. Then we define the set

$$\bar{H}_S(\rho, \mu_i, \Theta(\mathbf{n}, k)) = \bigcup_{j=1,\dots,\alpha} S_S(\mathrm{Proj}(\rho), \mu_i)\theta_j$$

for $\theta_1, \dots, \theta_\alpha \in \Theta(\mathbf{n}, k))$ and $\Theta(\mathbf{n}, k))$ the Herbrand schema of $\rho'$.

$\bar{H}_S(\rho, \mu_i, \Theta(\mathbf{n}, k))$ can be evaluated under any parameter assignment $\sigma$:

$$\sigma(\bar{H}_S(\rho, \mu_i, \Theta(\mathbf{n}, k))) = \bigcup_{j=1,\dots,\alpha} \sigma(S_S(\mathrm{Proj}(\rho), \mu_i))\theta_j$$

for $\theta_1, \dots, \theta_\alpha \in \sigma(\Theta(\mathbf{n}, k)))$ and $\Theta(\mathbf{n}, k))$ the Herbrand schema of $\rho'$.

**Example 9.2.6** Consider the regular proof schema $\rho_0(n)$ from Example 7.1.8 (it is called $\rho_1^R$ there), and its projection schema $\mathrm{Proj}(\rho_0(n)) =$

$$\cfrac{\begin{matrix}(\mathrm{Proj}(\hat{\rho}(n, V_S(n)) \\ \{V_S(0) \leftarrow \rho'\}, \{(1, r)\})) \\ S(n)\end{matrix} \qquad \begin{matrix}(\varphi(n)) \\ \vdash (P(\hat{f}(c,n)) \to P(g(c,n))) \to (P(c) \to P(g(c,n))), P(c) \land \neg P(\hat{f}(c,n))\end{matrix}}{\forall x(P(x) \to P(f(x))) \vdash (P(\hat{f}(c,n)) \to P(g(c,n))) \to (P(c) \to P(g(c,n))), \mathrm{CF}(\mathrm{Proj}(\rho_0(n)))}\ \land r$$

where $\varphi(n) =$

$$\cfrac{\cfrac{P(c) \vdash P(c) \qquad \cfrac{\cfrac{P(\hat{f}(c,n)) \vdash P(\hat{f}(c,n))}{\vdash P(\hat{f}(c,n)), \neg P(\hat{f}(c,n))}\ \neg_r \qquad P(g(c,n)) \vdash P(g(c,n))}{P(\hat{f}(c,n)) \to P(g(c,n)) \vdash P(g(c,n)), \neg P(\hat{f}(c,n))}\ \to_l}{\cfrac{P(c), P(\hat{f}(c,n)) \to P(g(c,n)) \vdash P(g(c,n)), P(c) \land \neg P(\hat{f}(c,n))}{P(\hat{f}(c,n)) \to P(g(c,n)) \vdash P(c) \to P(g(c,n)), P(c) \land \neg P(\hat{f}(c,n))}\ \to_r}\ \land_r}{\vdash (P(\hat{f}(c,n)) \to P(g(c,n))) \to (P(c) \to P(g(c,n))), P(c) \land \neg P(\hat{f}(c,n))}\ \to_r$$

and characteristic formula schema $G(n) = \mathrm{CF}(\rho_0(n))$ as defined in Example 9.1.3:

$$G(n) = H(n) \land P(c) \land \neg P(\hat{f}(c,n))$$

(note that $H(n) = \mathrm{CF}(\hat{\rho}(n, V_S(n)), \{(1, r)\}))$, where

$$H(0) = \neg P(\hat{f}(X^{\rho'}(0), 0)) \lor P(\hat{f}(X^{\rho'}(0), 0)).$$

and $H(n) =$

$H(p(n)) \land$

$$(\neg P(X^{\rho}(n)) \vee P(X^{\rho}(n))) \wedge (\neg P(\hat{f}(X^{\rho}(n), p(n))) \vee P(\hat{f}(X^{\rho}(n), n))).$$

Let $\mu_0$ be the occurrence of the formula $\forall x(P(x) \to P(f(x)))$ in the end-sequent of $\mathrm{Proj}(\rho_0(n))$. From Example 9.2.3 we know that

$$\begin{aligned} S_S(\mathrm{Proj}(\rho_0(n)), \mu_0(n)) = & \{x \leftarrow \hat{f}(X^{\rho}(n), p(n))\} \cup \\ & \{x \leftarrow \hat{f}(X^{\rho}(n-1), p(n-1))\} \cup \\ & \ldots \cup \\ & \{x \leftarrow \hat{f}(X^{\rho}(\bar{1}), p(\bar{1}))\}. \end{aligned}$$

The refutation schema of $G(n)$ is

$$\dfrac{\dfrac{(\hat{\rho}(X, n, \bar{0}))}{\vdash P(\hat{f}(c, n))} \qquad \dfrac{\dfrac{\dfrac{\dfrac{\vdash G(n)}{\vdash H(n) \wedge P(c) \wedge \neg P(\hat{f}(c, n))}}{\vdash P(c) \wedge \neg P(\hat{f}(c, n))}\,\wedge_{r2}}{\vdash \neg P(\hat{f}(c, n))}\,\wedge_{r2}}{P(\hat{f}(c, n)) \vdash}\,\neg_r}{\vdash}\,res$$

where $\hat{\rho}(X, n, l)$ is

$$\begin{aligned} D(\hat{\rho}) = \{\hat{\rho}(X, n, l) =_d \ & \underline{\text{if}}\ n = \bar{0}\ \underline{\text{then}}\ V_S(0, l) \\ & \underline{\text{else}}\ \rho(X, n, l)\{V_S(p(n), s(l)) \leftarrow \hat{\rho}(X, p(n), s(l))\}\}. \end{aligned}$$

$\rho(X, n, l) =$

$$\dfrac{\dfrac{\dfrac{V_S(p(n), s(l))}{\vdash P(\hat{f}(c, p(n)))} \qquad \dfrac{\dfrac{\dfrac{(\hat{\rho}_1(n, l))}{V_H(n, l)}}{\vdash H(n)}}{P(\hat{f}(X^{\rho}(n), p(n))) \vdash P(\hat{f}(X^{\rho}(n), n))}}{\vdash P(\hat{f}(c, n))}\,res\{X^{\rho}(n) \leftarrow c\}}{V_S(n, l)}$$

and $\hat{\rho}_1(k, l)$ is

$$\begin{aligned} D(\hat{\rho}_1) = \{\hat{\rho}_1(k, l) =_d \ & \underline{\text{if}}\ l = \bar{0}\ \underline{\text{then}}\ V_H(k, 0) \\ & \underline{\text{else}}\ \rho_1(k, l)\{V_H(s(k), p(l)) \leftarrow \hat{\rho}_1(s(k), p(l))\}\}, \end{aligned}$$

for $\rho_1(k, l) =$

$$\dfrac{\dfrac{\dfrac{\dfrac{V_H(s(k), p(l))}{\vdash H(s(k))}}{H(k) \wedge (\neg P(X^{\rho}(n)) \vee P(X^{\rho}(n))) \wedge (\neg P(\hat{f}(X^{\rho}(n), p(n))) \vee P(\hat{f}(X^{\rho}(n), n)))}}{\vdash H(k)}}{V_H(k, l)}$$

In the end, we have to replace all the initial sequents that are links of the form $V_S(\bar{0}, m)$ and $V_H(m, \bar{0})$ for $1 \leq m \leq n$ by derivations from $G(n)$ of $\vdash P(\hat{f}(c, \bar{0}))$ respectively $\vdash H(n)$: $V_S(\bar{0}, m)$ is replaced by $\tau_1 =$

$$\dfrac{\dfrac{\dfrac{\vdash G(n)}{\vdash H(n) \wedge P(c) \wedge \neg P(\hat{f}(c, n))}}{\vdash P(c) \wedge \neg P(\hat{f}(c, n))}\ \wedge_{r_2}}{\vdash P(c)}\ \wedge_{r_1}$$

and $V_H(m, \bar{0})$ is replaced by $\tau_2 =$

$$\dfrac{\dfrac{\vdash G(n)}{\vdash H(n) \wedge P(c) \wedge \neg P(\hat{f}(c, n))}}{\vdash H(n)}\ \wedge_{r_1}$$

To compute the Herbrand schema $\Theta^*(n)$, we have to regularize the refutation schema above. However, we can eliminate $\hat{\rho}_1$ (by pruning), and the remaining refutation schema is regular: $\Theta^*(n)$ is given by the s-substitution $\Theta(n) = \{X^{\rho}(n) \leftarrow c\}$: $D(\Theta^*) =$

$$\{\Theta^*(n) =_d \ n = 0 \ \underline{\text{then}}\ \emptyset \ \ \underline{\text{else}}\ \Theta(n) \circ \Theta^*(p(n))\}.$$

Hence,

$$\begin{aligned} \bar{H}_S(\rho_0(n), \mu_0(n), \Theta^*(n)) = \{x \leftarrow \hat{f}(c, p(n))\}\cup \\ \{x \leftarrow \hat{f}(c, p(n-1))\}\cup \\ \ldots\cup \\ \{x \leftarrow \hat{f}(c, p(\bar{1}))\}. \end{aligned}$$

For some numerals $\alpha$ this means that we obtain the instances

$$c, f(c), f(f(c)), \ldots, f^{\alpha-1}(c).$$

The Herbrand schema for $\mu_i$ of a proof schema $\rho$ evaluates under all parameter assignments $\sigma$ to the Herbrand instances for $\mu_i$ of $\sigma(\rho)$.

**Theorem 9.2.2** *Let $\rho_0$ be a regular proof schema of a skolemized prenex end-sequent and $\sigma$ a parameter assignment. Then*

$$\sigma(\bar{H}_S(\rho_0, \mu_i, \Theta(\mathbf{n}, k))) = \bar{H}(\sigma(\rho_0), \mu_i, \rho),$$

*where $\bar{H}(\sigma(\rho_0), \mu_i, \rho)$ is as in Definition 3.2.6.*

***Proof*** By definition

$$\sigma(\bar{H}_S(\rho_0, \mu_i, \Theta(\mathbf{n}, k))) = \bigcup_{j=1,\dots,\alpha} \sigma(S_S(\text{Proj}(\rho_0), \mu_i))\theta_j$$

for $\theta_1, \dots, \theta_\alpha \in \sigma(\Theta(\mathbf{n}, k)))$ and $\Theta(\mathbf{n}, k))$ the unification schema of the refutation schema $\rho'$ of $\text{CF}(\rho_0)$.

$\sigma(\Theta(\mathbf{n}, k)))$ is a $\text{RPL}_0^\Psi$ refutation of the instantiated characteristic formula schema $\sigma(\text{CF}(\rho_0))$ which by Theorem 9.1.1 is the characteristic formula of $\sigma(\rho)$. Thus, $\{\theta_1, \dots, \theta_\alpha\}$ for $\theta_1, \dots, \theta_\alpha \in \sigma(\Theta(\mathbf{n}, k)))$ is equal to a set $\{\theta_1, \dots, \theta_\alpha\}$ for $\theta_1, \dots, \theta_\alpha \in \Sigma(\rho)$, where $\rho$ is the refutation $\sigma(\rho')$. Therefore,

$$\sigma(\bar{H}_S(\rho_0, \mu_i, \Theta(\mathbf{n}, k))) = \bigcup_{j=1,\dots,\alpha} \sigma(S_S(\text{Proj}(\rho_0), \mu_i))\theta_j$$

for $\theta_1, \dots, \theta_\alpha \in \Sigma(\rho)$. By Proposition 9.2.1,

$$\sigma(S_S(\text{Proj}(\rho_0), \mu_i)) = S(\sigma(\text{Proj}(\rho_0)), \mu_i),$$

and thus

$$\sigma(\bar{H}_S(\rho_0, \mu_i, \Theta(\mathbf{n}, k))) = \bigcup_{j=1,\dots,\alpha} S(\sigma(\text{Proj}(\rho_0)), \mu_i)\theta_j$$

for $\theta_1, \dots, \theta_\alpha \in \Sigma(\rho)$.

# Chapter 10
# Schematic Reasoning in GAPT

Rather than developing a system from scratch, we integrated our interactive system for schematic proof analysis with schematic `CERES` within the software package GAPT (General Architecture for Proof Theory) [16]. The GAPT system provides numerous algorithms for the analysis, transformation, and construction of proofs in various formal calculi and has been used for `CERES` based proof analysis. GAPT is implemented in Scala and licensed under the GNU General Public License. The software is available under https://logic.at/gapt. GAPT also provides an interface for importing proofs from most major theorem provers and exporting proofs and other structures in TPTP format and its own internal JSON based format. A good reference summarizing the features of GAPT (*sans* the schematic components) is [16] and the GAPT documentation. Initially GAPT started as an implementation of the `CERES` method particularly suited for the analysis of Fürstenberg's proof of the infinitude of primes [4].

The schematic extension of GAPT depends on the formal structures developed in [15] and the improvements presented in [31]. One of the most interesting benefits of the interactive features which have been added to GAPT is the representation of the cut-structure of a proof schema as an inductive definition. This feature is based on the constructions developed in [31]. Similar frameworks for dealing with schematic proof objects as inductive definitions exists in systems such as COQ [9] and Isabelle [35] but without the proof analytic and transformation support provided by GAPT.

In this chapter we will present the current GAPT implementation and some of its features. The following proof schema will be our canonical example for illustrating the schematic features of the GAPT system.

**Example 10.0.1** Consider a proof schema

$$\mathcal{D}\colon \{(\delta, \rho(\delta, \vec{n}, 0), \rho(\delta, \vec{n}, n+1))\},$$



A. Leitsch et al., *First-Order Schemata and Inductive Proof Analysis*, Computer Science Foundations and Applied Logic, https://doi.org/10.1007/978-3-032-05741-9_10

where the end-sequent is

$$S(\delta) = P(a), \forall x(P(x) \to P(f(x))) \vdash P(\hat{if}(n))$$

using

$$\mathcal{E} = \begin{cases} \hat{if}(0) = a \\ \hat{if}(n+1) = f(\hat{if}(n)). \end{cases}$$

$\rho(\delta, \vec{n}, 0)$ is the derivation of the base case $S(\delta)\{n \leftarrow 0\}$:

$$\dfrac{\dfrac{P(a) \vdash P(a)}{P(a), \forall x(P(x) \to P(f(x))) \vdash P(a)}\; w:l}{P(a), \forall x(P(x) \to P(f(x))) \vdash P(\hat{if}(0))}\; \mathcal{E}$$

and $\rho(\delta, \vec{n}, n+1)$ is the derivation of the step case $S(\delta)\{n \leftarrow n+1\}$ :

$$\dfrac{\dfrac{\dfrac{\dfrac{(\nu_1)}{\forall x(P(x) \to P(f(x))) \vdash \forall x(P(x) \to P(f(x)))} \qquad (\nu_2)}{P(0), \bigwedge_{i=0}^{n}(P(i) \to P(i+1)), P(n+1) \to P(n+2) \vdash P(n+2)}\; cut}{P(a), \forall x(P(x) \to P(f(x))), \forall x(P(x) \to P(f(x))) \vdash P(\hat{if}(n+1))}}{P(a), \forall x(P(x) \to P(f(x))) \vdash P(\hat{if}(n+1))}\; c:l$$

where $\nu_1$ and $\nu_2$ are

$$\dfrac{\dfrac{\dfrac{\dfrac{P(\alpha) \vdash P(\alpha) \qquad P(f(\alpha)) \vdash P(f(\alpha))}{P(\alpha) \to P(f(\alpha)), P(\alpha) \vdash P(f(\alpha))}\; \to: l}{P(\alpha) \to P(f(\alpha)) \vdash P(\alpha) \to P(f(\alpha))}\; \to: r}{\forall x(P(x) \to P(f(x))) \vdash P(\alpha) \to P(f(\alpha))}\; \forall: l}{\forall x(P(x) \to P(f(x))) \vdash \forall x(P(x) \to P(f(x)))}\; \forall: r$$

$$\dfrac{\dfrac{\dfrac{S(\delta\colon P(a), \forall x(P(x) \to P(f(x))) \vdash P(\hat{if}(n)) \qquad P(f(\hat{if}(n))) \vdash P(f(\hat{if}(n)))}{P(a), P(\hat{if}(n)) \to P(f(\hat{if}(n))), \forall x(P(x) \to P(f(x))) \vdash P(f(\hat{if}(n)))}\; \to: l}{P(a), P(\hat{if}(n)) \to P(f(\hat{if}(n))), \forall x(P(x) \to P(f(x))) \vdash P(\hat{if}(n))}\; \mathcal{E}}{P(a), \forall x(P(x) \to P(f(x))), \forall x(P(x) \to P(f(x))) \vdash P(\hat{if}(n+1))}\; \forall: l$$

respectively and $\alpha$ is the eigenvariable of the strong quantifier rule in $(\nu_1)$.

## 10.1 Constructing Schematic Proofs

The existing proof construction mechanisms of the GAPT system still heavily relies on the syntactic constructions of Scala, i.e. GAPT does not have a dedicated input format. Nonetheless, GAPT contains a simple tactics language called *gaptic* for the construction of proofs providing a comfortable bottom-up development of proofs similar to the tactic languages found in proof assistants such as Coq and Isabelle. In this section we take the proof schema of Example 10.0.1 and formalize it using the GAPT system. This proof schema can be found in the examples directory of the current GAPT release as

```
FunctionIterationSchema.scala.
```

Also, in the same directory, one can find a refutation schema of the characteristic formula of Example 10.0.1 which is referred to as the

```
FunctionIterationRefutation.scala.
```

More complex examples can be found in this directory, as for instance the proof schema discussed in [14]. From the GAPT terminal one can load these examples by importing the objects as follows:

```
import examples.FunctionIterationSchema.
```

An overview of gaptic and the basic functionality of the GAPT terminal can be found in the GAPT User Manual.[1]

First of all note that schematic proofs are dependent on several core packages of GAPT.

```
import gapt.expr._
import gapt.proofs.context.update._
import gapt.proofs.gaptic._
import gapt.proofs.Sequent
```

These are the minimum imports needed for schematic proof construction. For those not familiar with the GAPT system, the *context* stores all information needed for the construction of a proof. As described before, a schematic proof is an ordered set of pairs of proofs of the same end-sequent, modulo parameter instantiation. The schematic extension of GAPT generalizes this concept to proof tuples with the same end-sequent, modulo parameter instantiation, allowing the use of inductive definitions more complex than the natural numbers. To enforce the shared end-sequent condition the system allows for *proof name declarations*. Figure 10.1 illustrates which symbols and definitions have to be added to the context within the proof object for the FunctionIterationSchema.

Note that the distinction between the inductive sort

```
Context.InductiveType("nat", hoc"0:nat",hoc"s:nat>nat")
```

[1] https://www.logic.at/gapt/downloads/gapt-user-manual.pdf.

```
object FunctionIterationSchema extends TacticsProof {
// individual sort
      ctx += Context.Sort( "i" )
// numeric sort
      ctx += Context.InductiveType( "nat", hoc"0 : nat",
                hoc"s : nat>nat" )
//symbol definitions
      ctx += hoc"f:i>i"
      ctx += hoc"a:i"
      ctx += hoc"P: i>o"
//inductive symbol definition
      ctx += PrimRecFun( hoc"if:nat>i>i",
                "if 0 x = x", "if (s y) x = (f (if y x))" )
//Proof names
      ctx += hoc"phi: nat>nat"
//Proof definitions
      val esPhi = Sequent(Seq( hof"!x(-P(x)|P(f(x)))",
                     hof"P(a)"),Seq(hof"P(if(n,a))"))
      ctx += Context.ProofNameDeclaration( le"phi n", esPhi )
         . . .
```

**Fig. 10.1** Proof schema object in GAPT

and the first-order term sort

Context.Sort("i")

must be made explicit, as well as the constants associated with the proof. Moreover, the definition of $\hat{\text{if}}(n+1, a)$ from Example 10.0.1 is added as a primitive recursive function

PrimRecFun(hoc"if:nat>i>i", "if 0 x = x",

"if (s y) x = (f (if y x))").

Now we can construct the base and step case proof associated with the proof definition of *phi* as illustrated in Fig. 10.2.

Note that the proof name has a function type of $nat > nat$. While it acts as a place holding constant, for proper integration into the system it has been implemented as a lambda term and thus, takes arguments and has a return type.

The base and step case proofs associated with phi are referred to as phiSc and phiBc. We add these proofs to the context using

ProofDefinitionDeclaration.

```
val esPhiSc = Sequent(
    Seq("Ant_1" -> hof"!x (-P(x) | P(f(x)))",
        "Ant_0" -> hof"P(a)"),
    Seq( "Suc_0" -> hof"P(if(s(n),a))" ) )
val phiSc = Lemma( esPhiSc ) {
    cut( "cut", hof"!x (-P(x) | P(f(x)))")
    allR( "cut", fov"A" )
              . . .
    ref( "phi" )
    unfold( "if" ) atMost 1 in "Suc_0"
    trivial
  }
ctx += Context.ProofDefinitionDeclaration(
      le"phi (s n)", phiSc )
```

```
val esPhiBc = Sequent(
    Seq("Ant_1" -> hof"!x (-P(x) | P(f(x)))",
        "Ant_0" -> hof"P(a)" ),
    Seq("Suc_0" -> hof"P(if(0,a))" ) )
val phiBc = Lemma( esPhiBc ) {
    unfold( "if" ) atMost 1 in "Suc_0"
    trivial
  }
ctx += Context.ProofDefinitionDeclaration(
    le"phi 0", phiBc)
}
```

**Fig. 10.2** Construction of base and step case proofs

Taking a close look at the code in Fig. 10.2, one will notice that phiBc is added to the context with a lambda expression le"phi 0" and phiSc with le"phi (s n)"; these are the cases of the inductive definition handled by the proof. It is not necessary for the cases to match the inductive definition nor for all cases to be covered. An example of such a schema would be the NdiffSchema found in

/gapt/examples/schema/NdiffSchema.scala.

The last point we would like to address concerning proof construction related to schema is the additional tactic added to the gaptic language, i.e. the reference tactic, and the extensive use of the unfold tactic. The reference tactic used in phiSc, ref("phi"), allows one to reference a proof, including the proof itself (self reference). The system will check to see if the reference is valid, i.e. there exists a proof name matching the goal the reference was called on. The unfold tactic, while not particular to schema, is used for unfolding the primitive recursive definitions such as $\widehat{\text{if}}(n+1, a)$. It essentially implements the equational theory $\mathcal{E}$.

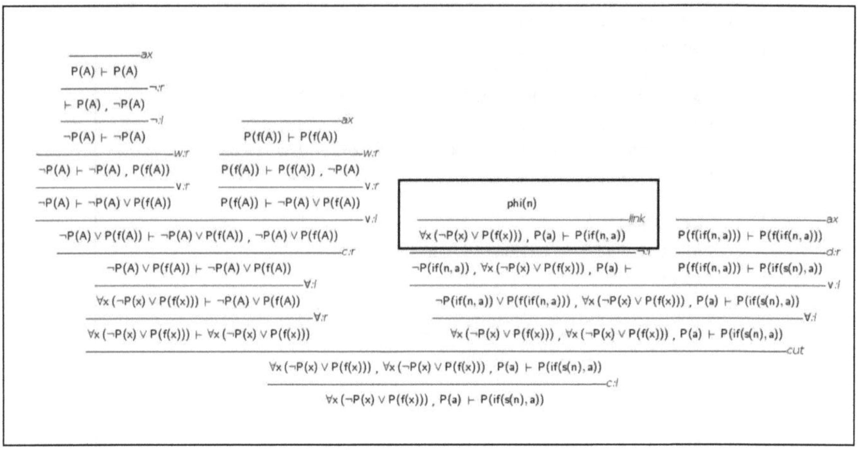


**Fig. 10.3** Prooftool output of step case proof. Notice the link to the proof phi(n)

The proofs phiBc and phiSc can be displayed in Prooftool (which is a viewer for proofs and other elements implemented in GAPT) using the command

```
prooftool(FunctionInterationSchema.phiBc)
```

(see Fig. 10.3). Also, we can instantiate the proof schema of phi with an arbitrary value;

```
instantiateProof.Instantiate(le"phi (s(s 0)"))
```

instantiates the proof schema with the value 2.

For more details, we refer the reader to a tutorial on schematic proof construction which can be found in the directory

```
/gapt/examples/schema/Schema Tutorial.
```

## 10.2 Schematic Proof Analysis

Proof schemata, when instantiated, are essentially **LK**-proofs and thus, any of the proof analytic tools and methods of GAPT can be applied. However, as described in Chap. 9 there are also tools specifically designed for uninstantiated proof schemata, especially those which contain cuts. For example, it is possible to extract schematic

```
//imports the Context
object FunctionIterationRefutation extends
  TacticsProof(FunctionIterationSchema.ctx){
//Produces a schematic struct from  Context
  val SCS: Map[CLS, ( Struct, Set[Var] )] =
      SchematicStruct("phi").getOrElse(Map())
//Produces a schematic characteristic formula
  val CFPRN = CharFormPRN( SCS )
//Constructs primitive recursive definitions
  of the characteristic formula
  CharFormPRN.PR( CFPRN )
                        . . .
}
```

**Fig. 10.4** Preamble for refutation schema construction

characteristic formulas or clause sets, i.e. a recursive representation of the cut-structure of the given proof schema. These objects are produced from an intermediate representation referred to as the *struct* of the proof.

In the previous section we provided a short introduction to proof schema construction in GAPT by building the FunctionIterationSchema. Using this same example we provide an introduction to the schematic proof analysis capabilities of GAPT. This exposition can be found in the examples directory as

FunctionIterationRefutation,

see Fig. 10.4.

Other than these few extra commands at the beginning of the object file, the construction of the refutation is quite similar to the construction of a proof schema. Essentially, $F \vdash$, where $F$ is the schematic characteristic formula, is proved. All these commands do is add the names and normalizers for the characteristic formula to the context:

```
gapt> FunctionIterationSchema.ctx.names
res0: Iterable[String] = List(s,if, =,f,a,P,0,phi,o,nat,i)
gapt> FunctionIterationRefutation.ctx.names
res1: Iterable[String] = List(phiSFAFF, phiSTATF, phiSFATF,
        s, f, =,f,a,P,0,phi,o,nat,i)
```

The names associated with the characteristic formula

phiSFAFF, phiSTATF, and phiSFATF

are automatically generated and are a combination of a proof name, phi, and a sequent of booleans, S(uccedent) F A(ntecedent) F F for phiSFAFF. The booleans mark cut ancestors. Once the schema of the characteristic formula is constructed an expansion proof for any instance can be produced (see Fig. 10.5).

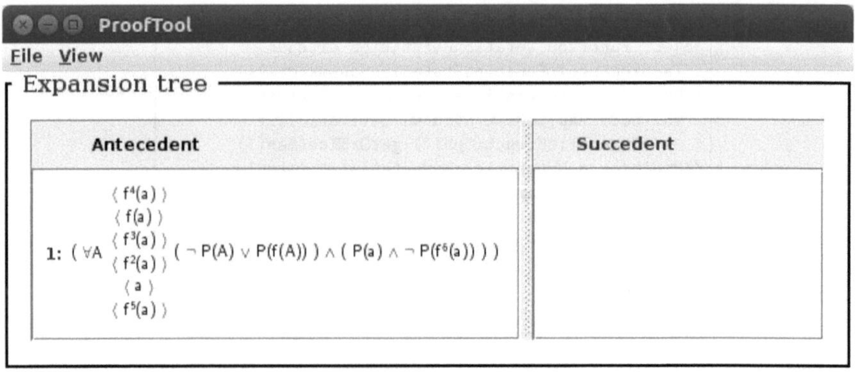


**Fig. 10.5** Expansion proof of instance 5 of the FunctionIterationRefutation schema

## 10.3 A Complex Example: The Eventually Constant Schema

In [12] a proof analysis of the so-called *Eventually Constant Schema* using the method of [15] was presented by manually producing instances of the proof and its cut-structure using various theorem provers. This interactive process has been streamlined using the GAPT system and schemata of both the proof

EventuallyConstantSchema

and the characteristic formula

EventuallyConstantRefutation

can be found in

/gapt/examples/schema/Schema.

The EventuallyConstantSchema (or ECS for short) is a formal proof of the following statement:

**Proposition 10.3.1** *Given a total monotonically decreasing function* $f : \mathbb{N} \rightarrow \{0, \cdots, n\}$, *for* $n \in \mathbb{N}$, *there exists an* $x \in \mathbb{N}$ *such that for all* $y \in \mathbb{N}$, *where* $x \leq y$, *it is the case that* $f(x) = f(y)$.

***Proof*** If the range only contains 0 then the theorem trivially holds. Let us assume it holds for a co-domain with $n$ elements and show that it holds for a co-domain with $n + 1$ elements. If for all positions $x$, $f(x) = n$ then the theorem holds, else if at some $y$, $f(y) \neq n$ then from that point on $f$ cannot map to $n$ because the function is monotonically decreasing, thus, $f$ will only have $n$ elements in its co-domain and the theorem holds in this case by the induction hypothesis.

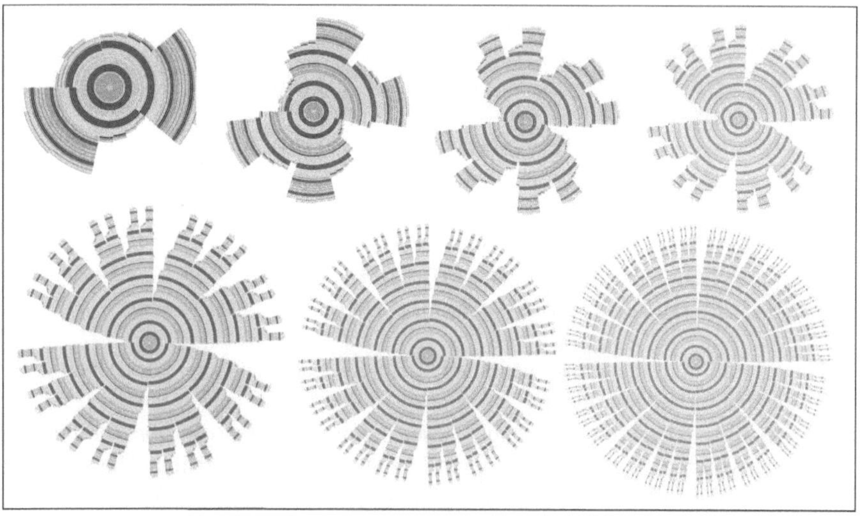

**Fig. 10.6** Instances 1–7 of the EventuallyConstantRefutation using the sunburst view of prooftool. Instance seven has 128 branches

The proof schema contains a sequence of $\Sigma_2$ cuts forcing the function to be monotone decreasing. A nice property of this statement is that both its proof schema and refutation schema have the same recursive complexity [12] in terms of finite representation, making it a perfect example to illustrate the method. Though, the schema grows linearly and the refutation exponentially (Fig. 10.6).

Concerning the cut formula, it is essentially the case distinction made in the stepcase of our mathematical proof. It has the following form:

$$\exists x \forall y \left( ((x \le y) \to n+1 = f(y)) \vee f(y) < n+1 \right).$$

The formal statement of the end-sequent has a $\exists\forall$ quantifier prefix like the cut formula:

$$\forall x (\bigvee_{i=0}^{n+1} i = f(x)), \forall x \forall y \Big(x \le y \to f(y) \le f(x)\Big) \vdash \exists x \forall y (x \le y \to f(x) = f(y)).$$

Both the CERES method and schematic CERES method were designed for proofs without *strong quantifiers* in the end-sequent. To get around this problem the proofs have to be Skolemized, hence the $g(\cdot)$ in the formal proof entered in the GAPT system (representing a Skolem function).

Given that the characteristic formula of the ECS is fairly complex we present it here as a recursively defined clause set. More details can be found in [12]. Notice that $C4(x, y, i, k)$ and $C4'(x, y, i, k)$ formalize the case distinction present in the cut formula.

$$\begin{aligned}
C1(x,k) &\equiv \vdash x(k) \leq x(k)\\
C2(x,k) &\equiv \vdash x(k) \leq g(x(k))\\
C3(x,i,k) &\equiv i = f(x(k)), i = f(g(x(k))) \vdash\\
C4(x,y,i,k) &\equiv y(k) \leq x(k), f(y(k)) < i+1 \vdash\\
&\quad f(x(k)) < i, i = f(x(k))\\
C4'(x,y,i,k) &\equiv y(k) \leq x(k+1), f(y(k)) < i+1 \vdash\\
&\quad f(x(k+1)) < i, i = f(x(k+1))\\
C5(x,k) &\equiv f(x(k)) < 0 \vdash\\
C6(x,k) &\equiv f(g(x(k))) < 0 \vdash\\
C7(x,k) &\equiv 0 \leq x(k) \vdash f(x(k)) < n, f(x(k)) = n
\end{aligned}$$

The refutation of the above recursively defined clause set can be described in the proof schema formalism (the implementation of the refutation uses the base and step case formalism as described in Chap. 7.3, see below). We use $\delta_0$ as label for the main refutation, which calls the proof schema $\delta$ containing the recursion. The schemata are based on the $\omega\iota$-theory $T^+$. The base case of $\delta_0$ is given by $\rho(\delta_0, 0) =$

$$\dfrac{\dfrac{\begin{matrix}(\pi_0)\\ f(\hat{g}(0,0)) = 0 \vdash\end{matrix} \qquad \begin{matrix}(\pi_1)\\ \vdash f(\hat{g}(0,0)) < 0,\\ f(\hat{g}(0,0)) = 0\end{matrix}}{\vdash f(\hat{g}(0,0)) < 0}\,res \qquad f(\hat{g}(0,0)) < 0 \vdash}{\vdash}\,res$$

where $\pi_0 =$

$$\dfrac{\dfrac{\begin{matrix}(\pi_2)\\ \vdash f(\hat{g}(1,0)) = 0\end{matrix}}{\vdash f(g(\hat{g}(0,0))) = 0}\,\text{unfold} \qquad \begin{matrix}f(X'(0,0)) = 0,\\ f(g(X'(0,0))) = 0 \vdash\end{matrix}}{f(\hat{g}(0,0)) = 0 \vdash}\,res(\Theta_{\pi_0})$$

for $\Theta_{\pi_0} = \{X'(0,0) \mapsto \hat{g}(0,0)\}$ and $\pi_2 =$

$$\dfrac{f(\hat{g}(1,0)) < 0 \vdash \qquad \dfrac{\dfrac{\vdash Z(0,0) \leq g(Z(0,0)) \qquad \begin{matrix}0 \leq X(0,1),\\ \vdash f(X(0,1)) < 0,\\ f(X(0,1)) = 0\end{matrix}}{\begin{matrix}\vdash f(g(0)) < 0,\\ f(g(0)) = 0\end{matrix}}\,res(\Theta_{\pi_2})}{\begin{matrix}\vdash f(\hat{g}(1,0)) < 0,\\ f(\hat{g}(1,0)) = 0\end{matrix}}\,\text{fold}}{\vdash f(\hat{g}(1,0)) = 0}\,res$$

for $\Theta_{\pi_2} = \{Z(0,0) \mapsto 0, X(0,1) \mapsto g(0)\}$ and $\pi_1 =$

$$\dfrac{\dfrac{\vdash Z(0,0) \leq Z(0,0) \qquad \begin{matrix}0 \leq X(0,0)\\ \vdash f(X(0,0)) < 0,\\ f(X(0,0)) = 0\end{matrix}}{\vdash f(0) < 0, f(0) = 0}\,res(\Theta_{\pi_1})}{\vdash f(\hat{g}(0,0)) < 0, f(\hat{g}(0,0)) = 0}\,\text{fold}$$

where $\Theta_{\pi_1} = \{Z(0, 0) \mapsto 0, X(0, 0) \mapsto 0\}$.

The step case is $\rho(\delta_0, n + 1) =$

$$\dfrac{\dfrac{\begin{array}{c}(\tau_0)\\ f(\hat{g}(0, 0)) = n + 1 \vdash\end{array} \qquad \begin{array}{c}(\tau_1)\\ \vdash f(\hat{g}(0, 0)) < n + 1,\\ f(\hat{g}(0, 0)) = n + 1\end{array}}{\vdash f(\hat{g}(0, 0)) < n + 1}\,res \qquad \begin{array}{c}(\delta, (\mathbf{n + 1, 0}))\\ f(\hat{g}(0, 0)) < n + 1 \vdash\end{array}}{\vdash}\,res$$

where $\tau_0 =$

$$\dfrac{\dfrac{\begin{array}{c}(\tau_2)\\ \vdash f(\hat{g}(1, 0)) = n + 1\end{array}}{\vdash f(g(\hat{g}(0, 0))) = n + 1}\,T^+ \qquad \begin{array}{c}(axiom)\\ f(X'(n + 1, 0)) = n + 1,\\ f(g(X'(n + 1, 0))) = n + 1 \vdash\end{array}}{f(\hat{g}(0, 0)) = n + 1 \vdash}\,res(\Theta'_{\tau_0})$$

where $\Theta'_{\tau_0} = \{X'(n + 1, 0) \leftarrow \hat{g}(0, 0)\}$ and $\tau_1 =$

$$\dfrac{\dfrac{\begin{array}{c}(axiom)\\ \vdash Z(n + 1, 0) \leq Z(n + 1, 0)\end{array} \qquad \begin{array}{c}(axiom)\\ 0 \leq X(n + 1, 0) \vdash\\ f(X(n + 1, 0)) < n + 1,\\ f(X(n + 1, 0)) = n + 1,\end{array}}{\vdash f(0) < n + 1, f(0) = n + 1}\,res(\Theta'_{\tau_1})}{\vdash f(\hat{g}(0, 0)) < n + 1, f(\hat{g}(0, 0)) = n + 1}\,T^+$$

where $\Theta'_{\tau_1} = \{Z(n + 1, 0) \leftarrow 0, X(n + 1, 0) \leftarrow 0\}$ and $\tau_2 =$

$$\dfrac{\begin{array}{c}(axiom)\\ f(\hat{g}(1, 0)) < n + 1 \vdash\end{array} \qquad \dfrac{\dfrac{\begin{array}{c}(axiom)\\ \vdash Z(n + 1, 0) \leq f(Z(n + 1, 0))\end{array} \qquad \begin{array}{c}(axiom)\\ 0 \leq X(n + 1, 1) \vdash\\ f(X(n + 1, 1)) < n + 1,\\ f(X(n + 1, 1)) = n + 1,\end{array}}{\vdash f(f(0)) < n + 1, f(f(0)) = n + 1}\,res(\Theta'_{\tau_2})}{\vdash f(\hat{g}(1, 0)) < n + 1, f(\hat{g}(1, 0)) = n + 1}\,T^+}{\vdash f(\hat{g}(1, 0)) = n + 1}\,res$$

for $\Theta'_{\tau_2} = \{Z(n + 1, 0) \leftarrow 0, X(n + 1, 1) \leftarrow f(0)\}$. This concludes the definition of the refutation schema $\delta_0$. It remains to define $\rho(\delta, (0, k))$ and $\rho(\delta, (n + 1, k))$.

$\rho(\delta, (0, k)) = f(\hat{g}(k, 0)) < 0 \vdash$.

Note that $f(\hat{g}(k, 0)) < 0 \vdash$ is an axiom.

$\rho(\delta, (n + 1, k)) =$

$$\dfrac{\dfrac{\begin{array}{c}(\tau_3)\\ f(\hat{g}(k, 0) = n,\\ f(\hat{g}(k, 0)) < n + 1 \vdash\end{array} \qquad \dfrac{\begin{array}{c}(axiom)\\ S\end{array} \qquad \begin{array}{c}(axiom)\\ S'\end{array}}{\begin{array}{c}f(X(n, k)) < n + 1 \vdash\\ f(X(n, k)) < n, f(X(n, k)) = n\end{array}}\,res(\Theta_1)}{f(\hat{g}(k, 0)) < n + 1 \vdash f(\hat{g}(k, 0)) < n}\,res(\Theta_2) \qquad \begin{array}{c}(\delta, (\mathbf{n, k}))\\ f(\hat{g}(k, 0)) < n \vdash\end{array}}{f(\hat{g}(k, 0)) < n + 1 \vdash}$$

where $S$ is $\vdash X'(n,k) \leq X'(n,k)$ and $S'$ is

$$Y(n,k) \leq X(n,k),\ f(Y(n,k)) < n+1 \vdash f(X(n,k)) < n,\ f(X(n,k)) = n,$$

$\Theta_1 = \{X'(n,k) \leftarrow X(n,k), Y(n,k) \leftarrow X(n,k)\}$, $\Theta_2 = \{X(n,k) \leftarrow \hat{g}(k,0)\}$, and $\tau_3 =$

$$\dfrac{\dfrac{(\tau_4)}{f(\hat{g}(k,0)) < n+1 \vdash f(g(\hat{g}(k,0))) = n} \qquad \dfrac{(axiom)}{A}}{f(\hat{g}(k,0) = n,\ f(\hat{g}(k,0)) < n+1 \vdash}\ res(\Theta_3)$$

for $S = f(X''(n,k+1)) = n,\ f(g(X''(n,k+1))) = n \vdash, \Theta_3 = \{X''(n,k+1)) \leftarrow \hat{g}(k,0)\}$, and $\tau_4 =$

$$\dfrac{\dfrac{\dfrac{(\delta,(\mathbf{n},(\mathbf{k}+\mathbf{1})))}{f(\hat{g}(k+1,0)) < n \vdash}}{f(g(\hat{g}(k,0))) < n \vdash}\ T^+ \qquad \dfrac{\dfrac{(axiom)}{\vdash X'(n,k) \leq g(X'(n,k))} \qquad \dfrac{(axiom)}{S}}{f(X'(n,k)) < n+1 \vdash f(g(X'(n,k))) = n,\ f(g(X'(n,k))) < n}\ res(\Theta_4)}{f(\hat{g}(k,0)) < n+1 \vdash f(g(\hat{g}(k,0))) = n}\ res(\Theta_5)$$

where $S$ is

$$Y(n,k) \leq X(n,k+1),\ f(Y(n,k)) < n+1 \vdash f(X(n,k+1)) = n,\ f(X(n,k+1)) < n,$$

and $\Theta_4 =$

$$\{Y(n,k) \leftarrow X'(n,k), X(n,k+1) \leftarrow g(X'(n,k))\},$$

and $\Theta_5 = \{X'(n,k) \leftarrow \hat{g}(k,0)\}$.

Note that $\rho(\delta,(n,k))$ admits proof recursion defined in Definition 7.2.5. So in the formalism defined in Chap. 7.2 we can define an inductive closure $\hat{\rho}_\delta(n,k)$ from $\rho_\delta(n,k)$ which stands for $\rho(\delta,(n,k))$.

The proofs of the

EventuallyConstantRefutation

illustrate the exponential growth, see Fig. 10.7. This point was also addressed in [12].

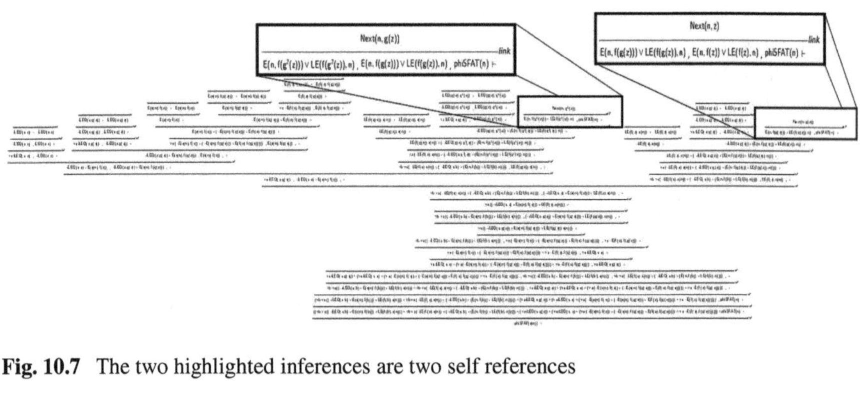


**Fig. 10.7** The two highlighted inferences are two self references

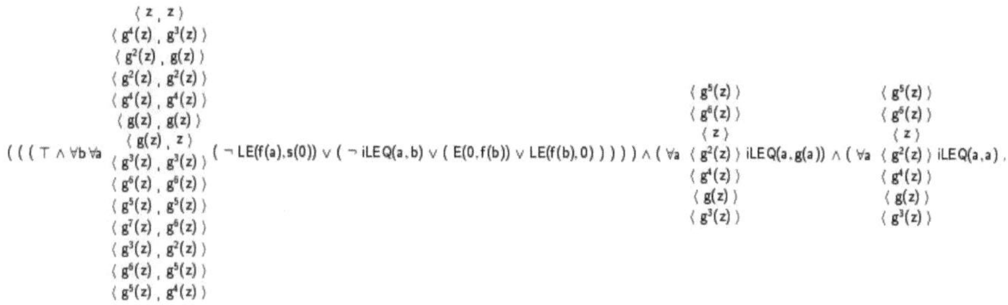


**Fig. 10.8** Part of the expansion proof of the ECS schema refutation instance 7 (the ECS schema has a large expansion proof and thus we present the most significant part)

As was already done for the

FunctionIterationRefutation,

we can also produce an expansion proof for the

EventuallyConstantRefutation,

which provides similar information as the substitutions discussed in [12], see Fig. 10.8.

# Chapter 11
# Open Problems and Future Research

The schematic CERES method has yielded significant results in proof theory, particularly in the analysis of proof schemata and the extraction of Herbrand systems. However, several challenges remain unresolved, leaving promising investigations for further exploration. This chapter outlines current open problems and concrete plans for future development, with particular emphasis on the completeness of schematic CERES, applications to arithmetic systems, and the ongoing extension of the GAPT system.

**Completeness of Schematic** CERES

One of the central unresolved problems in the theory of schematic proof analysis is the completeness of the schematic CERES method. While the method has demonstrated considerable power in extracting Herbrand systems from proof schemata, it remains an open question whether it is complete. Given a proof schema, the projection schema and the characteristic formula schema can be computed automatically. It remains an open question whether there always exists a schematic refutation of the characteristic formula schema–which is needed to compute a Herbrand system. Therefore, the completeness problem for schematic CERES reduces to the completeness of refutation schemata for characteristic formula schemata. Moreover, we plan to give a characterization of inductive proofs which can be analyzed by schematic CERES, that is the class of proofs having characteristic formula schemata which can be refuted by the schematic resolution calculus.

**Applications to Primitive Recursive Arithmetic (PRA)**

Existing work has demonstrated that the proof schema formalism simulates a class of proofs in **PRA** (see Chapter 8 for details). In future research we plan to extend these results, by also considering proofs involving quantification. The development of a schematic CERES method for **PRA**-proofs would constitute a major advancement. Indeed, together with a completeness result for schematic CERES, this implies a realization of Herbrand's theorem for an expressive fragment of formal number theory.



A. Leitsch et al., *First-Order Schemata and Inductive Proof Analysis*, Computer Science Foundations and Applied Logic, https://doi.org/10.1007/978-3-032-05741-9_11

### Extension of the GAPT System

The GAPT (General Architecture for Proof Theory) system has been instrumental in supporting both theoretical investigations and experimental applications of `CERES`. While the current GAPT implementation allows the extraction of a schematic characteristic formula and the construction of a projection schema, the construction of the refutation schema of the schematic characteristic formula cannot be automated. Continued development of GAPT remains a high priority. Some concrete directions are:

1. Extending the proof data base: The GAPT system provides a set of proofs which can be used for testing the schematic `CERES` method (one example is the ordered infinitary pigeonhole principle). However, in particular for inductive proofs the set of proofs in the current GAPT proof data base has to be extended in order to perform a substantial analysis and to obtain realistic results from experiments.
2. Schematic proof analysis: Automate the construction of the schematic Herbrand substitutions from projection and refutation schemata, and finally, the construction of the combined (final) Herbrand system of the proof schema.
3. User interface and automation: Developing user-friendly interfaces for proof schema construction and `CERES` transformation, with the goal of lowering the barrier to entry for non-expert users.

The ultimate goal would of course be to obtain a (at least semi-) automated analysis of mathematical proofs with induction (e.g. the pigeonhole principle or Fürstenberg's proof).

# References


1. Andrews PB (1971) Resolution in type theory. J Symb Log 36(3):414–432
2. Aravantinos V, Caferra R, Peltier N (2011) Decidability and undecidability results for propositional schemata. J Artif Intell Res 40:599–656
3. Aravantinos V, Echenim M, Peltier N (2013) A resolution calculus for first-order schemata. Fundam Informaticae 125(2):101–133
4. Baaz M, Hetzl S, Leitsch A, Richter C (August 2008) Spohr H (2008) Ceres: an analysis of Fürstenberg's proof of the infinity of primes. Theoret Comput Sci 403(2–3):160–175
5. Baaz M, Hetzl S, Leitsch A, Richter C, Spohr H (2008) Ceres: an analysis of Fürstenberg's proof of the infinity of primes. Theoret Comput Sci 403(2–3):160–175
6. Baaz M, Leitsch A (2000) Cut-elimination and redundancy-elimination by resolution. J Symb Comput 29(2):149–177
7. Baaz M, Leitsch A (2006) Towards a clausal analysis of cut-elimination. J Symb Comput 41(3):381–410
8. Baaz M, Leitsch A (2011) Methods of cut-elimination, vol 34. Springer Science & Business Media
9. Bertot Y, Castran P (2010) Interactive theorem proving and program development: Coq'Art the calculus of inductive constructions, 1st edn. Springer Publishing Company, Incorporated
10. Boolos G, Burgess JP, Jeffrey RC (2007) Computability and logic, 5th edn. Cambridge University Press, Cambridge
11. Brotherston J, Simpson A (2011) Sequent calculi for induction and infinite descent. J Log Comput 21(6):1177–1216
12. Cerna DM, Leitsch A (2016) Schematic cut elimination and the ordered pigeonhole principle. In: IJCAR, vol 9706. Lecture notes in computer science. Springer, pp 241–256
13. Cerna DM, Leitsch A, Lolic A (2021) Schematic refutations of formula schemata. J Autom Reason 65(5):599–645
14. Cerna DM, Lettmann MP (2017) Integrating a global induction mechanism into a sequent calculus. In: TABLEAUX, vol 10501. Lecture notes in computer science. Springer, pp 278–294
15. Dunchev C, Leitsch A, Rukhaia M, Weller D (2013) Cut-elimination and proof schemata. In: International Tbilisi symposium on logic, language, and computation. Springer, pp. 117–136
16. Ebner G, Hetzl S, Reis G, Riener M, Wolfsteiner S, Zivota S (2016) System description: Gapt 2.0. In: International joint conference on automated reasoning. Springer, pp. 293–301
17. Fürstenberg H (1955) On the infinitude of the primes. Am Math Mon 62:353
18. Fürstenberg H, Weiss B (1978) Topological dynamics and combinatorial number theory. J d'Analyse Mathematique 34(1):61–85




A. Leitsch et al., *First-Order Schemata and Inductive Proof Analysis*, Computer Science Foundations and Applied Logic, https://doi.org/10.1007/978-3-032-05741-9

19. Gentzen G (1934-35) Untersuchungen über das logische Schließen. Mathematische Zeitschrift 39:176–210, 405–431
20. Gentzen G (1935) Untersuchungen über das logische Schließen I. Math Z 39(1):176–210
21. Gentzen G (1936) Der Wiederspruchfreiheit Der Reinen Zahlentheorie. Math Ann 112:493–565
22. Girard JY (1987) Proof theory and logical complexity, vol 1. Biblopolis, Napoli
23. Gonthier G (2008) Formal proof-the four-color theorem. Notices AMS 55(11):1382–1393
24. Herbrand J (1930) Recherches sur la théorie de la démonstration. PhD thesis, Université de Paris
25. Hetzl S, Leitsch A, Weller D, Paleo BW (2008) Herbrand sequent extraction. In: AISC/MKM/calculemus, vol 5144. Lecture notes in computer science. Springer, pp 462–477
26. Kreisel G (1951) On the interpretation of non-finitist proofs-part i. J Symb Log 16(4):241–267
27. Kreisel G (1952) On the interpretation of non-finitist proofs-part ii. J Symb Log 17(1):43–58
28. Kreisel G (1958) Mathematical logic: What has it done for the philosophy of mathematics? In: Proceedings of the international congress of mathematicians, pp 164–173
29. Leitsch A, Lolic A (2019) Extraction of expansion trees. J Autom Reason 62(3):393–430
30. Leitsch A, Lolic A, Mahler S (2024) On proof schemata and primitive recursive arithmetic. In: Bjørner NS, Heule M, Voronkov A (eds) LPAR 2024 complementary volume, Port Louis, Mauritius, 26–31 May 2024, vol 18. Kalpa Publications in Computing, EasyChair, pp 117–130
31. Leitsch A, Peltier N, Weller D (2017) CERES for first-order schemata. J Log Comput 27(7):1897–1954
32. Lolic A (2020) Automated proof analysis by CERES. PhD thesis, Technical University of Vienna
33. Martelli A, Montanari U (1982) An efficient unification algorithm. ACM Trans Program Lang Syst 4(2):258–282
34. McDowell R, Miller D (2000) Cut-elimination for a logic with definitions and induction. Theoret Comput Sci 232(1–2):91–119
35. Nipkow T, Wenzel M, Paulson LC (2002) Isabelle/HOL: a proof assistant for higher-order logic. Springer, Berlin, Heidelberg
36. Robinson JA (1965) A machine-oriented logic based on the resolution principle. J ACM (JACM) 12(1):23–41
37. Schöning U, Pruim R (1998) The equivalence problem for loop (1)-and loop (2)-programs. Gems of theoretical computer science, pp 25–36
38. Sørensen MH, Urzyczyn P (2006) Lectures on the curry-Howard isomorphism, 2nd edn. Elsevier
39. Takeuti G (1987) Proof theory, 2nd edn. North Holland
40. Van der Waerden BL (1927) Beweis einer Baudetschen Vermutung. Nieuw Archiv Wiskunde 15(2):212–216

# Index




A. Leitsch et al., *First-Order Schemata and Inductive Proof Analysis*, Computer Science Foundations and Applied Logic, https://doi.org/10.1007/978-3-032-05741-9

MIX
Papier aus verantwortungsvollen Quellen
Paper from responsible sources
FSC® C105338

If you have any concerns about our products,
you can contact us on
**ProductSafety@springernature.com**

In case Publisher is established outside the EU,
the EU authorized representative is:
**Springer Nature Customer Service Center GmbH**
**Europaplatz 3, 69115 Heidelberg, Germany**

Printed by Libri Plureos GmbH
in Hamburg, Germany